Advances in Intelligent Systems and Computing

Volume 179

Editor-in-Chief

Prof. Janusz Kacprzyk
Warsaw
Poland

For further volumes:
http://www.springer.com/series/11156

Miloš Kudělka · Jaroslav Pokorný
Václav Snášel · Ajith Abraham
Editors

Proceedings of the Third International Conference on Intelligent Human Computer Interaction (IHCI 2011), Prague, Czech Republic, August, 2011

Springer

Editors

Miloš Kudělka
Faculty of Electrical Engineering and Computer Science
Technical University of Ostrava
Ostrava-Poruba
Czech Republic

Václav Snášel
Faculty of Electrical Engineering and Computer Science
Technical University of Ostrava
Ostrava-Poruba
Czech Republic

Jaroslav Pokorný
Faculty of Mathematics and Physics
Charles University
Prague
Czech Republic

Ajith Abraham
Machine Intelligence Research Labs (MIR Labs)
Scientific Network for Innovation and Research Excellence
Auburn, WA
USA

ISSN 2194-5357 ISSN 2194-5365 (electronic)
ISBN 978-3-642-31602-9 ISBN 978-3-642-31603-6 (eBook)
DOI 10.1007/978-3-642-31603-6
Springer Heidelberg New York Dordrecht London

Library of Congress Control Number: 2012943358

Printed on acid-free paper

Springer is part of Springer Science+Business Media (www.springer.com)

Preface

The Intelligent Human Computer Interaction conferences bring together scientists, engineers, computer users, and students to exchange and share their experiences, new ideas, and research results about all aspects (theory, applications, and tools) of intelligent methods applied to human–computer interaction and to discuss the practical challenges encountered and the solutions adopted. The previous two IHCI events (2009, 2010) were held in Allahabad, India. The Third Intelligent Human Computer Interaction (IHCI 2011) conference was held from August 29 to 31, 2011, in the historical city center of Prague, Czech Republic. The conference was organized by the VŠB—Technical University of Ostrava, Czech Republic, and the Charles University in Prague, Faculty of Mathematics and Physics, Czech Republic.

This book has five parts. The first part is focused on user interface and interaction, and the second and fifth parts present papers from the area of knowledge discovery. Other parts are focused on technological and theoretical background and applications and tools supporting the human–computer interaction.

Christopher J. Martin et al. present a preliminary study as a part of research aimed at designing and implementing an agent which detects negative emotions in a human user and expresses its own emotional reaction. Based on the theory of Situational Awareness, Andrea Pandurino et al. illustrate how five major "interface demons" are particularly relevant for RIA engineering and undermine an effective dialogue between users and RIA interfaces. Julius T. Nganji et al. propose the use of the Image Description Assessment Tool (IDAT), a Java-based tool containing some proposed heuristics, for assessing how well an image description matches the real content of the image on the Web. Šárka Zehnalová et al. are focused on engaging more interactive ways to image search based on four typical scenarios and introduce their own user interface of the Xingle testing system where all the scenarios are implemented. To improve the classification accuracy, Pavel Bobrov et al. propose to use Independent Component Analysis for μ-rhythm identification in data corresponding to motor imagery task performance during Brain–Computer Interface training and operation.

Vivek Kumar Singh et al. present an experimental work on evaluating hard and soft flat clustering algorithms for categorizing text documents. Erqiang Zhou and Bing Wu show that the Natural Language Interface can greatly reduce the efforts for writing or understanding queries on information systems. Siham Boulaknadel and Fadoua Ataa Allah present preliminary results of experiments with a corpus for Standard Amazighe language. Hussein Soori et al. describe very simple rules for stemming Arabic words.

Garnett Wilson et al. describe a novel fuzzy logic system designed to meet the real-world demand for providing intelligent ranking to large repositories of documents previously encoded with non-fuzzy (crisp) metadata. The goal of Manoj kumar Rajagopal et al. is to capture style from real human motion so that it can be rendered with a virtual agent that represents this human. Steven Vercruysse and Martin Kuiper use JavaScript and the new HTML5 canvas element to create a software module, GraphVis, for the online visualization and exploration of large relationship networks. Hong-Chi Shih et al. propose a Reduce Identical Event Transmission Algorithm (RIET) for Wireless Sensor Networks. Azeem Lodhi et al. are focused on providing representational support for business processes and they describe extensions of an existing modeling language. Omar S. Soliman et al. propose an automatic localization and boundary detection of retina images using basic filters to support ophthalmologists for the detection and diagnosis of harmful eye diseases. The study of Mohamed Achraf Dhouib et al. aims to make a profit of the technological diversity of several Decision Support Systems used to detect distress situations.

Takashi Matsuhisa investigates a communication logic based on the multimodal logic S5n, and it is treated from a multi-modal logical point of view. Radim Nedbal proposes a fully declarative language for encoding preferences conditional to the current state of the world represented as a relation database instance. Ivo Lašek and Peter Vojtáš evaluate various approaches to a user profile modeling for news recommendations. Petr Gajdoš and Jan Platoš describe the advantages of a Graphics Processor Unit (GPU) in parallel computation of Self-Organizing Network including a comparison with a multi-threaded CPU. Alexander A. Frolov et al. introduce a new Expectation-Maximization method which maximizes the likelihood of a Boolean Factor Analysis solution.

Anupam Srivastava et al. present a generic interactive framework based on human cognition where the system can learn continuously from the Internet and from its interaction with users. Aboul Ella Hassanien et al. introduce a hybrid scheme that combines the advantages of pulse coupled neural networks (PCNNs) and support vector machine in conjunction with type-II fuzzy sets and wavelets to enhance the contrast of the original images and feature extraction. Štěpán Minks et al. are focused on the strength extraction between authors from the DBLP database based on their association. Miloš Kudělka et al. propose a novel method for detecting local communities in social networks and describe an experiment based on student data.

We would like to express our sincere gratitude to the program committee, local organizing committee, and the numerous reviewers who helped us evaluate the papers and make IHCI 2011 a very successful scientific event. Gratitude is also expressed to Professor Janusz Kacprzyk, editor of the series publishing this book as well as to the Springer team for its excellent work. We hope that every reader will find this book interesting and inspiring.

Contents

Part I
User Interface and Interaction

Towards an Affective Self-Service Agent

Christopher J. Martin, Jacqueline Archibald, Leslie Ball and Lloyd Carson

Abstract While emotional intelligence plays a key role in facilitating satisfactory interactions between people, its application is both underexplored and underexploited in human–computer interaction. Self-service technology is increasingly being incorporated by goods and service providers, however user satisfaction is still less than ideal. Studies have been carried out in which an affective embodied agent has been found to reduce frustration in users of interactive computer systems. This paper presents a preliminary study as a part of research aimed at designing and implementing an agent which detects negative emotions in a human user and expresses its own emotional reaction with the aim of improving the user's mood and therefore their level of satisfaction in the context of a self-service interaction. We describe a study to determine customer facial expressions using facial Action Units (AUs) during interactions with self-service supermarket checkouts. Our preliminary results indicate that AU 23 and AU 24 were displayed with particular frequency.

1 Introduction

It has been suggested that a minimum level of emotional intelligence—the ability to recognise emotions in others and to express them oneself—is required for successful social functioning [31]. Emotional intelligence is a form of social

C. J. Martin (✉) · J. Archibald · L. Ball · L. Carson
University of Abertay, Bell Street, Dundee, UK
e-mail: c.martin@abertay.ac.uk

J. Archibald
e-mail: j.archibald@abertay.ac.uk

L. Ball
e-mail: l.ball@abertay.ac.uk

L. Carson
e-mail: l.carson@abertay.ac.uk

M. Kudělka et al. (eds.), *Proceedings of the Third International Conference on Intelligent Human Computer Interaction (IHCI 2011), Prague, Czech Republic, August, 2011*, Advances in Intelligent Systems and Computing 179, DOI: 10.1007/978-3-642-31603-6_1,

intelligence which can benefit many aspects of a human-human interaction including the perceived quality and reliability of information, the ease of social interaction and cooperation between parties [24, 25, 31]. Affective computing, the field of research which includes giving computers the ability to simulate emotional experience, emotional expressiveness and to detect and respond to emotion in human users [26], is becoming increasingly popular in the quest to create more satisfying and effective human–computer interactions, as systems which simulate emotional intelligence are likely to enjoy the benefits described above [24–27].

Self-Service Technology (SST) is being adopted in an increasing number of locations including supermarkets, rail stations and airports. Anecdotally, people report frustration when using these systems. Meuter et al. [22] found that over 36 % of dissatisfaction with SST interactions was caused by poor design, while Pujari found that 26 % of dissatisfaction was due to customer service problems [28]. Of these, less than 5 % of causes of dissatisfaction were attributed to the customer [22, 28]. SSTs would benefit from being made easier and more satisfying to use, since frustration caused by using a system can negatively affect the customer's feelings towards the retailer, leading to negative behavioural intentions and low retailer interest. Conversely, an increase in the perceived level of quality of the service provided by SSTs has been found to increase customer satisfaction and have a positive effect on customers' behavioural intentions [14, 19]. It has also been suggested that the perceived affective quality of a system influences its perceived usefulness and perceived ease of use, which in turn positively influence the user's behavioural intention to use the system [41].

This paper describes the first step in a larger project which aims to create a fully affective user interface system to be used in a self-service interaction with the aim of detecting and reducing negative emotions, such as anger and frustration, in the user. A literature review is presented then we report the methods and preliminary results of a filming study carried out in a supermarket in order to obtain video data of participants' emotion. Future work is proposed detailing how we will analyse the footage and how it will influence future applications.

2 Literature Review

To date, literature suggests that affective agents are not widely used in SSTs, and so present an opportunity for improving the usability of these systems. Current SSTs are unable to detect the emotions of customers. We propose that a system which is able to adapt its behaviour based on the emotions of a user will lead to greater satisfaction and improved behavioural intention in users.

Agents are virtual personas which are present during an interaction, communicating via any combination of printed text, spoken dialogue and graphical representation (known as an avatar). To improve human–computer interactions when shopping online, agents are sometimes employed to assist the user [1, 11, 21]. Affective agents employ characteristics such as limited emotional awareness and expressiveness, examples of which are the agents on the IKEA [11] and Marks and Spencer [21] websites. Typically, affective agents consist of an embodiment (an avatar) accompanying a chatbot-style interface and may express emotional facial reactions to the human–computer dialogue such as smiling when the user thanks the agent for its help [11] or looking confused if it does not understand the user's request [21].

The use of affective agents has been trialled in attempts to reduce frustration in computer-based tasks, however these have relied on "Wizard of Oz" functionality, where a human controls various parts of the system to give the illusion of fully automatic behaviour. Klein et al. [16] carried out a study in which participants played a game for a chance to win a cash prize. The game was designed to periodically freeze the player's on-screen character, reducing their chance of winning the prize, thereby inducing frustration. Players played one round of the game and, depending on the experimental condition, were given the opportunity to describe their level of frustration and possibly receive empathetic feedback. They were then offered the chance to play the game again. In conditions where the system gave empathetic feedback on the user's reported level of frustration, participants were found to play for longer than when no feedback was given. Another study was carried out by Jaksic et al. [13] which required participants to interact with a faulty shopping website. Faults included shopping cart errors such as items not being added or the contents of the cart being lost. An affective agent was employed to offer apologies and reassurance in response to negative emotional facial expressions. The study found that empathetic virtual agents could reduce frustration levels in moderately frustrated participants but increased frustration levels in highly frustrated participants. Agents have also been found to improve the effectiveness of advice giving systems. Berry et al. [2] analysed the Greta agent [29], an embodied affective agent which gives advice about eating disorders and empathetic feedback in a text dialog with a user. Empathetic feedback was manipulated to be either consistent or inconsistent with accepted social norms, for example an empathetically consistent agent would show happiness if the user stated that they were happy whereas an empathetically inconsistent agent would display sadness or anger. The embodied agent was perceived as more trustworthy than a video of a human helper, more helpful than a voice only agent and equally as helpful as a human. It was also liked more than a voice only agent. Advice given by the agent was remembered better when delivered by an emotionally consistent agent than by an emotionally neutral or emotionally inconsistent agent. Wang et al. [39] used an embodied affective agent to support a human user during a learning experience. The agent supported the user, commenting on their moods and complementing their mood with appropriate emotional reactions. The agent encouraged the user to keep working during a learning task and an empathetic agent was able to convince more subjects to attempt more exercises than a non-empathetic agent.

Facial expressions are a good input signal for computer-based emotion detection as they are the most effective method (either alone or in combination with vocal inflection) of expressing social and affective signals [24, 34, 38] and the six basic emotions (happiness, anger, fear, sadness, surprise and disgust) are universally displayed with facial expressions [7]. Another benefit is that, unlike other biometric measures such as galvanic skin response or heart rate, observation requires no physical contact, only a simple webcam type video camera.

Several attempts have been made to allow computers to recognise emotions from facial expressions [4, 8, 9, 12, 17, 18, 20, 23, 32, 36, 37, 40] using a variety of techniques. One method involves recognising facial Action Units (AUs): fine-grained descriptors of the movements of facial muscles which create facial expressions [15]. It is expected that emotional facial expressions will be displayed by users during interaction with SSTs as it has been found that 70 % of computer users encounter problems which cause them to physically or verbally abuse their computer [35]. Facial muscle activation has also been found to correlate positively with task difficulty [3], therefore we intend to use facial emotional expressions, captured on video, as input to our system. We will use Haar-like feature extraction [8, 18, 32, 40] due to its ease of training and implementation.

3 Proposed Research

To date, literature suggests that affective agents are not widely used in SSTs, and so present an opportunity for improving the usability of these systems. Current SSTs are unable to detect the emotions of customers. We propose that a system which is able to adapt its behaviour based on the emotions of a user will lead to greater satisfaction and improved behavioural intentions in users. Our proposed system will consist of three components: an input, process and output component. Our plan is that the input component will take video input from a camera and will detect, in real-time, a negative emotion from the user's facial expression. The processing component will make a decision on how best to behave based on the user's detected emotion. This behaviour will aim to reduce the strength of the user's negative emotion. The output component will orchestrate the non-verbal behaviour dictated by the processing component in an emotionally expressive avatar representing a human face. This behaviour will involve facial expressions, head gestures and body gestures. We aim to develop this system and investigate whether it is capable of reducing negative emotions in SST users.

This paper focuses on the initial stages of building the input system. In order to train an emotion detection system, emotional facial expression data need to be gathered. Databases of examples of emotional facial expressions exist [10], however these often focus on the "basic six" emotions (happiness, sadness, anger, fear, surprise and disgust) described by Ekman et al. [7] and such stimuli are often acted, rather than spontaneous. We felt it important to discover exactly which emotions are displayed during self-service interactions, as often people's range of everyday emotions are more complex and varied than the "basic six". To determine the range of emotions exhibited by SST users, we carried out a study in which we filmed and analysed the facial expressions of shoppers using self-scan checkouts in a supermarket. Once we know which emotions are expressed during SST interactions, we can identify whether negative emotions are expressed and, if so, use them to train a facial emotion recognition system.

A filming study was carried out to gather emotional data for the purpose of training the input component in our proposed system. A description of the method and preliminary results follow.

4 Method

Our study was carried out in a supermarket located in a city centre following ethical approval from both the supermarket management and the University of Abertay School of Computing and Engineering Systems Research Ethics Sub-Committee. The session took place on a Sunday afternoon between 1410 and 1710. Sunday was chosen as management advised that this was the least busy day of the week and minimum customer disruption was a high priority of the researchers. Further, we felt that customers who had queued for relatively short times might be more willing to take part. We also wanted to minimise disruption to increase the chance of being allowed to carry out further filming sessions, should they be required. The volume of customers using the self-service checkouts varied considerably over the course of the study, from no customers at all through to every checkout being in use and a queue of people waiting to pay. A video camera (a Canon Mini DV camcorder) was attached to a tripod which was positioned behind the self-service checkout and pointed at the user's face, with a distance of approximately 1 m between the camera lens and the participant's face. The height of the tripod was adjusted to ensure participants' faces were fully visible in the shot and were roughly centred in the scene. Figure 1 shows an image captured from one of the videos of a participant using the self-service checkout. Participants were recruited by approaching people in the queue, explaining the purpose of the study and asking whether they would mind being filmed scanning their shopping. Typically, the researcher asked "Would you mind being filmed while you scan your shopping? It's for research being carried out by Abertay University". Participants' heads and shoulders were filmed as they scanned and paid for their shopping. Thirty participants took part in the study, 17 male and 13 female. In an attempt to increase the appeal of taking part in the study, it was decided not to ask participants any questions, either personal or relating to their shopping experience, so that taking part added as little time as possible to the participant's shopping experience and to help them feel less like their privacy was being invaded. Although it would have been interesting to gather demographic data, information about participants' technological background and their current and previous experiences with SSTs, this study's main aim was simply to collect data on emotional facial expressions from as large a group of participants as possible. Customers with very few items (three or less) were not approached as it was felt that the amount of their time taken up with the explanation and informed consent signing was disproportionately large compared to the time it would take to scan their items.

Fig. 1 A still from a video of a participant using the self-service checkout

The footage was viewed by the researcher and facial expressions were manually identified. Facial expressions were judged to be activation of facial AUs which deviated from the participant's neutral facial expression. The footage was then broken into clips—shorter sequences of footage—of either facial expressions or neutral faces. The neutral face clips were discarded and the facial expression clips were repeatedly viewed by the researcher. Facial expressions exhibited as a result of interacting with a member of staff, a companion or the researcher filming the interaction were also discarded as they were not elicited by the SST device. At the peak of each expression (all AUs at maximum activation), the researcher noted the activated AUs in a spreadsheet and the number of occurrences of each AU was calculated (Fig. 2). It is possible that some AUs were exhibited by the participants but not observed by the researcher due to factors including the resolution of the recorded footage, the lighting in the scene and participant features such as hair cuts, glasses, hats, scarves and facial hair. The resolution of the video footage was 320×240 pixels.

5 Preliminary Results

Thirty interactions were videoed and subsequently analysed. The average length of an interaction was 302 s, with a minimum duration of 87 s and a maximum duration of 595 s. In total, 16 different facial AUs were identified in the footage. Facial Action Units were developed by Ekman and Friesen [6] as a way of coding the individual muscle movements involved in creating facial expressions. Figure 2 shows the number of occurrences of each AU observed in the videos.

A chi-square test for goodness-of-fit indicated that the frequency of occurrence of AU categories significantly deviated from the expected distribution, $x^2(15, \text{n} =$

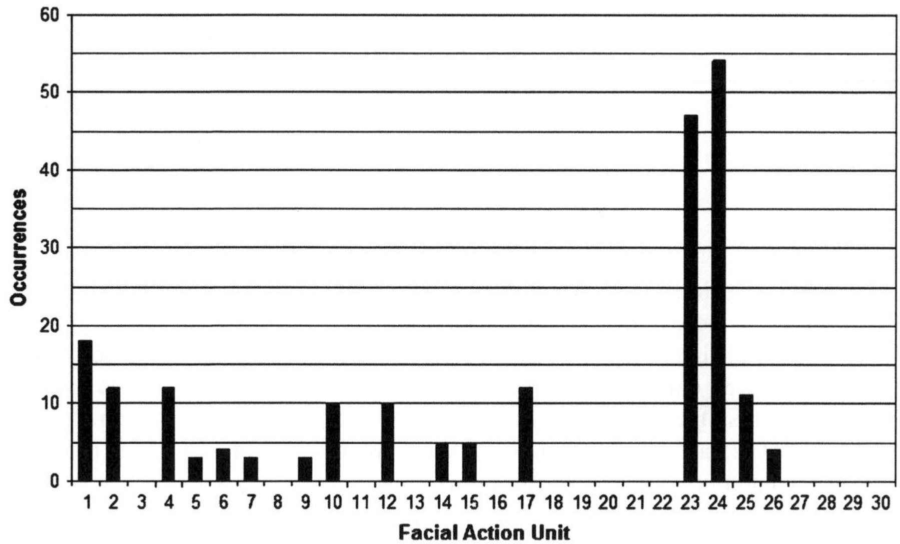

Fig. 2 Occurrences of Facial Action Units [6] observed in SST users

213) = 261.066, $p < 0.001$. Two facial behaviours, AU 23 and AU 24, occurred at least twice as frequently as any other AU.

6 Conclusions and Further Research

In this first phase of research we have obtained a collection of video clips which show a range of facial expressions displayed spontaneously by people using self-service checkouts at a supermarket. AU 23 and AU 24 (and possibly AU 1) were observed particularly frequently. When displayed in combination, AU 23 (lip tighten) and AU 24 (lip press) have been linked to anger [17, 20, 30], while AU 23 alone is also associated with this emotion [5, 33]. AU 1 refers to the inner portion of the brows being raised and is linked to sadness [5, 20], surprise [5, 20] and fear [20]. Figure 3 shows an example of AU 23 and AU 24 in combination. We hypothesise that their appearance in our videos is due to some negative emotional state elicited in the SST users.

A second filming session utilising a hidden camera (subject to ethical approval) may give more accurate information on participants' moods and emotions, as the study described in this paper had participants volunteering to take part. This could have affected results as the type of people who volunteered may have been less busy and therefore less stressed or may have been easier going in general. Also, people with a small number of items were not approached, however if people with few items

Fig. 3 The image shows a participant from our study exhibiting a combination of AU 23 and AU 24

anticipate a short interaction which turns out to be a long, difficult interaction they may experience more intense negative emotions.

The next stage in this research is to test the reliability of the categorisation of the facial Action Unit data, as our initial findings were obtained by having the footage analysed by a single coder with no formal training in the Facial Action Coding System. If AU 23 and AU 24 (or another AU or combination of AUs which is linked to negative emotion) are confirmed as occurring with particular frequency we will train a Haar-like feature extraction system to detect activation of these AUs in videos of human faces.

We will develop a system which consists of an emotion detection module (the Haar-like feature extractor), decision making logic and an emotionally expressive virtual agent. The system will detect a negative emotion, indicated by the activation of relevant AUs, and trigger the agent to perform some non-verbal affective behaviour with the aim of improving the user's emotional state. It is hypothesised that such a system will lead to more satisfying interactions with SSTs.

The behaviour exhibited by the agent in order to alleviate negative emotions will be multi-modal, involving body gesture, facial expressions and printed and spoken dialogue. Before a decision is made more research must be carried out into suitable facial expressions to express. Should the agent look annoyed, sharing the user's perceived mood? Or could anger be interpreted as the agent being angry at the user because they made a mistake? Should the agent express happiness, with the aim of the user being influenced by the agent's mood and thereby improving their own? Or will this contrasting emotion result in an agent as inane and inappropriate as the Microsoft paperclip? Many participants expressed clearly positive facial expressions (smiles and laughter) when they interacted with a member of staff and it was interesting to note that even participants who appeared to be greatly frustrated, stressed and impatient while using the self-scan machine quickly expressed happiness and even laughter when a shop assistant came to their aid and interacted with them in a friendly manner. Based on this observation, we intend to initially investigate friendliness and positive emotional expressions as a means of improving a user's mood. It is important for us to keep in mind that an agent which, on a technical level, works as intended may not be accepted by a user and therefore may be unsuccessful in its purpose of improving a human–computer interaction. We may evaluate our affective system as working correctly, however when implemented in a real-world scenario our affective agent may be seen as an unwanted distraction by the user, as was the case with the infamous Microsoft Office Paper Clip.

Acknowledgments The researchers would like to thank the supermarket for allowing us to carry out the filming session. We would also like to thank the participant who kindly allowed us to use their image in this paper.

References

1. ASDA: ASDA VA. http://groceries.asda.com/asda-estore/index.jsp?utm_source=homepage&utm_medium=cta&utm_campaign=startShoppingDeflt# . Cited 14 Feb 2011
2. Berry, D.C., Butler, L.T., de Rosis, F.: Evaluating a realistic agent in an advice-giving task. Int. J. hum.-comput. Stud. **63**(3), 304–327 (2005)
3. Branco, P., Firth, P., Encarnacao, L.M., Bonato, P.: Faces of emotion in human-computer Interaction. In: CHI'05 Extended Abstracts on Human Factors in Computing Systems (CHI'05), New York, 2005
4. Buenaposada, J.M., Munoz, E., Baumela, L.: Recognising facial expressions in video sequences. Pattern Anal. Appl. **11**(1), 101–116 (2008)
5. Carrol, J.M., Russell, J.A.: Facial expressions in Hollywood's portrayal of emotion. J. Pers. Soc. Psychol. **72**(1), 164–176 (1997)
6. Ekman, P., Friesen, W.V.: Facial Action Coding System: A technique for the measurement of facial movement. Consulting Psychologists Press, Palo Alto (1978)
7. Ekman, P.: Universals and cultural differences in the judgments of facial expressions of emotion. J. Pers. Soc. Psychol. **53**(4), 712–717 (1987)
8. Essa, I.A., Pentland, A.P.: Coding, analysis, interpretation, and recognition of facial expressions. IEEE Trans. Pattern Anal. Mach. Intell. **19**(7), 757–763 (1995)
9. Gunes, H., Piccardi, M.: Fusing face and body gesture for machine recognition of emotions. In: IEEE Workshop on Robot and Human Interactive Communication, ROMAN, 2005
10. HUMAINE Association: Databases. http://emotion-research.net/wiki/Databases. Cited 28 April 2011
11. IKEA, IKEA Help Center: http://193.108.42.79/ikea-us/cgi-bin/ikea-us.cgi (2011). Cited 14 Feb 2011
12. Ioannou, S.V., Raouzaiou, A.T., Tzouvaras, V.A., Mailis, T.P., Karpouzis, K.C., Kollias, S.D.: Emotion recognition through facial expression analysis based on a neurofuzzy network. Neural Netw. **18**(4), 423–435 (2005)
13. Jaksic, N., Branco, P., Stephenson, P., Encarna, L.M.: The effectiveness of social agents in reducing user frustration. In: CHI '06 Extended Abstracts on Human Factors in Computing Systems. ACM: Montral, Quebec, 2006
14. Johnson, A.J., Pepperman, K., Quist, L.: The role of emotion in self-service technologies: How can the U-Scan system be improved to positively influence users' emotions and shopping experiences. http://74.125.155.132/scholar?q=cache:6zCQBXL0EjIJ:scholar.google.com/+%22self+service%22+checkout+anger\&hl=en\&assdt=2001 (2008). Cited 15 Dec 2010
15. Kanade, T., Cohn, J.F., Tian, Y.: Comprehensive database for facial expression analysis. In: Fourth IEEE International Conference on Automatic Face and Gesture Recognition, Grenoble, 2000
16. Klein, J., Moon, Y., Picard, R.W.: This computer responds to user frustration. In: CHI '99 Extended Abstracts on Human Factors in Computing Systems. ACM: Pittsburgh, 1999
17. Kotsia, I., Pitas, I.: Facial expression recognition in image sequences using geometric deformation features and support vector machines. IEEE trans. Image Process. **16**(1), 172–187 (2007)
18. Kumano, S., Otsuka, K., Yamato, J.: Pose-invariant facial expression recognition using variable-intensity templates. Int. J. Comput. Vis. **83**(2), 178–194 (2009)
19. Lin, J., Hsieh, P.: The role of technology readiness in customers' perception and adoption of self-service technologies. Int. J. Serv. ind. manag. **17**(5), 497–517 (2006)

20. Lucey, P., Cohn, J.F., Kanade, T., Saragih, J., Ambadar, Z., Matthews, I.: The extended Cohn-Kanade dataset (CK+): a complete dataset for action unit and emotion-specified expression. In: Computer Vision and Pattern Recognition Workshop on Human-Communicative Behaviour, 2010
21. Marks and Spencer: Ask Hanna. http://money.marksandspencer.com/help/overview (2011). Cited 14 Feb 2011
22. Meuter, M.L., Ostrom, A.L., Roundtree, R.I., Bitner, M.J.: Self-service technologies: understanding customer satisfaction with technology-based service encounters. J. Mark. **64**(3), 50–64 (2000)
23. Michel, P., El Kaliouby, R.: Real time facial expression recognition in video using support vector machines. In: ICMI'03, Vancouver, November 2003. ACM (2003)
24. Pantic, M.: Affective computing. In: Pagani, M. (ed.) Encyclopedia of Multimedia Technology and Networking, 1, pp. 8–14. Idea Group Reference, Hershey (2005)
25. Pantic, M., Sebe, N., Cohn, J.F., Huang, T.: Affective multimodal human–computer interaction. In: Proceedings of the 13th Annual ACM International Conference on Multimedia. ACM: Hilton, Singapore, 2005
26. Picard, R.W.: Affective Computing. MIT Press, Cambridge (1997)
27. Picard, R.W.: Toward machine emotional intelligence: analysis of affective physiological state. IEEE Trans. Pattern Anal. Mach. Intell. **23**(10), 1175–1191 (2001)
28. Pujari, D.: Self-service with a smile? Self-service technology (SST) encounters among Canadian business-to-business. Int. J. Serv. Ind. Manag. **15**(2), 200–219 (2004)
29. de Rosis, F., Pelachaud, C., Poggi, I., Carofiglio, V., de Carolis, B.: From Greta's mind to her face: modelling the dynamics of affective states in a conversational embodied agent. Int. J. hum.-Comput. Stud. **59**(1–2), 81–118 (2003)
30. Rozin, P., Lowery, L., Imada, S., Haidit, J.: The CAD triad hypothesis: a mapping between three moral emotions (contempt, anger, disgust) and three moral codes (community, autonomy, divinity). J. Pers. Soc. Psychol. **76**(4), 574–586 (1999)
31. Salovey, P., Mayer, J.D.: Emotional intelligence. Imagination Cogn. Pers. **9**, 185–211 (1990)
32. Satiyan, M., Nagarajan, R.: Recognition of facial expression using Haar-like feature extraction method. In: International Conference on Intelligent and Advanced Systems (ICIAS). Kuala Lumpur, 2010
33. Schmidt, R.M., Ruth, C.R., Walters, C.M., Robinson, T.B., Shiyanov, P.A., Goodyear, C.D., Funke, M.A., McIntire, L.K.: Facial expressions of emotion and the assessment of performance. Air Force Research Laboratory, 711th Human Performance Wing, Human Effectiveness Directorate, Wright-Patterson Air Force Base, OH: Wright-Patterson Air Force Base, Ohio, 2010
34. de Silva, L.C., Ng. P.C.: Bimodal emotion recognition. In: Proceedings of Fourth IEEE International Conference on Automatic Face and Gesture Recogntion. Grenoble, 2000
35. Stone, D.: User Interface Design and Evaluation. Open University, UK (2005)
36. Tian, Y., Kanade, T., Cohn, J.F.: Recognizing action units for facial expression analysis. IEEE Trans. Pattern Anal. Mach. Intell. **23**(2), 97–115 (2001)
37. Valstar, M., Pantic, M.: Fully automatic facial action unit detection and temporal analysis. In: Conference on Computer Vision and Pattern Recognition Workshop (CVPRW'06). IEEE (2006)
38. Vinciarelli, A., Pantic, M., Bourlard, H., Pentland, A.: Social signal processing: state-of-the-art and future perspectives of an emerging domain. In: 16th ACM international Conference on Multimedia. Vancouver, British Columbia, Canada 2008
39. Wang, C., Chen, G., Liu, C., Liu, B.: Design an empathic virtual human to encourage and persuade learners in e-learning systems. In: Proceedings of the First ACM International Workshop on Multimedia Technologies for Distance Learning. ACM: Beijing, (2009)
40. Yang, P., Liu, Q., Metaxas, D.N.: Boosting coded dynamic features for facial action units and facial expression recognition. In: IEEE Conference on Computer Vision and Pattern Recognition, Minneapolis, 2007
41. Zhang, P., Li, N.: The importance of affective quality. Commun. ACM **48**(9), 105–108 (2005)

Balancing Complex Page Modeling and Usability for Rich Internet Applications

Andrea Pandurino, Luca Mainetti, Davide Bolchini and Roberto Paiano

Abstract The growth of Rich Internet Applications (RIAs) calls for new conceptual tools that enable designers and web engineers to model and keep under control the design complexity unleashed by innovative interaction (with increasing communication potential) and carefully consider the impact of the design decisions on the optimal flow of the user experience. Based on the theory of Situational Awareness, in this paper we illustrate how 5 major "interface demons" are particularly relevant for RIA engineering and undermine an effective dialogue between users and RIA interfaces. From this analysis, we propose a set of conceptual design primitives (Rich-IDM) to enable designers and web engineers to characterize the complex components of RIA models and take design decisions which meet both usability and communication requirements.

1 Introduction

Rich Internet Applications (RIAs) are an increasingly common family of interactive systems that is pervasive in many web domains, from e-commerce to branding sites, to social networking, and other Web 2.0 services. With respect to traditional web

A. Pandurino (✉) · L. Mainetti · R. Paiano
Faculty of Engineering, Department of Innovation Engineering, Graphics and Software Architectures Lab, University of Salento, Via per Monteroni, 73100Salento, LE, Italy
e-mail: andrea.pandurino@unisalento.it

L. Mainetti
e-mail: luca.mainetti@unisalento.it

R. Paiano
e-mail: roberto.paiano@unisalento.it

D. Bolchini
School of Informatics at IUPUI, User Simulation and Experience Research Lab, UIndiana University, 535 W. Michigan St, Indianapolis, IN46202, USA
e-mail: dbolchin@iupui.edu

M. Kudělka et al. (eds.), *Proceedings of the Third International Conference on Intelligent Human Computer Interaction (IHCI 2011), Prague, Czech Republic, August, 2011*, Advances in Intelligent Systems and Computing 179, DOI: 10.1007/978-3-642-31603-6_2,

applications, low-latency interactivity, widget-rich pages, and desktop-like navigation among screens are some of the most apparent features of RIAs that are affecting not only our way to use, but also our paradigms to conceive, design and engineer the future generations of web applications. It is possible to identify at least two broad categories of meta-requirements or high-level driving forces behind the evolution and articulation of RIAs: *communication effectiveness* and *usability* [16]. On the one hand, by communication effectiveness we mean the set of requirements concerned with the key messages that the application wants to convey to its target audience. In this perspective, it can be argued that the highly-engaging, interaction-rich, and dynamic user experience afforded by modern RIAs is a potentially great enabling factor to meet these requirements, by it unleashes designs aimed at inducing pleasure, fun and even surprise. On the other hand, usability is traditionally concerned with the degree to which the application can be used efficiently and effectively by the intended target audience to achieve the expected tasks in specific contexts of use (ISO 9241).

Ensuring that users can make sense of the application features, understand what they can do and easily identify, locate and manipulate the means to do it are fundamental concerns for a successful user experience [7]. Highly effective applications from the communication perspective can be catastrophic from a usability standpoint; and highly usable sites can be so dry, crude and minimal that miss to make a positive, long-lasting impression on the target audience. Looking at this tension through the theoretical lens of Situation Awareness [2], in this paper we will analyze phenomena of poor usability of RIAs, and we will indicate how these issues can be addressed by adopting a fine-grained systematic modeling of the user experience design. We advocate that the notion of *user interface plasticity*, originally introduced by [6] as "the capacity of user interfaces to adapt, or to be adapted, to the context of use while preserving usability", and proposed by other authors [17] as a criterion to evaluate "design decisions... in terms of their influence on UI ergonomic quality, using Ergonomic Criteria", must be effectively extended to the interaction flow between the user and the application. In RIAs, this calls for new or updated methods to engineer the system's *information architecture*, correctly balancing technical concerns (rich interface widgets), and communication concerns (related to UX requirements). To argument our ideas, case studies will be shown in order to practically explain five well-known SA demons [2]: Attention Tunneling, Misplaced Saliency, Errant Mental Model, Requisite Memory Trap, and Data Overload. These case studies will correspond to five existing RIA applications, spanning form e-commerce domain to social software. To address these issues, we propose a suite of simple conceptual design primitives that can enable designers to avoid these pitfalls on RIA pages and yet to engineer complex interactions leveraging RIA technologies. The remainder of the paper is organized as follows: Sect. 2 is a systematic introduction to the poor RIA design problems using the Situational Awareness theory; Sect. 3 reports on key related work in the area of RIA engineering approaches and User eXperience (UX) requirements design; Sect. 4 gives a brief introduction to our methodological approach in order to address these issues; in Sect. 5 the conclusions summarize our key messages.

2 Interface Deamons for RIA Design

The new features of RIAs tend to exhibit potential "demons" that can negatively affect the usability of the interaction, and that can be vividly characterized by following Endsley's general principles of "poor" Situational Awareness (SA) [2]. In the following, we present each demon providing an example derived by the analysis of real RIA websites.

Attentional Tunneling. At the micro-interaction level, design features that mix static content (such as text and image), multimedia (such as animation, etc.), and new, dynamic and reactive interface widgets can divert the user's attention from the intended, unifying message of the page.

This unexpected effect that is increasingly common in RIA interfaces, is well known as a fundamental problem of *Attentional Tunneling* in the theory of SA [2]. It is mainly caused by the expanding number of interaction rules that highlight specific interface elements over others at any given time. A clear example is visible in the design of Hogan site (Fig. 1), a well-known Italian shoe brand (www.hoganworld.com). The whole site content revolves around the variety of shoe products manufactured by Hogan. The site design is based on a real world metaphor: *the street*. This metaphor is used as the *main information channel* to provide access to the entire site's content. The main site section is a panoramic, interactive 360-degree photo of a street. On this photo, there are some hot-spots (represented by specific objects) that allow the user accessing specific features of the site. This type of *exploratory interface* is quite common in RIAs. By definition, an exploratory interface is not necessarily self-explanatory and immediately intuitive. Users need to explore the panoramic photo to "discover" hot-spots (they are not explicit from the outset), and to memorize their location: link labels suddenly appear as users accidentally mouse over the various parts of the map, and disappear as the users move away. This interaction also implies that users encounter many redundant links (many parts of the map lead to the same place), and overall it is very difficult to get a clear sense of how much is available and what are all the navigation options for the users. Attentional Tunneling comes into play as this page design emphasizes so much this exploratory interface that users risk to be *exclusively* drawn to it, thus neglecting the other parts of the page, which provide primary navigation options. The main navigation, in fact, is the "Navigon" link (reported on the left of Fig. 1) that is presented as a minor, hidden, and deemphasized option, which is communicated as equivalent in importance as secondary, "service" links, such as Contacts. Finally, once drawn to explore the map, there is no way for users to be contextually alerted of complementary (more traditional and comprehensive) navigation options, enabling them to move out of the exploratory channel. How to address this design issue to avoid usability problems and still leverage the communication potential of attractive interface components?

Misplaced Saliency. Moving from the level of the access structures (how to navigate to locate and reach contents) to the actual content base of the application, recurring Attentional Tunneling can easily bring users to misplace (or mis-locate) the "saliency" of the overall message of the page content. This problem can be

Fig. 1 www.hoganworld.com, homepage (Oct. 2010)

summarized by the implicit question of the user: What is the designer showing me now? What is the main intent of this page? A clear example of *Misplaced Saliency* is visible on the site of Gilette (www.gilette.com). The main goal of the Gilette site is to illustrate information about the shaving products. In detail, the example is focused on the page that shows the features of the razor "Gilette Fusion Power" (Fig. 2). In the middle of the page, the picture of the razor has several hot-spots that highlight each one a specific feature. It is clear that the designer needs to address the user attention on the razor (present in the third section) in order to encourage the user to click the hot spot. Instead, the second section, which provides the animated shape of the razor testimonial (3D avatar), results doubtless more salient for the user. Although, the animation through its movements tries to spur the user to access the razor, it remains overly prominent over the content itself. This dynamics is also compounded by the fact that the avatar suggests that he is going to do something, and this calls the user for waiting upon his actions. What is the relationship between this animated avatar and the content?

Errant Mental Model. At the general design level, users may have difficulty in capturing the underlying conceptual model of the designers—which should ideally match the user's mental model—thus giving birth to so called *Errant Mental Model* [2]. An example of this demon is visible in the homepage of the site Scrapblog (www.scrapblog.com). The homepage has the goal of providing a complete vision of the site offers and the access structures to them. The homepage (showed in Fig. 3) is dominated by the Sect. 2, which is composed by a set of pictures that alternate to the user (carousel), and that highlight the site features. When the page is showed, the user can act on two buttons placed in the second and third section, and labeled in the same way "start creating". The interface makes immediately clear the semantic relation between the two buttons because: (i) they have the same label; (ii) they have a similar look&feel aspect; and (iii) the button of the third section does not have its

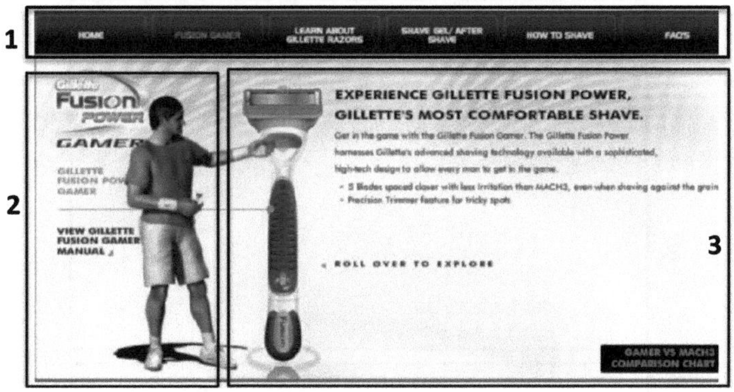

Fig. 2 www.gilette.com. Showcasing the features of the fusion power razor (Oct. 2010)

own description, but it use the explanation of the button of the second section. This relation defines a mental model that connects the two buttons through their goals, their look&feel aspects and their positions. In a first moment, the homepage seems well structured and does not present usability problems but, after few seconds, the carousel updates the central part of the page, showing a new picture (Fig. 4) that contains a new button in Sect. 2.

The change of the pictures is not compliant with the mental schema defined before by the user. The new couple of buttons, compared with the previous one, has the same graphical aspect, the same position, and label with similar meaning ("Start Creating" and "Get Started Now"). But the goals of the two buttons are different: "Get Started Now" allows accessing the specific site features for the kids, while "Start Creating" allows creating a generic product (as in the first version). This mismatch between the first and second versions of the page is an example of *Errant Mental Model* demon. The carousel makes a modification of the functionalities of the page elements, without a correct change of the context, such as a change of the position or of the look&feel. What is the perceived relationship between the several elements of the page? How can these relationships be modeled in order to delivery a correct message to the site visitor?

Requisite Memory Trap. The massive use of animations (such as sliding windows), that trigger continuous changes of the interface, greatly stresses user working memory by forcing users to recall their specific position in the local and global information architecture, the affordance of specific controls, and where they are located. This effect is traditionally called *Requisite Memory Trap* demon [2]. An example of this demon can be found in the homepage of the site of communication carrier Verizon. The homepage that we do not report for lack of space, shows the features of the offered services. The page content is not so dense, but the page is long and requires the use of the mouse scroll to be completely viewed. The page presents various mechanisms that allow users to change (hide/unhide) completely the provided information without page reloads: there are 6 areas but 17 several contents that can

Fig. 3 www.scrapbook.com. The view of the homepage when the user accesses the site (Oct. 2010)

Fig. 4 www.scrapblog.com. The second image of the carousel of the homepage of Scrapblog (Oct. 2010)

contain other dynamically showed elements. The information showed *one-shot* to the user is few if compared to the global quantity of information contained in the page. But, to have a complete schema of the information, the user could access all the hidden elements, which are mutually exclusive. This situation is directly related to the Required Memory Trap demon. What is the role of mutual interactions between the page elements in the information delivery? How can the designer check the user required cognitive workload to understand the showed information?

Data Overload. As related effect, the current trend to use RIA enhanced interface to compress the information in a single page view is prone to cause information overload in the dialogue with the users. This is known as *Data Overload* demon [2].

In the analysis of this demon, it is necessary to take in care that the concept of Data Overload in a web environment is related not only to the number of line of data presented to the user at the same time. This concept must be seen as the capability of the user to create a schema of the information compliant with his/her background. In a standard application (focused on specific information contents) the quantity of information puts this capability in difficulty while in a web application the user becomes in crisis because he/she has to manage many little pieces of heterogeneous information. In both cases, even if for different reasons, it is possible to recognize the Data Overload demon. An example of this demon is present in the homepage of the site of the stylists Dolce and Gabbana (www.Dolce&Gabbana.com). The site homepage contains a set of big pictures that present in a single view all the images about collections, restaurants, events, and shows that are not semantically connected. The effect for the user is an overload of heterogeneous and not focused contents. How can the designer manage heterogeneous information delivered to the user?

In this analysis of potential problems of RIA user experiences the emerging fundamental issue is that the chances that users get lost in this complex, interactive dialogue are very high.

3 Related Work

The research on RIA conceptual modeling has followed two main tracks: the first one is related to the modeling of data and business logic on the client-side, the second one is related to the modeling of enhanced user interface elements [12]. Many design methodologies that are helpful in the modeling of the traditional web application have been extended in order to model the new features of the RIA. OOHDM design framework proposes a design process in which the interface and its behavior are explicitly analyzed and examined [18]. The Rich Internet application User eXperience model [15] (RUX-model) is a significant methodological advance in RIA interface model. RUX-model through its abstract interface level describes the aspects of the interface common to all the RIA technologies. WebML [4] adopts all the RUX-Model level related to the interface aspects while UWE (UML-based Web Engineering [11]) extends its connection rules between the used presentation model and the RUX abstract interface meta-model. The approach offered by OOH4RIA [13] is based on a model-driven process that specifies the artifacts to model a complete RIA for a Google Web Toolkit (GWT) [9].

Besides these approaches, there is a research area that is focused on the specific user interface features. Model Driven Engineering (MDE) approach is been pushed [6] to the limit in order to reconcile high-level modeling techniques with low-level programming to model the advanced features of modern application. Moreover, many studies are exploring new issues: (i) the needs to have a systematic approach for characterizing UI tuning in terms of quality in use along context of use variations [17]; (ii) how human-computer interaction can be based on discourse modeling, even without employing speech or natural language [10]. The system AUGUR [8] provides context-aware interaction support for navigating and entering data in arbitrary form-based web applications. Formal methods are widely used to analyze human-computer interaction in order to prevent the gap between the interface and the user mental model [5]. Furthermore, the formal models of user interface designs are used as the basis for designing software evaluation studies [3].

According to the need of evaluating the interface in the early phase of the design and to support the designer to project correctly the communication aspects, we start our approach analyzing the interface of RIA through a set of demons of Situation Awareness [2]. Whereas all these approaches provide support for bridging the gap between the interface features of RIAs and methodologies for conceptual and technological development, they lack in characterizing the fluid, smooth and organic nature of the user interaction and navigation in RIA. To meet the challenge to model all the RIA aspects, we propose to extend this perspective to examine the connection between RIAs interface modeling (well connected to an underlying conceptual structure) and the requirements for the user experience.

4 The Rich-IDM Primitives

The new features of the RIA have changed radically the user experience. Thus, many design methodologies have requested specific extensions to manage correctly the new RIA features: webML [4] and UWE [11] are been extended through RUX [15]. These approaches formalize the technological aspects of RIA but they do not consider enough (in our opinion) the evaluation of changes in the user experience and how it evolves.

The importance of the user experience analysis has been already described through the *interface demons* caused by the incorrect use of the augmented RIA interface. Thus, it is clear that it is really important to have suitable methodological tools in which the single primitive has a well defined communicative semantics and allows to design the fluid, smooth and organic nature of interaction and navigation paradigms of RIA. On the basis of these needs, in the following, we present a methodological approach called Rich-IDM [14] based on Interactive Dialogue Model [1]. Interactive Dialogue Model is a dialogue-based design technique for model the communicative structure of information-intensive interactive applications. At first, we present briefly all the primitives of Rich-IDM considering their notation, semantics and user experience aspects (see Table 1). Then, for each primitive we describe its semantic and its specific features. In this analysis, we presume that the reader knows the basic concepts of IDM (available at [1]).

The RIA Page Element. In Rich-IDM, the minimum piece of information is called RIA-Page Element and it is managed as a unique block. In detail, the RIA-Page Element is defined as coherent atomic fragment of RIA page which displays a specific content, mapped to a dialogue act belonging to the IDM logical design. It could be specialized in: (i) Introductory RIA-page Element that has the main goal to introduce the specific content and, often, is related to an introductory dialogue act;

Table 1 Rich-IDM modeling primitives at a glance

Name	Notation	Design semantics	User experience concern
Content RIA-Page Element		A coherent, atomic fragment of RIA page which displays a content unit, as directly mapped from IDM content dialogue acts	Perceive the "rheme" (novel message) of the interactive communication flow
Introductory RIA-Page Element		A fragment of a RIA page which displays mechanisms to enable access to multiple instances of a topic, as directly mapped to a introductory dialogue act of the L-IDM design	Grasp the coherence of a group of topics and be able to start exploring them
Transition RIA-Page Element		The reification of an IDM transition dialogue act on the RIA page. It allows users to follow the semantic relation of two topics	Explore analog topic through associative, lateral paths
User experience core	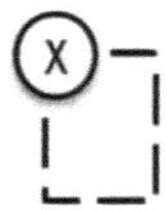	A connected composition of page elements, which communicates the semantic nucleus of what is offered to the user at a given moment	Focus the attention on the primary, central message of the page
RIA-Handle		An interaction affordance, which enables users moving within two or more page elements of the same user experience core	Activate micro-interactivity on content elements without leaving the context of the core of the page
Context view		A set of user experience cores that maintains navigational context, orientation, organic, and fluid transition between the cores	Keep peripheral attention on the secondary, framing elements of the page
Default element		Indicate the default RIA-Page Elements	–

(ii) the Content RIA-page Element that, mapping (one or more) content dialogue act, displays to the user the payload of the dialogue; and (iii) Transition RIA-page Element that shows the semantic information that links two topics. The relation of these primitives with the model of the information (defined in IDM) allows establishing a direct link with the RIA requirements. Thus, it is possible to evaluate in the early phase of the design the balance between the different types of delivery contents. For example, it is clear that the excess in a single page of Introductory RIA-Page Elements (that are used to access to heterogeneous contents) creates some difficulty for the user that has serious problem to understand the page structure and to establish the correct actions to do. This causes the "Data Overload" demon that could be prevented using Rich-IDM. Relating the primitives to their semantic, Rich-IDM allows balancing the type of the contents and to group semantically the several introductory acts the user can alternatively view.

RIA-Handle. RIAs propose to the user an interaction mechanism that is different from a classical web application. Thus, after the definition of the RIA-Page Elements, that are the interaction objects, it is necessary to define a new primitive that is able to model the mutual relation between the elements of the page. This new concept is the RIA-Handle, which main goal is to model all the dynamic aspects of the UX. The RIA-Handle is a directional relationship between the RIA-Page Elements involved in a user action. Its main focus is not on how the specific technologies manage the events, but on how the user can navigate with the application activating the transition to the target. The RIA-Handle captures the syntactic of RIA interaction, on top of the semantics modeled by the other elements. From the methodological point of view, the RIA-Handle allows designers to represent all the relations contained in information architecture.

A carefully use of the RIA-Handle helps the designer to manage correctly several interface demons. In particular, the number of the RIA-Handle which target is the same User Experience Core (see next paragraph) of the source can express a measure of the interface objects that change and modify the UX (such as in the "Requisite Memory Trap" demon); in fact, each handle acts on an object (the target) and could have effects also on the source (that, for example, could be hidden). Furthermore, the several kinds of the events (such as MouseClick, MouseOver) can measure the number of the possible kind of interactions managed by the interface, and thus it measures the elements of the dialogue. A great number of event types increase the complexity perceived by the user, who has to learn how to compose these events to reach a specific content or goal.

User Experience Core. In a standard web application, the user attention is focused on "a page". On the contrary, in RIAs many information elements (that could not be semantically directly connected) can be collapsed in the same page. Thus, the core of the dialogue is not directly related with the displayed elements. To resolve this need, a completely new concept called User Experience Core is defined. Its main goal is to model clearly the elements of the Rich-IDM design that must be the heart of the dialogue with the users. The designer must manage the elements contained in the User Experience Core carefully because their perception affects strongly the sense and the quality of the message delivered to the user. On this basis, the semantics of

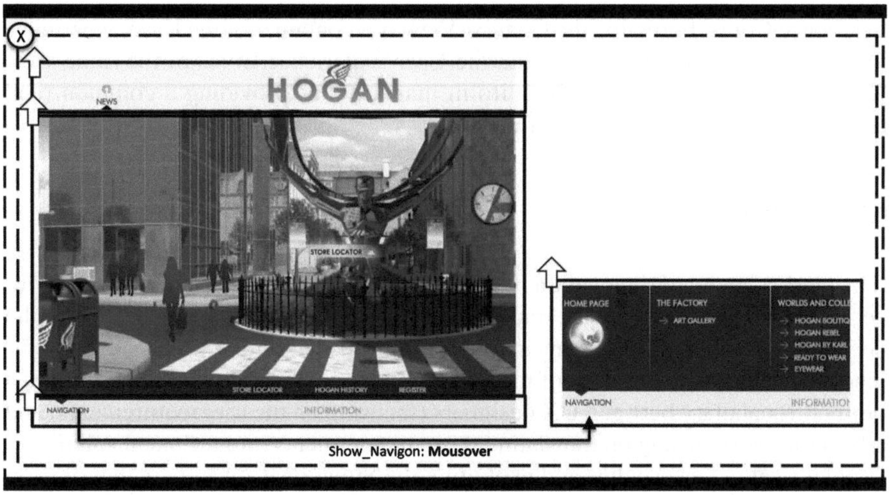

Fig. 5 The Rich-IDM model of the homepage of www.hoganworld.com (Oct. 2010)

the User Experience Core is to define the unit of perception of the dialogue. Formally, the User Experience Core is a container of the RIA-page Elements. At the start of the navigation, the default RIA-page Elements showed to the user is marked with the Default Element described in Table 1. The RIA-Handle mechanism is used to model the navigation between the User Experience Cores independently if they are (or not) part of the Context View (see the next paragraph). Many RIA interface demons are mitigate by a correct use of the User Experience Core. In fact, we have to consider that the User Experience Core is strictly focused on a specific subject of the dialogue, thus creating a semantic unit directly connected to the IDM model. From this point of view, the designer and the communication expert could cooperate to include in the User Experience Core all the contents needed to establish the dialogue with the user. For example (Fig. 5) we report the design of the homepage of the site of Hogan (already analyzed above). This model contains three Introductory RIA-Page elements that have the same semantic function to allow the access to the site features. But, the designer, probably to improve the attractive aspects, gives more emphasizes to the central section (where there is the panoramic images), while the other two sections are placed in a reduced space and are unhidden through the RIA-Handle "show_navigon". These graphical differences are in contrast with the semantics of the elements and cause the "Attentional Tunneling" demon that could be evaluated throught Rich-IDM methodology.

Context View. The look&feel of the RIA is more important and often the visualization aspects are used to define specific area of the application and delimitates related arguments. To capture these very important characteristics for the quality of the dialogue flow, designers need to define the User Experience Cores that must be shown in the same way to the user. The Context View allows to the designers to define a specific navigational context allowing harmoniously connecting related User

Experience Cores. Formally, the Context View is a container of the User Experience Core. Its name is due to the idea that the User Experience Cores of the same view context are shown to the user in uniform manner, thus providing a common (and stable) visualization environment to the user.

5 Conclusion

RIAs mark a landmark innovation in the way web engineering can unleash superior user interactions. To balance the trade-off between the potential sophistication of the user interface and the need to ensure proper usability, cognitive workload, and efficiency, we have proposed a set of high-level modeling constructs, which provides a common vocabulary between UX designers (focusing on the user requirements) and web engineers (focusing on translating those requirements into technical solutions). Through a set of case studies and exploiting the Situational Awareness theory in an original way, we have shown the potential of our approach in pointing out one direction to integrate a user interaction and usability perspective in the modeling of RIAs. In particular, our contribution is a new point of view in RIA presentation design, new are some Rich-IDM derived concepts if compared to related works, and novel is the cross-disciplinary application of the SA principles. In detail, we propose new primitives with strong semantics: the User Experience Core describes the communication semantic nucleus (offered to the user in a given moment) connecting the page elements; the Content RIA-Page Elements, Introductory RIA-Page Elements and the Transition RIA-Page Elements inherit their semantic from Interactive Dialog Model and allow the user to access the RIA information; the RIA-Handle enables users moving within two or more page elements allowing to model the dynamic aspects of the RIAs; and the Context View maintains navigational context, orientation, organic, and fluid transition between the User Experience Core. Moreover, through case studies, we demonstrated how these primitives could help designers avoiding poor Situational Awareness effects, as characterized by Endsley's principles. In short, we proposed a simple way to fill the gap between hypermedia design and user experience design for RIA, which is an open issue both from the web engineering point of view and the human computer interaction one.

As future research, we plan to define model metrics in order to provide UX designers and web engineers with means to derive usability measurements directly form Rich-IDM diagrams, so supporting them to discover poor usability situations.

References

1. Bolchini, D., Paolini, P.: Interactive dialogue model: a design technique for multi-channel applications. IEEE Trans. Multimed. **8**(3), 529–541 (2006)
2. Bolstad, C.A., Costello, A.M., Endsley, M.R.: Bad situation awareness designs: what went wrong and why. In: Proceedings of the 16th World Congress of the International Ergonomics Association, Netherlands (2006)

3. Bowen, J., Reeves, S.: Developing usability studies via formal models of UIs. In: Proceedings of the 2nd ACM SIGCHI Symposium on Engineering Interactive Computing Systems EICS'10, pp. 175–180. ACM, New York (2010)
4. Bozzon, A., Comai, S., Fraternali, P., Carughi, G.T.: Capturing RIA concepts in a web modeling language. In: Proceedings of the 15th International Conference on World Wide Web WWW'06, pp. 907–908. ACM, New York (2006)
5. Combefis, S., Pecheur, C.: A bisimulation-based approach to the analysis of human-computer interaction. In: Proceedings of the 1st ACM SIGCHI Symposium on Engineering interactive computing systems EICS'09, pp. 101–110. ACM, New York (2009)
6. Coutaz, J.: User interface plasticity: model driven engineering to the limit!. In: Proceedings of the 2nd ACM SIGCHI Symposium on Engineering Interactive Computing Systems EICS'10, pp. 1–8. ACM, New York (2010)
7. Durrani, S., Durrani, Q.S.: Applying cognitive psychology to user interfaces. In: Proceedings of the 1st International Conference on Intelligent Human Computer Interaction, pp. 156–168. Springer, India (2009)
8. Hartmann, M., Daniel, S., Muhlhauser, M.: AUGUR: providing context-aware inter-action support. In: Proceedings of the 1st ACM SIGCHI symposium on Engineering Interactive computing systems EICS'09, pp. 123–132. ACM, New York (2009)
9. Houben, G.J., Van der Sluijs, K., Barna, P., et al.: Chapter 10. Web Engineering: Modelling and Implementing Web Applications, HCI Series. Springer-Verlag, ISBN: 978-1846289224 (2007)
10. Kaindl, H.: Model a discourse and transform it to your user interface. In: Proceedings of the 2nd ACM SIGCHI Symposium on Engineering Interactive Computing Systems EICS'10, pp. 371–372. ACM, New York (2010)
11. Koch, N., Knapp, A., Zhang, G., et al.: UMLBased Web engineering: an approach based on standards. Web Engineering: Modelling and Implementing Web Applications, HCI Se-ries, vol. 12, No. 7, pp. 157–191. Springer-Verlag (2007)
12. Linaje, M., Sanchez-Figueroa, Preciado: Engineering rich Internet application user interfaces over legacy Web models. IEEE Internet Comput. **11**(6), 1089–7801 (2007)
13. Meliá, S., Gómez, J., Pérez, S., Diaz, O.: A model-driven development for GWT-based rich Internet applications with OOH4RIA. In: Proceeidngs of IEEE ICWE'08, pp. 13–23. New York (2008)
14. Pandurino, A., Bolchini, D., Mainetti, L., Paiano, R.: Rich-IDM: extending IDM to model rich Internet applications. In: Proceedings of 12th ACM International Conference on Information Integration and Web Based Applications and Services iiWAS2010, pp. 145–152 (2010)
15. Preciado, J.C., Linaje, M., Morales-Chaparro, R., et al.: Designing rich Internet applications combining UWE and RUX-method. In: Proceedings of IEEE ICWE'08, New York (2008)
16. Spagnolo, L., Bolchini, D., Paolini, P.: Beyond findability: search-enhanced information architecture for content-intensive RIAs. J. Inf. Archit. **2**(1), 19–36 (2010)
17. Serna, A., Calvary, G., Scapin, D.L.: How assessing plasticity design choices can im-prove UI quality: a case study. In: Proceedings of the 2nd ACM SIGCHI Symposium on Engineering interactive Computing Systems EICS'10, pp. 29–34. ACM, New York (2010)
18. Urbieta, M., Rossi, G., Ginzburg, J., Schwabe, D.: Designing the interface of rich Internet applications. In: Proceedings of the 2007 Latin American Web Conference, IEEE Computer Society, pp. 144–153. Washington, DC (2007)

Describing and Assessing Image Descriptions for Visually Impaired Web Users with IDAT

Julius T. Nganji, Mike Brayshaw and Brian Tompsett

Abstract People with visual impairments, particularly blind people face alot of difficulties browsing the web with assistive technologies such as screen readers, when websites do not conform to accessibility standards and are thus inaccessible. HTML is the basic language for website design but its ALT attribute on the IMG element does not adequately capture comprehensive image semantics and description in a way that can be accurately interpreted by screen readers, hence blind people do not usually get the complete description of the image. Most of the problems however arise from web designers and developers not including a description of an image or not comprehensively describing these images to people with visual impairments. In this paper, we propose the use of the Image Description Assessment Tool (IDAT), a Java-based tool containing some proposed heuristics for assessing how well an image description matches the real content of the image on the web. The tool also contains a speech interface which can enable a visually impaired individual to listen to the description of an image that has been uploaded unto the system.

J. T. Nganji (✉)
Distributed Reliable Intelligent Systems (DRIS) Lab,
Department of Computer Science,
University of Hull, Cottingham Road,
HullHU6 7RX, United Kingdom
e-mail: J.Nganji@hull.ac.uk

M. Brayshaw · B. Tompsett
Department of Computer Science, University of Hull,
Cottingham Road,
HullHU6 7RX, United Kingdom
e-mail: M.Brayshaw@hull.ac.uk

B. Tompsett
e-mail: B.C.Tompsett@hull.ac.uk

M. Kudělka et al. (eds.), *Proceedings of the Third International Conference on Intelligent Human Computer Interaction (IHCI 2011), Prague, Czech Republic, August, 2011*, Advances in Intelligent Systems and Computing 179, DOI: 10.1007/978-3-642-31603-6_3, © Springer-Verlag Berlin Heidelberg 2013

1 Introduction

The primary language for authoring web pages is the Hyper Text Markup Language (HTML). HTML is used to position resources, including images on web pages as well as provide a description of some media. Image description in HTML is achieved through the ALT and LONGDESC attributes of the IMG element. Whilst ALT only enables a short description of the image within the <IMG> tag as seen in Listing 1,

Listing 1 A sample <img> element with an ALT attribute

```
<p>
<img src = "images/man.jpg" alt = "a man in a corn field"/>
</p>
```

LONGDESC could be used for longer descriptions of the image within the <IMG>tag as represented in Listing 2 or could be used to specify the location of a description of the image, in the form of a URI as shown in Listing 3. However, it is not supported by some web browsers and some assistive technologies [1].

Listing 2 A sample <img> element with a longdesc attribute

```
<p>
<img src = "images/man.jpg" alt = "a man in a corn field"
longdesc = "imagedescription.html"/>
</p>
```

Assistive technologies such as screen readers interact with web pages to read the content to people with visual impairments. Nevertheless, most websites do not provide adequate text alternatives to images [1] hence images are not well interpreted to blind people.

Listing 3 A sample <img> element with a d link

```
<p>
<img src = "images/man.jpg" alt = "a man in a corn field"
longdesc = "imagedescription.html"/>[a href = "imagedescrip
tion.html" title = "a detailed description of the image"/>
D</a>
</p>
```

The Web Content Accessibility Guidelines (WCAG) 2.0,[1] a W3C recommendation of 11 December 2008 advises web content creators to provide text alternatives for any non-text content and this is also reiterated by [2]. According to WCAG 2.0 guidelines, *the objective of this technique is to create a text alternative that serves the same purpose and presents the same information as the original non-text content. As a result, it is possible to remove the non-text content and replace it with the text alternative and no functionality or information would be lost. This text alternative should*

[1] http://www.w3.org/TR/WCAG20/

not necessarily describe the non-text content. It should serve the same purpose and convey the same information. This may sometimes result in a text alternative that looks like a description of the non-text content. But this would only be true if that was the best way to serve the same purpose.

For screen readers to convey the correct information about an image to a visually impaired person, image description needs to be detail enough to provide a complete and true description. For instance, the image may have been taken in a specific location, during a specific season and may contain a number of objects (people, animals, etc.) of different sex (male, female) who may be displaying different emotions (smiling, frowning, etc.). All these in addition to other features of the image are things that are seen and appreciated by people with normal eye sight but which are hidden from visually impaired users.

The semantic web [3] is more meaningful and can enable machines to understand information and to communicate with each other in ways that have not been possible before. An ontology is "a specification of a representational vocabulary for a shared domain of discourse" [4]. Ontologies can facilitate semantic image description which can ease understanding of images on the web. Even with that, the ontology needs to contain comprehensive information about the image.

This paper therefore seeks to encourage web developers to present accurate and comprehensive image descriptions for visually impaired people by providing ten heuristics that will help in accomplishing this. The heuristics could be used as classes in an ontology to describe the image. The remainder of this paper is as follows: Sect. 2 looks at the problems associated with an inaccessible web for people with visual impairments, Sect. 3 reviews image description while Sect. 4 explains the considerations for a good image description. Section 5 on the other hand provides a tool for assessing image descriptions. Section 6 presents some limitations and future work. The paper concludes in Sect. 7 with a summary.

2 The Difficulties of Describing Images on the Web

For most people with normal eyesight, browsing the web is a painless task because most websites have been designed for people with normal vision. They can readily click on objects and follow links to other resources, even for websites that do not conform to accessibility and usability norms. This is, however, not the case for people with visual impairments, particularly those with congenital blindness who may have to rely on assistive technologies such as screen readers to browse through websites. Whilst the use of screen readers is helpful in browsing web content, they read out web content sequentially, and so could take a long time to reach the information required by the user thus [5] developed a multimodal interface to improve web accessibility for visually impaired people. When web designers do not adhere to web accessibility guidelines, visually impaired people browsing the web with screen readers will be excluded from accessing information since the assistive technology will be unable to interpret the content to the user [6]. Chiang et al. [7] also noted some problems people with visual impairments face when browsing the web, such as problems associated with linking from one document to another given that they cannot follow visual cues, and the increasing use of multimedia and interactive documents which are difficult for such people to access.

Providing an inaccessible website means discriminating against disabled people by denying them access to online services. Most governments today have legislation to prevent discrimination of disabled people, with prosecutions for those who do not adhere to such legislations. The Americans with Disabilities Act[2] (ADA) of 1990 requires employers to provide "reasonable accommodation" and mechanisms for "effective communication" to workers with disabilities [7] while in the UK, the Disability Discrimination Act[3] (DDA) of 1995 and 2005 requires service providers to make "reasonable adjustments" to their services to meet the needs of disabled people.

We noted earlier that screen readers are unable to describe an image to a user if no description is provided. However, in the event where a description is provided but is insufficient or wrong, the visually impaired user will still be unable to understand the image. This calls for good practice in image description and employment of alternative technology to improve understanding of the image. Consequently, the problems associated with image description by assistive technologies need to be tackled with cutting edge technology, which we will explore in Sect. 5.

3 Review of Image Description

Much work has been done on image analysis with focus on retrieval. In the mid-1980s, Sara Shatford [8] extended Erwin Panofsky's [9] *pre-iconographical, iconographical* and *iconological* model of describing an art work. Shatford further subdivided the three levels into *who, what, when* and *where*. Recent research has shown the lack of adequate image description [1] hence, [10] created a web mediator that automatically adds ALT tags based on an analysis of the image contents. Hollink et al. [11] developed a classification framework for classification of image descriptions by users and found that users preferred general descriptions rather than specific or abstract descriptions. Most research on media semantics has focused on semantically annotating media to facilitate search and retrieval e.g. [12, 13]. Nevertheless, there is a lack of research towards adequately describing images for assistive technologies to interpret to people with visual impairments.

4 Comprehensive Image Description

With the advent of the semantic web, ontologies have been employed in image annotation. The Resource Description Framework (RDF) [14] or the Web Ontology Language (OWL) [15] can be used to convey a meaningful description of an

---

2 http://www.ada.gov/

3 http://www.dwp.gov.uk/employer/disability-discrimination-act/

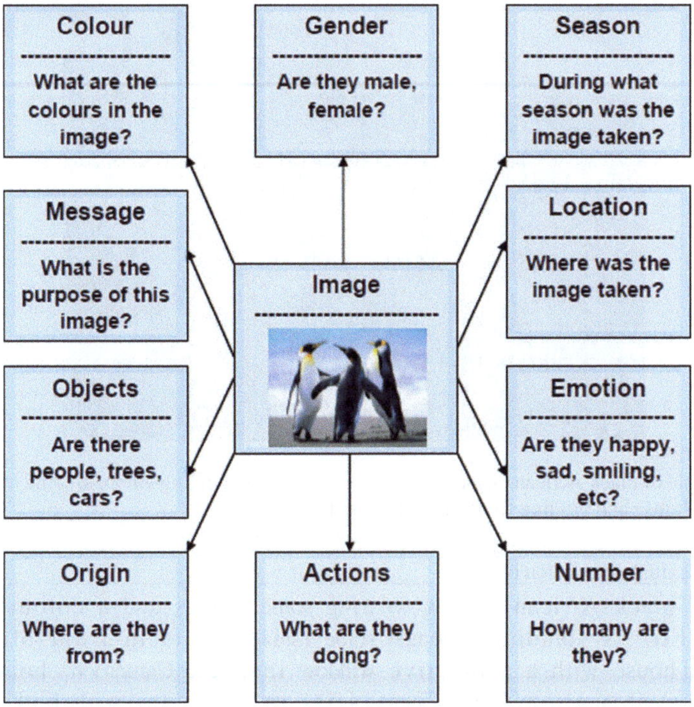

Fig. 1 Comprehensive image description heuristics

image because of their semantics. However, to comprehensively describe images for assistive technology interpretation, the image description needs to take into consideration some important heuristics that capture the full meaning and description of the image as represented in Fig. 1. This follows a *Who, What, When, Where* and *How* approach. For instance, *who* asks the questions relating to the people in the image, while *what* relates to other non-human objects including buildings, trees, automobile, etc. including their descriptions such as colour. *When* on the other hand asks questions related to time such as when the picture was taken (time, season, etc.) while *where* seeks to find out the location such as where the image was taken, the positions of various objects in the image, etc. *How* relates to actions, emotions, etc.

These heuristics could be used as classes in an ontology to describe images. An image's description can be represented on three levels: text, structured, and ontology-based structured similar to [16]. Considering Fig. 2, the text description could be as in Listing 4. The textual description serves to provide a description of the image in natural language, which could be understood by humans.

Fig. 2 A male of black African origin wearing some lenses is standing amongst some maize plants in front of a house and holding a camera with both hands

Listing 4 Textual description of Fig. 2

```
A male of black African origin wearing some lenses and a colourful
African shirt is standing amongst some maize plants in front of a non
plastered house with a protective window in Buea, Cameroon, holding
and staring at a green maize fruit still on the plant with both hands
```

Although the above description could be well understood by humans, machines might not readily understand this, causing assistive technology interpretation to be a difficult task. Semantic web technologies can overcome the limitations of information retrieval using keyword-based methods [17] as it makes information more meaningful to people by making it more meaningful to machines [18]. Similarly, web ontologies could be employed to overcome the challenges of describing images with HTML. Nevertheless, that is not the focus of this paper which seeks more to use a tool to encourage comprehensive image descriptions and to assess the accuracy of the descriptions.

5 Image Description Assessment

Tools have been developed to evaluate various contents of websites. The Healthcare Website Assessment Tool (HWAT) is one of such tools used to evaluate the quality of osteoporosis websites [19]. Similarly, to evaluate image descriptions, we propose the Image Description Assessment Tool (IDAT) which adopts ten heuristics earlier described with various quality indicators and weighting scores as represented in Table 1.

Each heuristic has a weighting score of 10 and the extent to which the image description conforms to that heuristic is thus scored on a maximum of 10. The total

Table 1 Image description assessment tool

Heuristic	Quality indicator	Maximum weighting score
Location	Where was the image taken? Does the description of the image adequately convey this information?	10
Objects	What are the objects in the image (people, trees, cars, buildings, etc.)? How well does the image description convey this information?	10
Gender	What is the gender of the people and the animals in the image? Are they male or female and does the description convey this message?	10
Colours	What are the various colours in the image? Do tree leaves colours show it is autumn, spring, summer, winter, or is it rainy or dry season and does the image description adequately convey this?	10
Emotions	What emotions (smiling, frowning, etc.) are expressed by people, animals, etc. in the image and does the image description adequately convey this information?	10
Numbers	How many of those objects are in the image and does the image description adequately convey this information?	10
Message	What is the image all about? How well does the image description adequately convey this message?	10
Origin	What is the origin of the people, animals, etc. in the image and does the image description convey this information?	10
Action	What is the state of motion of the objects in the image? Are they moving, still, etc. and are these well represented in the description?	10
Season	During what season was this image taken and if applicable, is this represented in the description?	10

Table 2 Image description with IDAT

Heuristic	Present or absent (1 or 0)	Weighting score (/ 10)	Maximum weighting score
Location	0	–	–
Objects	1	10	10
Gender	1	10	10
Colours	1	0	10
Emotions	0	–	–
Numbers	1	10	10
Message	1	5	10
Origin	0	–	–
Action	1	8	10
Season	0	–	–
Total		43	60

weighting score will thus be 100 for an image that contains a scenario with ten heuristics. People describe images differently and in their interpretation of images have various qualities or heuristics. It is therefore imperative to make provisions for this in calculating the accuracy of the description as in the following formula.

Let x_i be the weighting score on 10 for an individual heuristic and n be the total maximum weighting score.

Then the percentage accuracy (P) of the image description to the original image could be derived as follows:

$$P = \sum xi/n \times 100 \tag{1}$$

Let us consider the following description (see Fig. 2 caption) of an image which was taken during the rainy season in Buea, a small town in Cameroon.

To assess how close the description is to the original image, IDAT can be applied as in Table 1. When a heuristic is present in the description, 1 is recorded and a weighting score given on a scale of 10. Where a heuristic is absent in the description but the image clearly shows the presence of that heuristic, 1 is recorded and a score of 0 is given. On the other hand, where a heuristic is absent in the description and the image clearly shows that the heuristic cannot be applied to the description, 0 is recorded and no score is attributed. Based on these conditions, we can now assess the description of Fig. 2 as shown in Table 2.

Where the image was taken cannot be determined by looking at the image, hence no weighting score is given. The season has not been mentioned. The maize plant requires rainfall to grow during the rainy season but it is also possible to be grown during the dry season if it is well watered. That description scores zero for colours as the green colour of the plant and other dominant colours in the image have not been mentioned. The message is not 100 % accurate because the man in the image is not holding a camera. It follows from the equation that the percentage accuracy of

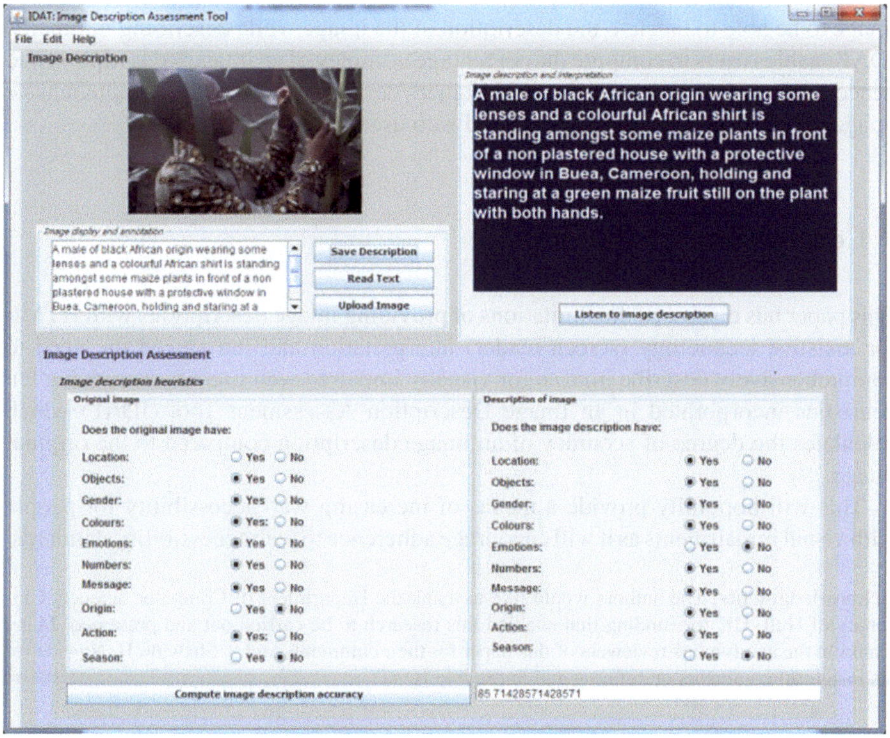

Fig. 3 IDAT showing some features including entering image description, listening to the description and calculating the image description accuracy

the description is

$$P = 43/60 \times 100 = 71.67 \tag{2}$$

The accuracy of the description is thus 71.67 %.
A screenshot of IDAT is seen in Fig. 3.

6 Limitations and Future Work

In assessing image description with IDAT, the tool relies on the assessor's ability to interpret the image, which in itself may be faulty. The need for a tool to automate the process has been raised and further research will investigate this.

IDAT implementation is nearly completed and once the process is over, user testing will be done and the results used to improve the tool. This system allows users to upload any image and to provide a description of the image with a text to speech output of the description as depicted in Fig. 3. It also allows the person describing the image to be able to listen to the description of the image while describing it. Finally, IDAT enables users to compute the percentage accuracy of an image's description and hence can be used to encourage web designers/developers to provide comprehensive image descriptions for visually impaired web users.

7 Conclusion

This paper has discussed the limitations of providing image descriptions with HTML for assistive technology (screen reader) interpretation and introduced the need to comprehensively describe images for visually impaired web users by proposing ten heuristics incorporated in an Image Description Assessment Tool (IDAT) which calculates the degree of accuracy of an image description compared to the original image.

This will hopefully provide a means of increasing web accessibility for people with visual impairments as it will encourage adherence to web accessibility standards.

Acknowledgments The authors would like to thank the Department of Computer Science, University of Hull, UK for funding that enabled this research to be carried out and presented. Many thanks to the anonymous reviewers of this paper for their comments and to Shawulu H. Nggada for his insightful comments on refining and improving IDAT.

References

1. Petrie, H., Harrison, C., Dev, S.: Describing images on the web: a survey of current practice and prospects for the future. In: Universal Access in HCI: Exploring New Dimensions of Diversity, Vol. 8, Proceedings of the 3rd International Conference on Universal Access in Human-Computer Interaction, 22–27 July 2005, Las Vegas, Nevada). New Jersey: Lawrence Erlbaum Associates, (2005)
2. Baguma, R., Lubega, J.T.: Web design requirements for improved web accessibility for the blind. In: Fong, J., Kwan, R, Wang, F.L. (eds.) Hybrid Learning and Education. Lect. Notes in Comput. Sci., pp. 392–403. Springer, Heidelberg (2008)
3. Berners-Lee, T., Hendler, J. and Lassila, O.: The Semantic Web, Scientific American, pp. 35–43 (2001)
4. Gruber, T.R.: A translation approach to portable ontology specifications. Knowledge Acquisition **5**(2), 199–220 (1993)
5. Yu, W., Kuber, R., Murphy, E., Strain, P., McAllister, G.: A novel multimodal interface for improving visually impaired people's web accessibility. Virtual Reality **9**, 133–148 (2006)
6. Shi, Y.: The accessibility of queensland visitor information centre's websites. Tour. Manag. **27**(2), 829–841 (2006)
7. Chiang, M.F., Cole, R.G., Gupta, S., Kaiser, G.E., Starren, J.B.: Computer and world wide web accessibility by visually disabled patients: Problems and Solutions. Surv. of Ophthalmol., vol. 50, no.4, (2005)
8. Shatford, S.: Analyzing the subject of a picture: a theoretical approach. Cataloging & Classif. q. **6**(3), 39–62 (1986)
9. Panofsky, E.: Studies in Iconology: Humanistic themes in the art of the renaissance, pp. 5–9. Harper & Row, New York (1972)

10. Keysers, D., Renn, M., Breuel, T.M.: Improving accessibility of HTML documents by generating image-tags in a proxy. In: Proceedings of the Ninth International ACM SIGACCESS Conference on Computers and Accessibility, pp. 249–250 (2007)
11. Hollink, L., Schreiber, G., Wielinga, B., Worring, M.: Classification of user image descriptions. Int. J. Hum. Comput. Stud. **61**(5), 501–626 (2004)
12. Yang, H.C., Lee, C.H.: Image semantics discovery from web pages for semantic-based image retrieval using self-organizing maps. Expert Syst. Appl. **34**(1), 266–279 (2008)
13. Elahi, N., Karlsen, R., Akselsen, S.: A context centric approach for semantic image annotation and retrieval, computationworld. Computation world: Future computing, service computation, cognitive, adaptive, content, patterns, pp. 66–668 (2009)
14. Lassila, O., Swick, R.: Resource description framework model and syntax specification. World Wide Web Consortium (1999). http://www.w3.org/TR/1999/REC-rdf-syntax-19990222. Accessed 10 Mar 2011
15. McGuinness, D.L., Van Harmelen, F.: Web Ontology Language Overview. World Wide Web Consortium (2004). http://www.w3.org/TR/owl-features/ Accessed 10 Mar 2011
16. Ruotsalo, T.: Methods and Applications for Ontology-Based Recommender Systems. Aalto University, Doctoral Diss (2010)
17. Castells, P., Fernández, M., Vallet, D.: An adaptation of the vector-space model for ontology-based information retrieval. IEEE Trans. Knowl. Data Eng. **19**(2), 261–272 (2007)
18. Ohler, J.: The Semantic web in education. Educause Q. **31**(4), 7–9 (2008)
19. Lewiecki, E.M., Rudolph, L.: AKiebzak, G. M., Chavez, J. R., Thorpe, B. M.: Assessment of osteoporosis-website quality. Osteoporos Int **17**, 741–752 (2006)

Image Search: A Story of One User Interface

Šárka Zehnalová, Zdeněk Horák and Miloš Kudělka

Abstract With the rapid development of information technology, the emphasis on the quality of user interfaces has been increasing recently, also with regard to mobile platforms, accessibility etc. In this paper we focus on engaging more interactive ways to image search. While observing and discussing with users about how they wish to proceed during search of images we detected four typical scenarios. We present all of them on concrete examples. We also describe what kind of image features our system works with and how we detect them. We introduce our own user interface of Xingle testing system where are all the mentioned scenarios implemented. The Xingle system works with about half a million images that were collected for testing purposes.

1 Introduction

Orientation in large sets of images is one of the challenges in the field of search technology. There are several interconnected issues. One of them is the search for semantics in images. The currently prevailing approach is based on the use of tags. There are support systems available that help users with image tagging by proposing tags that are as accurate as possible and appropriate for the search. Internet search engines on the other hand can obtain semantic information from the analysis of the context in which the images are located on the site. The second area is the

Š. Zehnalová (✉) · Z. Horák · M. Kudělka
FEI, VSB-Technical University of Ostrava, 17. listopadu 15,
708 33Ostrava-Poruba, Czech Republic
e-mail: zehnalova@inflex.cz

Z. Horák
e-mail: zdenek.horak@vsb.cz

M. Kudělka
e-mail: miloš.kudělka@vsb.cz

M. Kudělka et al. (eds.), *Proceedings of the Third International Conference on Intelligent Human Computer Interaction (IHCI 2011), Prague, Czech Republic, August, 2011*, Advances in Intelligent Systems and Computing 179, DOI: 10.1007/978-3-642-31603-6_4, © Springer-Verlag Berlin Heidelberg 2013

automated analysis of the visual features of the images. Detected features are useful for examination of the similarity of images or for the search in stock photo sites. General image search typically utilizes both approaches, the semantics based on tags is combined with the automatic features detection. Image retrieval techniques for content-based search and similarity search have been described and used in many projects. For example MUFIN project focuses on general similarity demonstrating it on MUFIN Image Search.[1] The third area is the way the user is supposed to formulate his query for the image, eventually how to subsequently specify the search target.

In the approach we took we also use both semantics based on tags as well as automatic feature detection. However, in our article the primary focus is on how to work efficiently with this information during searching. For the purposes of testing we use our own data storage and we have built our testing user interface. We focus on four typical scenarios that we have tracked down from the work of the users. We present these scenarios on simple examples. For testing purposes we use the dataset with about half a million images that we downloaded from Flickr.

In the next section we discuss related work. The third section introduces our image feature detection system and its capabilities. In the fourth section we explain its user interface in detail and illustrate the usage of the system (w.r.t. its user interface) on several real search scenarios and discuss the accessibility point of view.

2 Related Work

The amount of image data available on the Internet is increasing constantly. Popular photo sharing website Flickr[2] is hosting more than five billion images with about 4,000 new pictures being uploaded every minute. A lot of research has been done on improving image search engines. Most of the systems (Google Image Search,[3] Bing,[4] Yahoo![5]) offer keyword-based query formulation as a starting point. Textual Query Suggestion (TQS) techniques [7] were applied to support the users in their search. Clustering methods [2] were developed to offer organization of results, such as MediaFaces [5] used by Yahoo. The feature options such as color, image type (photo, clip art, …) and image size are available. However the UI of these engines still resemble traditional text search.

The capabilities of content-based and visual similarity search should be incorporated into the user interface. Design recommendations based on analysis of user behavior can be found in this paper [1]. They describe the differences between image and web search pointing out that image search is often more exploratory and often used purely for entertainment. Any user interface should support this. An innovative approach is used in Mind Finder [6] system. A query panel allowing interactive

[1] http://mufin.fi.muni.cz/imgsearch

[2] http://www.flickr.com

[3] http://www.google.com/imghp

[4] http://www.bing.com/?scope=images

[5] http://images.search.yahoo.com/images

Fig. 1 Description of particular UI components

sketching and tagging serves as an entry point. Our approach in this work is to combine search by keywords, features and example while keeping UI simple and intuitive (see Fig. 1).

3 Image Features

In this section we present features we are capable of detecting. Some of the features are related to the whole image only, but many of them can also be used to describe some parts of the image. As an example of feature related to the whole image can be considered the image type—we can detect, whether the presented image is a photo, clip-art or some kind of pencil sketch. Further details about the detection system can be found in our previous paper [3].

3.1 Color Features

According to used colors we are able to find out whether the image is gray-scaled, and if not, whether the image is toned into some specific hue. Also we can say if the image is light, dark or if the image is cool or warm.

The next group of features is color features. We want to describe the image in terms of colors in the same way as a human will, but it is not sufficient only to count the ratio of one color in the image or in some area of the image. A more complex histogram is also not enough. We should consider things like dithering, JPEG artifacts and the subjective perception of colors by people. Therefore, the color identification is not very straightforward. We have generated a reasonable group (approx. one thousand) of color prototypes. Each prototype has been classified by several human annotators.

The annotation results were then joined using simple voting system. The investigated pixel is classified using the color of the nearest prototype.

This process may seem complicated at first glance, but has been selected after several attempts as giving subjectively the best results. The colors we are currently able to identify are red, green, blue, yellow, turquoise, violet, orange, pink, brown, beige, black, white and gray. Using color spatial distribution, color histograms and below mentioned shapes recognition we are also able to detect the background color.

3.2 Shape Features

To obtain more precise information about the processed image, we have decided to employ an image segmentation technique. Using simple segmentation algorithms we were able to separate regions with same (or almost same) color. But to be able to index these shapes, we need to describe them. We have calculated the center of this shape and using this point and different angles we have sliced the shape into several regions. For each region, we have computed the maximum distance from the center. Following the changes of this distance (peaks, regularity) we are able to distinguish between different basic shapes (rectangle, circle, triangle, etc.).

Of course, this approach is not general. We use it only for bigger shapes and we ignore possible holes within the shapes. Because we use mostly downsampled versions of source images, we can guarantee the effectiveness of processing. And because we use high quality downsampling, our results are similar to a person's first glance. The shapes we are able to detect are: line, rectangle, circle, triangle and quad. Shapes that cannot be classified are used in two ways—for similarity search and for compound features.

3.3 Compound Features

By observing the way users evaluate the similarity between images, we have introduced several new compound features which are inferred from the already presented ones. Some images which were—according to many particular features (and even after prolonged human inspection)—similar, were at first glance marked by users as being not similar. We have investigated reasons for this behaviour and one of them was, that users can often very quickly evaluate the diversity of the image in terms of colors, shapes and image background and prefer this criterion at the first sight. Therefore we have constructed several diversity features specifying

- how many different colors in different regions are present in the image,
- how many different shapes (and what kind of shapes) are present in the image,
- how notable is the background and if it consists of one or more colors.

3.4 *Texture Features*

We can detect three kinds of image textures. To identify fine-detailed textures, we divide the image into small blocks and calculate histograms for each component of the color model. If the histograms are similar enough (in the sense of Manhattan distance), but the image contains some variance (to avoid trivial cases), the image probably contains fine texture. Coarse and grainy textures are detected using the concepts of Rotation Invariant Perimeter and Invariant Compactness. The first method allows us to calculate the length of the shape boundary in the low resolution space much more precisely. Using the Invariant Compactness we get the description of the small shapes (several pixels) which is invariant to scale, rotation and translation. The repetition of shapes with a similar description along with the size of these shapes gives us an indication on coarse or grainy texture.

4 Image Search System

In this section we describe our testing user interface for image search. The development of this UI went through several phases. We discussed user requirements before each phase, analyzed them and made changes accordingly.

So the scenarios that follow didn't originate from a concrete specification but were developed as a result of several week-long intensive discussions over the application usage. We have put special emphasis on clarity, directness and flexibility.

4.1 *UI Overview*

The user interface is divided into four basic parts and in several points resembles the classical search engine interface (see Fig. 1). The topmost part—header—contains noticeable system logo, so the user can quickly identify the page he is looking at. As in many search engines, the entry point for the system is the search box. This box is accompanied by an autocomplete function, which is in detail explained later in the text (Fig. 1/1).

Left column of the interface contains image features, which can be used to filter search results. All features are represented by corresponding icons. First group (Fig. 1/3) contains global image features, such as image ratio (portrait, landscape, square), image type (photo, clip-art, drawing) and image color scheme (full color, gray scale, color toned). Second group (Fig. 1/4) contains shape features (circle, ellipse, triangle, rectangle and line). In the third section (Fig. 1/5) the user can specify an arbitrary combination of colors which should be on the retrieved images. In the last group (Fig. 1/6) the user can—using slider—select requested background color. The upper part of this column contains one more thing—the image selected as an example image for similarity search (Fig. 1/2). When there is no such image selected (or uploaded by user), there is short information text encouraging users to use this functionality. This kind of criteria specifiers allows the users to quickly build up their mental models.

The biggest part (Fig. 1/9) of the user interface is reserved for the search results. The images are partly preloaded to ensure uninterrupted browsing experience. The search results are divided into several pages where the number of images per page is calculated dynamically using the screen resolution. Last image of each page is reserved for the Next page button as it is the most used button from the paging section. Hovering mouse over any image on the screen shows image detail dialog with larger image thumbnail, image tags, information about image author, etc. Clicking the image thumbnail select the image as an example image for similarity search.

The bottom-most part contains sporadically used information such as copyright information, paging links and additional search results information, such as number of retrieved images and total number of images in the dataset.

As the user selects requested tags and image features, he constructs the search filter (which is used to retrieve the images). As part of this filter is represented using icons, we have decided to summarize the selected filter using regular sentences, such as 'looking for similar picture containing yellow having black background' (Fig. 1/8). This feature helps the user to discover the capabilities of the system quickly. The row with this summary is displayed dynamically using an animation under the search box, so the user notices it immediately.

To involve the user in the system from the first visit, we have introduces two features—at the homepage, we present several popular tags, which can be clicked by the user and obtain interesting relevant images (Fig. 1/7). Some of these images are presented in the search results area from the very beginning.

All non-trivial processes (such as selecting image as an example image or longer running searches) are accompanied with an animation to ensure the user exactly knows what is happening.

As described in the previous section, we are able to detect a set of features. However, we allow the user to search for only some subset of them. The rest is used for similarity queries only. Many of these features are rather technically oriented, somehow complex or hard to be represented by simple icon. After several testing sessions we have found out, that these features were not used by the users at all. From the same reason we allow the user to specify one background color only, even if we are capable of detecting images with two or more background colors.

4.2 Technical Specification

The technical background of our image search system is service oriented. The user interface running in the user's web browsers is based on HTML/jQuery/JSON and is supported by server ASP.NET application that communicates with the data storage using web services. Internally we use the 'bag of words' [4] approach to retrieve and find similar images. The user interface have been developed (and heavily tested) to be running not only on desktop PC's, but also on mobile platforms, such as iOS and Android.

4.3 Search Scenarios

In the following sections we present four search scenarios realized using our image search system. These scenarios have been selected after tracking interaction between users and our system as the typical ones.

Scenario 1: I Know What I'm Looking For

In this scenario the user is looking for something specific. As an example we may consider a user interested in finding image of blue butterfly on green background.

1. The user starts typing 'butterfly' text to the search field. While typing, suggestions that lead to non empty search result are automatically offered (see Fig. 2/1 left). He/she chooses 'butterfly' and clicks on the search button. The results are shown almost instantly (see Fig. 2/1 right).
2. Next the user chooses 'green' background color from the slider. Again, the new set of results is shown instantly (see Fig. 3).
3. The last step is to check 'blue color' option. After that the final set of images is obtained (Fig. 2/3). Now the user may go on specifying the query in the same manner by adding more colors or selecting different background of image etc.

Scenario 2: I Like This One

In this scenario the user will use query by example approach. In the results area he/she see an image which is somehow similar to what he/she has in mind. In this case the user will not need to type.

1. The users hovers an image by moving mouse cursor over the picture. By clicking the detail of the picture it is selected as an example image (Fig. 3/1) and the similarity search begins automatically.
2. Next step is specifying the query by adding some keyword. To do that they move the mouse cursor over any image and click on one of the keywords in the detail. Selected keyword will appear in the search field (Fig. 3/2). Again, the search begins automatically.
3. The final set of result images is obtained and the user may continue to specify the query by adding any number of keywords in the same manner. All other feature options are also still available.

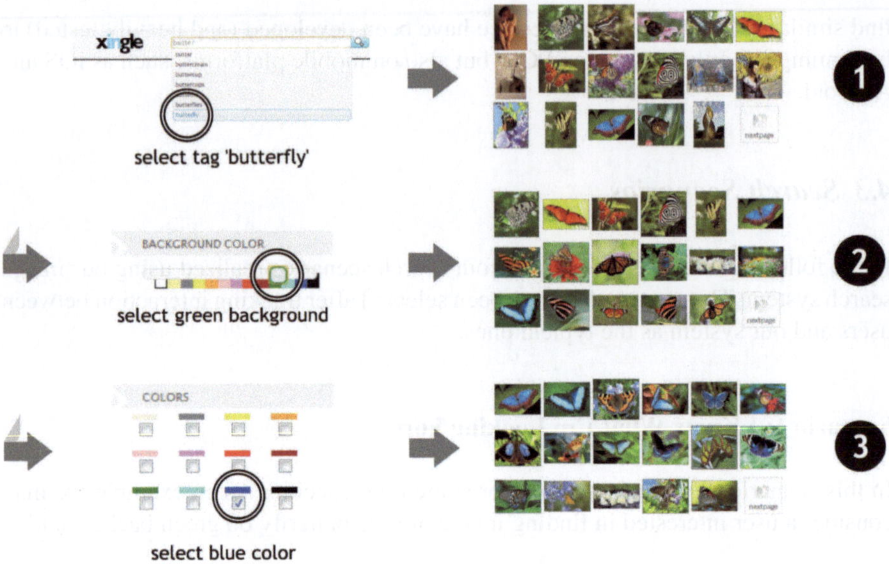

Fig. 2 Illustration of the *I know what I'm looking for* scenario

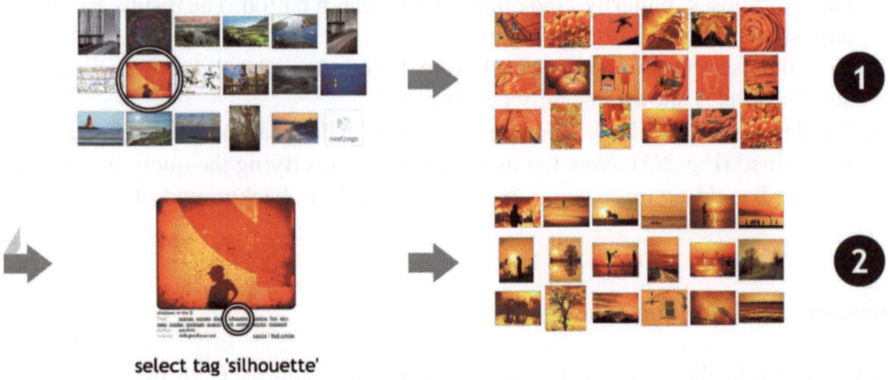

Fig. 3 Illustration of the *Query by example* scenario

Scenario 3: MY Example

In this scenario the user is looking for similar image to the one he/she has.

The requested image should be similar both visually and by content.

1. The user selects an image through 'upload your own' option—(Fig. 4/1). The similarity search begins automatically.
2. Next step is specifying the query by adding keyword describing the content. To do that the user can either type the keyword or select the keyword from the detail of some picture (Fig. 4/2 left). Again, the search begins automatically.
3. The final set of result images is obtained (Fig. 4/2 right).

Fig. 4 Illustration of the *Query by MY example* scenario

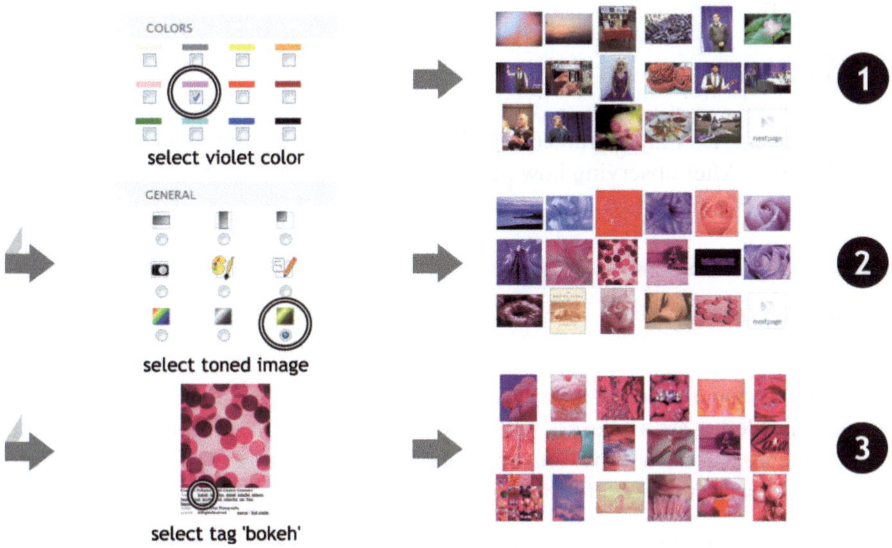

Fig. 5 Illustration of the *Inspire me!* scenario

Scenario 4: Inspire Me!

In this case the user has only a vague idea of what he/she wants. It may be a picture for decoration purposes or just a picture in some mood. As an example we may consider a user, whose only criterion is something violet.

1. The user checks 'violet color' option. The search begins automatically (Fig. 5/1).
2. Looking at results, the user decides to choose 'color-tone' option (Fig. 5/2). Again, the search begins automatically.
3. The user selects one picture to see its keywords. He/she picks 'bokeh' keyword to alter the search. The final set of result images is obtained (Fig. 5/3).

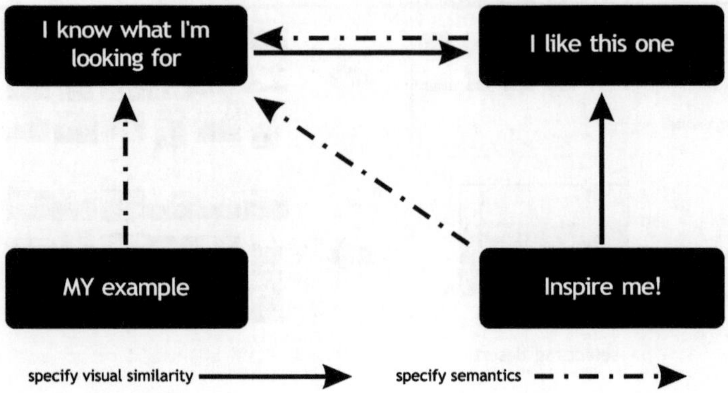

Fig. 6 Scenarios crossover

Scenarios Crossover

All of the scenarios may be mixed together. The user can combine them to support his/her search. After observing how people work, the most often used crossovers are illustrated on Fig. 6.

4.4 Accessibility

Several years ago, the accessibility was an important topic mostly from the point of view of disabled users and users with special needs. Today the topic has shifted to incorporate also users with various input devices and running platforms. Therefore we have focused on the two following issues:

1. We have tried to minimize the need to type by providing suggestions while typing the keywords. After first word is completed we provide the user with the new set for the second keyword etc. The suggested combination of keywords always leads to some search result. It is guaranteed that user will find something when picking keywords from the suggested ones. We also allow the user to pick any keyword from the detail of any image just by single click. To be more precise, we have identified 14 low level use cases, and more than 85 % of them can be solved without a keyboard.
2. Many users come from the platforms that are missing classical mouse-like pointing devices, but are accompanied by touch screens. Therefore our UI has been adapted to these devices—all buttons are big enough (and distant enough from each other) to be used by fingers and the role of Hover effects has been reduced.

5 Conclusion

In this paper we have described our image search system. We have focused heavily on the user interface, which was discussed with respect to particular image search scenarios. In the future we would like to detect new complex features in the images, therefore we will have to find its representations in the UI.

As opposed to all classical UI's for advanced image search (Google, Stock photo sites) we tried a more interactive approach. Having search launched automatically after any user action such as completing one keyword, clicking on any feature or changing settings, users can see the progress of their query. They can change the query easily by single click and see which option has narrowed their search. In case they don't look for one very specific picture, they may even get inspired in the process.

Acknowledgments This work was partially supported by SGS, VSB-Technical University of Ostrava, Czech Republic, under the grants No. SP2011/172.

References

1. Andre, P., Cutrell, E., Tan D., Smith, G.: Designing novel image search interfaces by understanding unique characteristics and usage. In: Proceedings of the 12th IFIP TC 13 International Conference on, Human-Computer Interaction, pp. 340–353 (2009)
2. Ding, H., Liu J., Lu, H.: Hierarchical clustering-based navigation of image search results. In: Proceeding of the 16th ACM international conference on Multimedia, pp. 741–744 (2008)
3. Horak, Z., Kudělka M., Snášel, V.: FCA as a tool for inaccuracy detection in content-based image analysis. In: IEEE International Conference on Granular Computing, pp. 223–228 (2010)
4. Lewis, D.: Naive (Bayes) at forty: the independence assumption in information retrieval. In: Machine Learning ECML-98, pp. 4–15 (1998)
5. van Zwol, R., Sigurbjornsson, B., Adapala, R., Garcia Pueyo, L., Katiyar, A., Kurapati, K., Muralidharan, M., Muthu, S., Murdock, V., Ng, P., et al.: Faceted exploration of image search results. In: Proceedings of the 19th international conference on World wide web, pp. 961–970 (2010)
6. Wang, C., Li Z., Zhang, L.: MindFinder: image search by interactive sketching and tagging. In: Proceedings of the 19th international conference on World wide web, pp. 1309–1312 (2010)
7. Zha, Z.J., Yang, L., Mei, T., Wang, M., Wang, Z., Chua, T.S., Hua, X.S.: Visual query suggestion: towards capturing user intent in internet image search. ACM Trans. Multimedia Comput. Commun. Appl. **6**, 1–19 (2010)

Brain Computer Interface Enhancement by Independent Component Analysis

Pavel Bobrov, Alexander A. Frolov and Dušan Húsek

Abstract Brain-Computer Interface is aimed as a direct communication pathway between human or animal brain and an external device. A reliable, accurate and fast identification of a being's intention based on EEG signal scanning is crucial part of the system. To improve the classification accuracy we propose to use Independent Component Analysis for μ-rhythm identification in data corresponding to motor imagery task performance during Brain-Computer Interface training and operation. We show that independent components corresponding to the μ-rhythm allow for higher classification accuracy comparing to raw EEG recordings usage.

1 Introduction

Brain-Computer Interface (BCI) in general is a direct communication pathway between human or animal brain and an external device. The EEG-based BCI considered operates in the following way. A set of mental tasks is chosen and each task is attributed to a particular command to an external device. Thus, to operate the device via the interface one has to perform these mental tasks. In case a device

P. Bobrov (✉)
Institute for Higher Nervous Activity and Neurophysiology of Russian Academy of Sciences and VSB Technical University of Ostrava, 17 listopadu 15/2172, 708 33Ostrava, Czech Republic
e-mail: p-bobrov@yandex.ru

A. A. Frolov
Mathematical neurobilogy of learning, Institute for Higher Nervous Activity, Neurophysiology of Russian Academy of Sciences, Butlerova 5a, Moscow, Russian Federation
e-mail: aafrolov@mail.ru

D. Húsek
Nonlinear Modelling, Institute of Computer Science, Academy of Sciences of the Czech Republic, Pod Vodárenskou Veži 2, Prague 8Prague, Czech Republic
e-mail: dusan@cs.cas.cz

M. Kudělka et al. (eds.), *Proceedings of the Third International Conference on Intelligent Human Computer Interaction (IHCI 2011), Prague, Czech Republic, August, 2011*, Advances in Intelligent Systems and Computing 179, DOI: 10.1007/978-3-642-31603-6_5, © Springer-Verlag Berlin Heidelberg 2013

movement is controlled, it seems natural for human to perform motor imagery (MI) tasks, i.e. imagine to be moving his extremities. Furthermore, what is perhaps even more important, it has been shown that such mental task execution can result in distinguishable EEG patterns (event-related synchronization and de-synchronization of the μ-rhythm [10]). Due to these two facts large part of BCI community founded their research on MI. This is why motor imagery based BCI is a well established technology nowadays, as demonstrated within such worldwide successful projects as those that arose in the Laboratory of Brain-Computer Interfaces Graz [9] or were solved under the Berlin Brain-Computer Interface mark [2].

Independent Component Analysis has been previously used for μ-rhythm extraction [7]. Here we go further and propose to use independent components corresponding to the μ-rhythm for MI recognition and classification accuracy than using raw EEG recordings. We hypothesize that this method allows for higher accuracy.

2 Method

To prove our hypothesis that identification of the μ-rhythm using Independent Component can help to achieve better results, we have set up proper experiments for such BCI performance evaluation. First, we set up standardized experimental protocol (Sect. 2.1). Then we had to chose a classification algorithm. This is described and discussed in Sect. 2.2. We also had to decide for proper version of ICA (Sect. 2.4). Next, we had to define method of measurement of classification accuracy (Sect. 2.3). All obtained experimental data were analyzed and classified off-line (Sect. 2.4) in two ways. In the first case, they were decomposed into two independent components using two different ICA algorithms for comparison (infomax and SOBI) and in the second case with no preprocessing before classification. Having compared obtained classification accuracies we show that our hypothesis is sound.

2.1 Experimental Protocol

During the experiment, every subject had to perform two MI tasks—namely—left and right hand contraction imagining, and relaxation. Each subject was trained to perform the tasks for five days. Each day two experiment sessions were held: training and testing. During the training session subject performed the tasks with no feedback, the session was used to train classifier. During the testing session classification results were provided to subject as a feedback, thus allowing to test both classifier and subject's ability to operate the BCI.

Each session was organized as follows. Subject was sitting in front of computer monitor which was used to cue him what task to perform. There were gray arrows, upper, left, and right, placed around white central circle which served to fix the subject's gaze. Changing the color of an arrow into green indicated that certain task is to be performed. The upper arrow corresponded to relaxation, left arrow corresponded to left hand movement imagining, and right arrow corresponded to right hand movement imagining. When MI tasks were performed during the testing session the feedback was provided by changing color of the central circle into green in case the result of classification coincided with the task cued to be performed. The feedback was off when subject relaxed.

Both sessions were divided into blocks. Each block included two 15 s periods when left hand movement had to be imagined, and two 15 s periods when right hand movement had to be imagined. These periodic sequences were randomly changed from block to block. Each MI period was preceded by 7 s relaxation period. Each period, either MI or relaxation, was preceded by 3 s transient interval when the corresponding arrow color was changed into blue. The training session consisted of two such blocks, while the testing session consisted of eight blocks.

EEG was recorded using 27 electrodes placed in Fz, F3, F4, Fcz, Fc3, Fc4, Cz, C1, C2, C3, C4, Cpz, Cp1, Cp2, Cp3, Cp4, Pz, P1, P2, P3, P4, Po3, Po4, Po7, Po8, O1, O2 positions. The data were digitized with 200 Hz sampling frequency and transferred to PC using NBL40 encephalograph (NeuroBioLab, Russia). Ten subjects participated in the experiment.

After all experiments had been carried out the data were processed off-line, separately for each subject and day of the study. Both training and testing session recordings were concatenated. Then, independent component analysis of the data was performed and components representing μ-rhythm were identified. After that accuracy of two MI task classification was estimated, based on the recordings made by all EEG channels and based on μ-rhythm independent components alone.

2.2 Classification

We have previously shown that the basic Bayesian classifier approach gives results comparable to those of much more complex ones [3, 5]. Let us suppose that there are L different mental tasks and for each mental task the EEG signal distribution is Gaussian with zero mean and non-singular covariance matrix C_i, $i = 1, \ldots, L$. Then, the probability to obtain EEG signal X under the condition that it corresponds to the ith mental task performing is $P(X \mid i) \propto e^{-\frac{V_i}{2}}$, where $V_i = X^{\mathrm{T}} C_i^{-1} X + \ln(\det(C_i))$. Following the Bayesian approach, the maximum value of $P(X \mid i)$ determines the class to which X belongs. Hence, the signal X is considered to correspond to performing the kth mental task as soon as $k = \mathrm{argmin_i}\{V_i\}$. The equality $X^{\mathrm{T}} C_i^{-1} X = \mathrm{trace}\left(C_i^{-1} X X^{\mathrm{T}}\right)$ implies that

$$V_i = \mathrm{trace}\left(C_i^{-1} X X^{\mathrm{T}}\right) + \ln(\det(C_i)). \quad (1)$$

It is easy to see that for such classifier training is sufficient to estimate the covariance matrices C_i for all mental states. In our experiments the training session data were used to obtain the matrices. In practice, all V_i are rather variable so we computed average value $\langle V_i \rangle$ according Formulae (1) for each 1 s epoch when estimating classification accuracy:

$$\langle V_i \rangle = \mathrm{trace}\left(CC_i^{-1}\right) + \ln\left(\det(C_i)\right), \tag{2}$$

where C denotes an epoch data covariance matrix computed as $\langle XX^{\mathrm{T}} \rangle$.

2.3 Classification Accuracy Estimation

In order to estimate the classifier performance, cross-validation procedure was used. Six of twenty 15 s periods of left hand imagining were randomly selected and were attributed to the training set, while remaining fourteen periods were attributed to the testing set. The periods of right hand imagining were split in the same manner. The classifier was trained using the training data set. The data of the testing set were split into epochs of 1 s length and each epoch was classified. One hundred of such splits were made.

As a result, a confusion matrix $P = (p_{ij})$ was obtained. Its element p_{ij} is an estimate of probability of the ith mental task performance to be recognized when the jth mental task is cued to be performed. Given the confusion matrix, three classification accuracy indexes were computed: average probability of correct recognition (p), mutual information between recognized states and presented instructions (g), and Cohen‘s kappa index (κ).

The indexes are calculated as

$$p = \mathrm{mean}_i(p_{ii}) \tag{3}$$

$$g = -\sum_{i,j} p_{ij}p_j \log_2\left(p_{ij}/p_{i0}\right) \tag{4}$$

$$\kappa = \frac{\sum_i p_{ii}p_{0i} - \sum_i [p_{0i}p_{i0}]}{1 - \sum_i [p_{0i}p_{i0}]}, \tag{5}$$

where p_{0j} is the probability of the jth instruction to be presented and $p_{i0} = \sum_j p_{ij}p_{0j}$ is the probability of the ith mental state to be recognized. The probabilities p_{0j} were estimated by dividing the number of epochs corresponding to the jth state by the number of all epochs.

2.4 Relevant Independent Components Identification

Independent Component Analysis algorithms seek to represent multi-channel signal $\mathbf{X}(t)$ recorded by N_c sensors as

$$\mathbf{X} = \mathbf{W}\xi(t), \tag{6}$$

where the ith column of weighting matrix $\mathbf{W}$ determines an impact of the ith independent component into the signal recorded by all channels, and variable $\xi_i(t)$ determines magnitude of the ith component at the moment t. The main assumption on the components ξ is that they are statistically independent. In our case, $N_c = 27$ so there were 27 components obtained.

Our experimental data were decomposed into independent components using two algorithms, infomax [6, 8] and SOBI [6] algorithms. More precisely, their implementations as available in EEGLAB toolbox [4] were used.

After ICA was performed independent components relevant to classification were identified, using the following "greedy" algorithm. The algorithm started with the set of all obtained components. At each step, a component was removed from the set so that the cross-validation procedure described in Sect. 2.4 would give the highest value of a particular index for set of the remaining components. We ran this algorithm based on p, g, and κ indices.

3 Results

There are two main results of the study. First one, the relevant component identification allows for comparing the classifier performance on the raw EEG recordings against its performance on the most relevant components alone. Second, our study shows that identification of the μ-rhythm using Independent Component Analysis can significantly enhance BCI performance. We would like discuss these results in more detail in Sect. 3.1 and Sect. 3.2.

3.1 Components Relevant to Classification

We considered the two components remaining in the end as those most relevant for classification since the removal of other components would either increase the accuracy or would not change it considerably. Typical classification index value dependence on the number of the independent components for one subject is shown in Fig. 1.

It turned out that the components corresponding to μ-rhythm in left and right hand area of motor cortex were identified as the most relevant to classification regardless of what index was used. These components will be referred to as MI components.

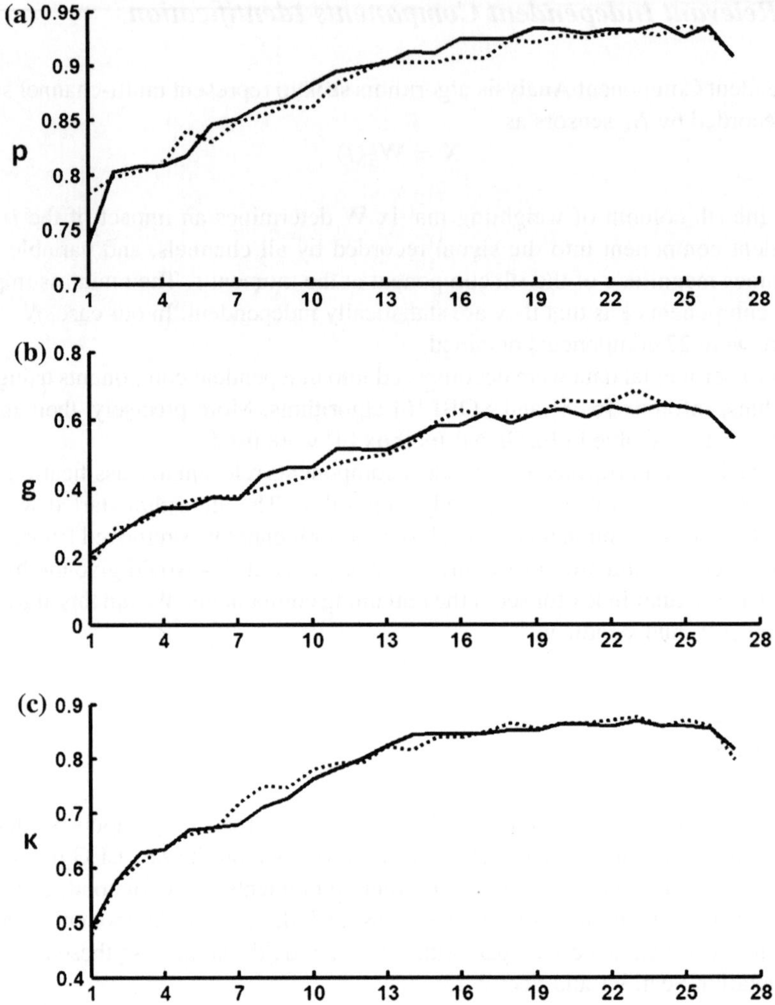

Fig. 1 Values of p (**a**), g (**b**), and κ (**c**) indexes in dependence of number of the independent components removed. Graphs are plotted based on the data of one of the subjects. The first *(leftmost)* values correspond to the case when no component was removed. Two components removed in the end were considered to be the most relevant to classification. *Thick lines* - infomax, *dotted lines* - SOBI.

In order to identify the component as related to μ-rhythm, we examined its power spectrum density based on the recordings corresponding to different mental task execution, as its impact into the signal on different channels which was given by absolute values of the corresponding mixing matrix column elements. Components, which had peaks in 8–13 and 20–30 Hz band, were contributing most into the signal recorded by either C3, Cp3, P3 or C4, Cp4, P4 channels were identified as MI

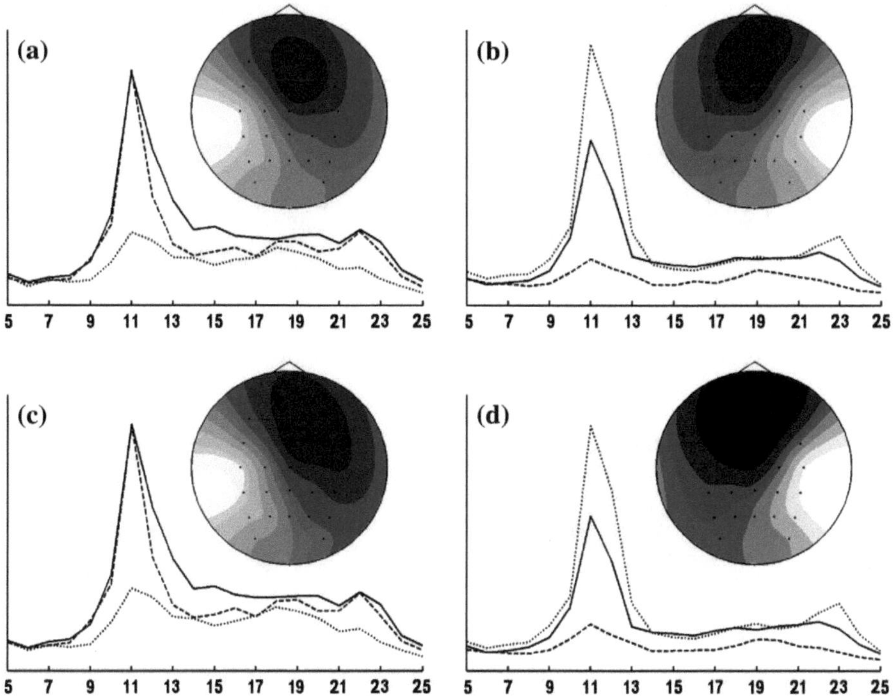

Fig. 2 Power spectrum densities and contribution into signal at the EEG electrodes MI components identified for one subject. *Lines* represent PSD evaluated based on recordings corresponding to relaxation *dashed*, left hand movement imagery *solid*, and right hand movement imagery *dotted*. **a**, **b** represent the components obtained using the infomax algorithm and **c**, **d** represent the components obtained using the SOBI algorithm

components. Such components could not be identified for two subjects who had, moreover, shown poor performance—so they were excluded from the data set. The components identified for the other subjects were quite the same. Figure 2 represents contribution distribution and power spectra of typical MI components identified for the subject whose data are represented in Fig. 1. The components represented on the left panels (a, c) are related to μ rhythm generated in right hand sensory-motor cortex area, while the components represented on the right panels (b, d) are related to μ rhythm generated in left hand sensory-motor cortex area. It is important that components obtained by infomax and SOBI algorithms which are based on different principles are almost the same. This fact supports the hypothesis on their physiological nature and, what is more, it also shows that SOBI, a much faster algorithm, can be used for μ-rhythm detection.

Figure 3 shows time-frequency structure of each component obtained by short-time Fast Fourier Transform. Reactions of μ-rhythm synchronization and de-synchronization are clearly visible. It is worth mentioning that we have observed

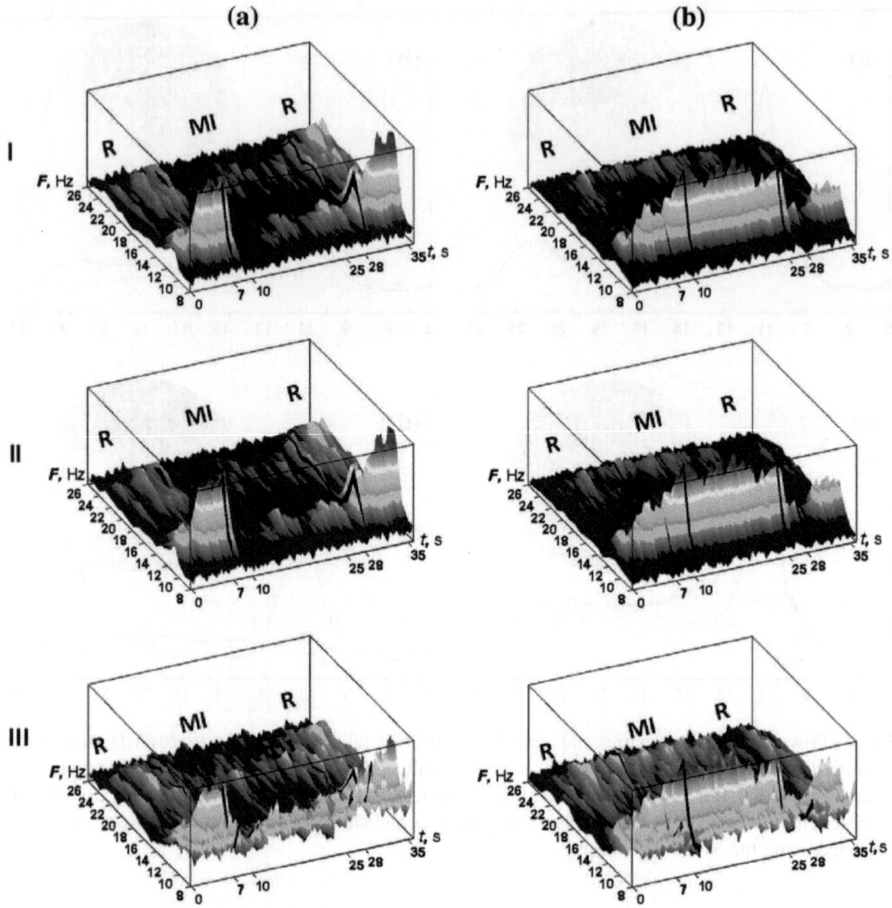

Fig. 3 Event related desynchronization (**Ia, IIa,** and **IIIa**) and synchronization (**Ib, IIb,** and **IIIb**) during one subject's right hand movement imagery. The picture was obtained by averaging SFFT coefficients over all MI periods (marked as **MI**) together with preceding and following rest periods (marked as **R**). **I** correspond to MI component obtained using the infomax algorithm, panels **II** correspond to MI component obtained using the SOBI algorithm, and **III** correspond to signal recorded by C3 electrode.

the reactions considering raw signal recorded by C3 electrode only in two subjects, suggesting necessity of using such tools as ICA for μ-rhythm detection.

Another result of the relevant component identification is that it allows for comparing the classifier performance on the raw EEG recordings against its performance on the most relevant components alone.

3.2 *Classification Accuracy*

Since MI components detected using two algorithms were the same, we will concentrate on the results of MI task classification for the SOBI components. Average values of classification accuracy indexes over all subjects and experiment days, obtained after raw data processing, are $p = 0.62 \pm 0.05$, $g = 0.10 \pm 0.04$, and $\kappa = 0.25 \pm 0.08$. Maximum values of the indexes are $p = 0.87$, $g = 0.45$, and $\kappa = 0.75$. These indexes turned out to be significantly higher (two-sample t-test, $P < 0.001$ for all indexes) after only two MI components were processed using cross-validation procedure. Average values in the later case became $p = 0.70 \pm 0.05$, $g = 0.20 \pm 0.08$, and $\kappa = 0.42 \pm 0.09$, with maximum $p = 0.93$, $g = 0.65$, and $\kappa = 0.86$.

4 Conclusion

The study shows that identification of the μ-rhythm using Independent Component Analysis can significantly enhance BCI performance. Since it is possible to use quite fast SOBI algorithm for μ-rhythm identification, then weighting matrix can be obtained after short training session along with classifier parameters. Whether the weighting matrix changes along with the subject training is a question which requires further investigation.

Acknowledgments This paper has been partly elaborated in the framework of the IT4Innovations Centre of Excellence project, reg. no. CZ.1.05/1.1.00/02.0070, supported by Operational Programme 'Research and Development for Innovations' funded by Structural Funds of the European Union and state budget of the Czech Republic and partly supported by the projects AV0Z10300504, GACR P202/10/0262, 205/09/1079.

References

1. Broy, M.: Software engineering—from auxilary to key technologies. In: Broy, M., Dener, E. (eds.) Software Pioneers, p. 13. Springer, Heidelberg (2002)
2. Blankertz, B., Dornhege, G., Krauledat, M., Muller, K., Curio, G.: The non-invasive Berlin brain-computer interface: fast acquisition of effective performance in untrained subjects. NeuroImage **37**, 539–550 (2007)
3. Bobrov, P., Frolov, A., Cantor, C., Fedulova, I., Bakhnyan, M.: brain-computer interface based on generation of visual images. PLoS ONE **6**(6), e20674 (2011). doi:10.1371/journal.pone.0020674
4. Delorme, A., Makeig, S.: EEGLAB: An open source toolbox for analysis of single-trial EEG dynamics including independent component analysis. J. Neurosci. Methods **134**, 9–21 (2004)
5. Frolov, A., Husek, D., Bobrov, P.: Comparison of four classification methods for brain computer interface. Neural Netw. World **21**(2), 101–192 (2011)
6. Hyvarinen, A., Karhunen, J., Oje, E.: Independent component analysis. Willey, New York. pp.480 (2001)

7. Kachenoura, A., Albera, L., Senhadji, L., Comon, P.: ICA: a potential tool for BCI systems. IEEE Signal Process. Mag. **25**(1), 57–68 (2008)
8. Makeig, S., Bell, A.J., Jung, T.-P., Sejnowski, T.J.: In: Touretzky, D., Mozer, M., Hasselmo, M. (eds.) Advances in neural information processing systems vol.8, pp.145–151. MIT Press, Cambridge, (1996)
9. Neuper, C., Scherer, R., Wriessnegger, S., Pfurtscheller, G.: Motor imagery and action observation: modulation of sensorimotor brain rhythms during mental control of a braincomputer interface. Clin. Neurophysiol. **120**, 239–247 (2009)
10. Pfurtscheller, G.: EEG event-related de-synchronization (ERD) and event related synchronization (ERS). In: Niedermeyer, E., Lopes da Silva, F.H. (eds.) Electroencephalography: basic principles, clinical applications and related fields, 4th edn. pp. 958–967 . Williams and Wilkins.Baltimore, (1999)

Part II
Knowledge Discovery I

Evaluating Hard and Soft Flat-Clustering Algorithms for Text Documents

Vivek Kumar Singh, Tanveer Jahan Siddiqui and Manoj Kumar Singh

Abstract Document clustering refers to unsupervised classification (categorization) of documents into groups (clusters) in such a way that the documents in a cluster are similar, whereas dissimilar documents are assigned in different clusters. The documents may be web pages, blog posts, news articles, or other text files. A popular and computationally efficient clustering technique is flat clustering. Unlike hierarchical techniques, flat clustering algorithms aim to partition the document space into groups of similar documents. The cluster assignments however may be hard or soft. This paper presents our experimental work on evaluating some hard and soft flat-clustering algorithms, namely K-means, heuristic k-means and fuzzy C-means, for categorizing text documents. We experimented with different representations (*tf, tf.idf, Boolean*) and feature selection schemes (with or without stop word removal and with or without stemming) on some standard datasets. The results indicate that *tf.idf* representation and the use of stemming obtains better clustering. Moreover, fuzzy clustering obtains better results than K-means on almost all datasets, and is also a more stable method.

V. K. Singh
Department of Computer Science, Banaras Hindu University (BHU), Varanasi, India
e-mail: vivek@bhu.ac.in

T. J. Siddiqui
Institute of Applied Physics and Technology, University of Allahabad, Allahabad, India
e-mail: jktanveer@yahoo.com

M. K. Singh (✉)
DST-Centre for Interdisciplinary Mathematical Sciences (DST-CIMS),
Banaras Hindu University (BHU), Varanasi, India
e-mail: mks_kjist@yahoo.co.in

M. Kudělka et al. (eds.), *Proceedings of the Third International Conference on Intelligent Human Computer Interaction (IHCI 2011), Prague, Czech Republic, August, 2011*, Advances in Intelligent Systems and Computing 179, DOI: 10.1007/978-3-642-31603-6_6,

1 Introduction

Document clustering refers to organizing the documents into various clusters such that similar documents belong to same cluster and dissimilar documents belong to different cluster. Traditionally, clustering algorithms have been used for clustering data patterns that comprised of numerical values of very low dimensionality. However, with the availability of large amount of unstructured text on the web, clustering algorithms have found useful applications in search engines. Document clustering helps in effectively navigating the web documents. The basic idea in clustering text document is to use their term frequency statistics to group the documents having similar terms in one cluster. The terms (also referred as features) act as indicators of topic/ contents of a document. The most commonly used representation of documents is vector space model which represents each document as feature vector in a multi-dimensional space. The number of dimensions corresponds to total number of unique document attribute in the collection.

Existing document clustering algorithms can be broadly categorized as *flat* and *hierarchical* or *hard* and *soft* [1–3]. Flat clustering partitions the document space into different clusters. Hierarchical clustering creates a hierarchy of clusters where a document of lower level cluster is also a member of the corresponding higher level cluster. The hard and soft algorithms on the other hand vary in their degree of document assignment to a cluster. Hard clustering computes a hard assignment—each document is a member of exactly one cluster. In soft clustering algorithms a document's assignment is a distribution over all clusters, i.e., a document has fractional membership in several clusters in contrast to Boolean memberships in hard clustering.

This paper presents our experimental work on performance evaluation of hard and soft flat-clustering algorithms. We have implemented the K-means, its heuristic variant and the fuzzy C-means algorithms. K-means is a hard flat-clustering algorithm whereas fuzzy C-means is a soft flat-clustering algorithm. K-means and fuzzy C-means are one of the most widely used flat-clustering algorithms because of their simplicity and computational efficiency. While many past experiments have addressed the performance evaluation of some clustering algorithms, the comparison and analysis presented in this paper are different. We have evaluated the performance of these flat clustering algorithms (both hard and soft partitioning) with different representation and feature selection schemes. We collected certain standard datasets such as Reuters 21,578, Classic and 20 Newsgroups and formulated different test cases from them. We processed the documents in the dataset to produce document vectors and obtained experimental results. We have also computed both internal and external evaluation criteria such as RSS, purity and F-measure. The rest of the paper is organized as follows. Section 2 describes the clustering problem, different approaches and evaluation measures. Section 3 describes our experimental setup. Section 4 presents the results obtained and Sect. 5 provides a short summary of the experimental results and their relevance.

2 Document Clustering

Flat clustering (hard) can be defined as follows: Given (i) a set of documents D$=(d_1, d_2, \ldots, d_N)$, (ii) desired number of clusters, K, and (iii) an objective function that evaluates the quality of clustering; the clustering task is to compute an assignment $\gamma : D \rightarrow \{1, \ldots, K\}$ that minimizes (or in certain cases maximizes) the objective function [4]. The objective function is usually the distance or similarity between documents. The objective function in K-means is to minimize the average distance between documents and their centroids or, equivalently, to maximize the similarity between documents and their centroids. The clustering problem thus requires computing similarity between documents in order to assign them to a particular cluster.

Hard clustering algorithms may be exhaustive, assigning each document to some cluster, or non-exhaustive, where some documents may not be assigned to any cluster. K-means and Expectation-Minimization (EM) are examples of flat clustering algorithms [5]. The documents in hard clustering are assigned exclusively to one cluster. Soft Clustering algorithms (fuzzy clustering) use membership values (inspired by fuzzy set theory) to associate degree of membership of a document to a particular cluster. The membership values are in the interval [0,1]. Fuzzy clustering produces a clustering but not a partition. Documents in fuzzy clustering belong to more than one cluster. The degree of membership of a document to a particular cluster specifies how close the document is to the cluster [6]. A hard clustering can nevertheless be obtained from fuzzy clustering by taking a threshold of the membership function value, i.e., a document may be believed to be part of that cluster for which its membership value is greater than a threshold value (say 0.7). Fuzzy C-means [7] is a popular flat soft clustering algorithm.

2.1 Representing Text Documents

Vector Space Model [4] is one of the most commonly used representation schemes for text documents. It represents documents as vectors of features representing terms that occur within them. These vectors are represented in a multi-dimensional space in which each dimension corresponds to a distinct term in the entire document set. Usually each feature takes some numerical value which is usually some function of frequency of terms, e.g. *tf, tf.idf or Boolean* presence [8]. Term Frequency (*tf*) is a count of how often a term appears within the document. Words that appear often have high *tf* value. The *idf* is a measure of inverse document frequency of the term. If the total number of documents in the set is *N*, and df_t be the number of documents that contain a term *t*, then the *idf* for the term *t* is computed as in Eq. 1:

$$idf_t = log(N/df_t). \tag{1}$$

2.2 Feature Selection and Similarity Computation

The text documents to be clustered usually contain a large number of terms and the total number of distinct terms in the entire document set is often very large. This presents severe computational challenges. Moreover, not all terms may be equally important in terms of being representative term of a document. For example, stop words like 'the', 'is', 'am', 'are', 'who' etc. occur very frequently in almost all documents and also do not carry representative information about the topic of a document. Another key observation that can help in reducing the number of terms and also improve clustering results is that terms often occur in various forms in documents still representing the same semantic concept. For example words like 'computer', 'computing', 'computational', 'computes' and 'computed' all are derivatives of the root 'comput'. This can be used to further reduce the number of distinct terms in the document space by representing all forms of a word by its root. This process is commonly known as stemming. It can help in reducing the dimensionality and also improving the clustering quality. Sometimes a term thresholding is also used to include only those terms that occur a specified number of times in the document. The reduced vectors used for computing clusters are termed as feature vectors. Every entry in the feature vector can thus be any of the three possible measures: *tf*, *tf.idf* or *Boolean presence*.

Once the feature set to be used for clustering is selected, we need a mechanism to compute similarity between feature vectors of two documents. There are various similarity measures that can be used for this purpose. One common choice is the Cosine Similarity measure. The Cosine Similarity of documents d_1 and d_2 can be computed by using Eq. 2.

$$Cosine\ Similarity(d_1, d_2) = \frac{FV(d_1).FV(d_2)}{|FV(d_1)||FV(d_2)|} \tag{2}$$

Here, the numerator represents the dot product of the feature vectors $FV(d_1)$ and $FV(d_2)$, and the denominator is product of their Euclidean lengths. This measure will give a minimum value of 0 for vectors that do not have any terms in common and a maximum value of 1 means to vectors that share all the terms. Hence, it can be used to measure similarity between two documents. The documents to be placed in the same cluster are those which have high cosine similarity among themselves.

2.3 K-means and Heuristic K-means Algorithms

The K-means algorithm is one of the most popular flat clustering algorithms, well suited to clustering large document datasets due to its relatively low computational cost. This algorithm assigns every document to one of the K clusters. Ideal cluster in K-means is a sphere with the centroid as its center of gravity. The objective in K-means is to minimize the average distance of documents from their cluster centers, where a cluster center is defined as the mean or centroid of the documents in a cluster. The centroid, μ, of the documents in a cluster, ω, is computed as in Eq. 3.

$$\mu(\omega) = \frac{1}{|\omega|} \sum_{x \in \omega} \overline{x} \tag{3}$$

The K-means algorithm begins by selecting K random seeds in the document space. These K points are assumed to represent centers (centroid) of the K initial clusters. The algorithm then proceeds to compute the distance (or similarity) of each document from all the K points. These distance values are used to assign every document to one of the K clusters. A document is assigned to a cluster which is closest to it, or in other words, to the cluster whose centroid has the smallest distance from the documents, out of all such K centroids. Once all documents are assigned to one of the K clusters, the centroids of all the K clusters is recomputed to represent the new centre of the changed cluster. Thereafter, this process is iterated, where in each iteration the distance of every document from all the cluster centroids is calculated and the document is assigned to the cluster whose centroid is closest to it. This iterative process is repeated until cluster assignment converges or up to a fixed number of iterations. The K-means algorithm often uses a residual sum of squares or RSS (squared distance of each document vector from its cluster centroid summed over all vectors) as a terminating criterion. This measure also serves as an internal quality measure and is described in Sect. 2.5. The K-means algorithm is known to converge, however the clustering quality is sensitive to initial seeds (the K random points used). Moreover, there is no guarantee that the K-means algorithm will always achieve the global minimum (measured in terms of RSS value).

Due to sensitivity of the algorithm to the initial K seeds, certain heuristics for seed selection may be used to improve the K-means results. The biggest disadvantage of the K-means algorithm is that it may converge to a local minimum if the K random seeds chosen are not appropriate. To minimize the sensitivity of the K-means algorithm to initial seed selection, different heuristics are tried [8]. These heuristics may include: (i) excluding outliers from the seed set; (ii) trying out multiple starting points and choosing the clustering with the lowest cost; and (iii) obtaining seeds from another method such as hierarchical clustering. However, these variants add some computational cost to the algorithm. Another robust variant that often works well for a large variety of document distributions is to select i (e.g. i = 10) random document vectors for each cluster and use their centroid as the seed for this cluster. We have used this scheme in our experimental formulation to improve K-means clustering results.

2.4 Fuzzy C-means Algorithm

Unlike hard clustering methods, fuzzy clustering does not provide hard clusters but returns a degree of membership of each document for every cluster. Hard clustering algorithms generate partitions of documents whereas fuzzy clustering outputs clustering but not a partition. Hence, in fuzzy clustering each cluster is a fuzzy set of all the patterns. A popular fuzzy clustering algorithm is Fuzzy C-means (FCM). It is similar in approach to K-means except that it produces a membership matrix, which contains the degree of membership of a document to all the clusters. If there are N documents and K clusters, then it begins by selecting a N $\times$ K membership matrix U. Every element $u_{i,j} \in [0, 1]$ of this matrix represents the degree of membership of a document x_i in cluster c_j. Thereafter, using U the value of a fuzzy criterion function associated with each partition is obtained. One possible fuzzy criterion function is given in Eq. 4

$$E^2(\chi, U) = \sum_{i=1}^{N} \sum_{k=1}^{K} u_{i,j} \|x_t - c_k\|^2 \tag{4}$$

where, c_k is the kth fuzzy cluster center. After computing the criterion function, documents are reassigned to clusters to reduce criterion function value and the matrix U is recomputed. This process is repeated until the algorithm converges (the entries in U do not change significantly). The ordered pairs (i, u_j) in each cluster represent the ith document and its membership value to the cluster u_j. Larger membership values indicate higher similarity. A hard clustering result can also be obtained from FCM by thresholding the membership value. FCM is a stable and robust method that is quite insensitive to initialization. At the same time it is a simple and computationally efficient algorithm.

2.5 *Performance Evaluation Measures*

The objective functions in most of the clustering algorithms try to achieve high intra-cluster similarity (documents within a cluster are similar) and low inter-cluster similarity (documents from different clusters are dissimilar). This is often viewed as an internal criterion of clustering quality. The residual sum of squares (RSS), defined as the squared distance of each vector from its centroid summed over all vectors, is a measure of internal criterion. The goal is to minimize the RSS value. It is equivalent to minimizing average squared distances. The RSS computation is done as in Eq. 5.

$$RSS = \sum_{k=1}^{K} RSS_k; \qquad \text{where} \quad RSS_k = \sum_{x \in \omega} |\overline{x} - \overline{\mu}(\omega_k)|^2 \tag{5}$$

However, good scores on internal criterion (like RSS) do not always turn out to be an accurate and effective measure of clustering quality, as we will have best RSS value when every document is a singleton cluster. An alternative to the internal criterion is therefore direct evaluation (an external criterion) of the clustering result in the application of interest. There are five popular external criteria used for direct evaluation the clustering quality. These are $Purity$, $Entropy$,

Normalized Mutual Information, *Rand Index* and *F-measure*. We describe here three of these important measures that are frequently used.

To compute *Purity*, each cluster is assigned to the class which is most frequent in the cluster, and then the accuracy of this assignment is measured by counting the number of correctly assigned documents and dividing by N. For example, if a cluster contains 12 documents of **x** type, 27 documents of **y** type and 03 documents of **z** type, then **y** being the most frequent class is taken as the actual class of that cluster. This is done for all the clusters. *Purity* is then calculated by evaluating how many documents are assigned to the correct class, as in Eq. 6

$$Purity(\Omega, C) = \frac{1}{N}\sum_{k} max_j(|w_k \cap c_j|) \tag{6}$$

where, $\Omega = \{w_1, w_2, \ldots, w_k\}$ is the set of clusters and $C = \{c_1, c_2, \ldots, c_j\}$ is the set of classes. Here, w_k denotes the set of documents in the cluster w_k and c_j denotes the set of documents in the class c_j. Bad clustering has *Purity* close to '0' and a perfect clustering has *Purity* value '1'. High *Purity* is thus easy to achieve when the number of clusters is large. The *Purity* becomes 1 if each document gets its own cluster. Therefore, *Purity* may also be not a good measure of clustering quality.

The Rand Index (RI) [9] and F-measure [10, 11], instead of using information theoretic approach, takes an alternative approach. They both use True-Positive (TP), True-Negative (TN), False-Positive (FP) and False-Negative (FN) decisions from Information Retrieval domain. A TP decision assigns two similar documents to the same cluster and a TN decision assigns two dissimilar documents to dissimilar clusters. FP and FN are two erroneous assignment decisions possible. A FP decision assigns two dissimilar documents to same cluster and a FN decision assigns two similar documents to different clusters. RI is then computed as a measure of percentage of correct decisions, as in Eq. 7

$$RI = \frac{TP + TN}{TP + FP + FN + TN} \tag{7}$$

RI is thus a measure of accuracy. The F-measure however avoids giving similar weight to FP_s and FN_s since separating similar documents is sometimes worse than putting pairs of dissimilar documents in the same cluster. The F-measure therefore penalizes FN_s more strongly than FP_s by selecting a value $\beta > 1$, as in Eq. 8.

$$P = \frac{TP}{TP + FP}, \quad R = \frac{TP}{TP + FN}, \quad F_\beta = \frac{(\beta^2 + 1)PR}{\beta^2 P + R} \tag{8}$$

Here, P refers to precision and R refers to recall. The value of β has to be carefully chosen and is usually done based on past observations.

3 Experimental Work

We have implemented K-means, heuristic K-means and Fuzzy C-means algorithms for clustering text documents. We obtained certain standard datasets, described in Sect. 3.1, and represented the documents in them using vector space model. We explored with different feature selection schemes. The algorithm implementations are then run on the processed standard datasets to obtain clustering results. The results of different implementations have been evaluated through internal and external criteria.

3.1 Datasets

We obtained the experimental data from three standard datasets, namely Reuters-21578 [12], Classic 7095 [13], and 20 Newsgroups [14]. Reuters-21578 is a collection of 21,578 newswire articles, originally collected and labeled by Carnegie Groups, Inc. and Reuters, Ltd. in the course of developing the CONSTRUE text classification system. The articles are assigned classes from a set of 118 topic categories. Reuters-21578 is basically a dataset for text classification. We have extracted two smaller sample data from the Reuters-21578 for our cluster evaluation. Classic dataset is another well known benchmark dataset that has a total of 7,095 documents. Classic dataset has two versions Classic 3 and Classic 4. Classic 4 comprises of a total of 7,095 documents from the document collection of CACM, CISI, CRAN and MED. Classic 3 includes only CISI, CRAN and MED. Classic 4 has 3,204 documents from CACM, 1,460 from CISI, 1398 from CRAN and 1033 from MED. 20 Newsgroups is another widely used text classification collection originally collected by Ken Lang. It consists of 1,000 articles from each of the 20 Usenet newsgroups. The name of the newsgroup is regarded as the category of the article. After removal of some duplicate entries, it contains a total of 18,941 articles. We have obtained a total of six datasets of varying sizes (500–1,049 documents). The dataset used in our implementation is described in Table 1.

3.2 Document Representation and Feature Selection

We have preprocessed the documents and transformed them into vector representations. This involved removing non-term tags, such as HTML and other non-relevant tags. The label information of documents in some datasets has been read while preprocessing. We have used different kinds of representation as well as feature selection schemes. The different runs of the clustering algorithms have used *tf*, *tf.idf* and Boolean presence to represent term related information of a document. While selecting entries of the term vector we have explored with removing and non-removal of stop words, and performing and not performing stemming. A customized stop words

Table 1 Dataset used in the experimental work

Dataset	Source	No. of documents	No. of classes
Reu-01	Reuters-21578	1,000	05
Reu-02	Reuters-21578	1,049	10
Cla-01	Classic 4 (7095)	500	04
Cla-02	Classic 4 (7095)	1,000	04
20New-01	20 Newsgroups	500	20
20New-02	20 Newsgroups	1,000	20

list has been used to remove stop words and the Porter suffix stripping algorithm [15] have been used for performing stemming. Hence due to the various combinations of choices, a particular algorithm has been run on one dataset in twelve variations; three for representation (*tf*, *tf.idf* & *Boolean*), two for stop words (*with and without stop words*) and two for stemming (*with and without stemming*). A total of 12 runs each for on 06 different datasets for all the 03 algorithms thus involved a minimum of 216 runs. Moreover, since K-means, Heuristic K-means and Fuzzy C-means, all the algorithms involve stochastic elements; we had to make at least 03 runs for each of the different combinations and choose the best result, resulting in a total of 648 runs.

3.3 Implementing K-means, Heuristic K-means and Fuzzy C-means

We have implemented K-means and Heuristic K-means in JAVA and the Fuzzy C-means algorithm in MATLAB. The document preprocessing and feature vector computation has however been implemented in JAVA for all the three algorithms. This was done primarily due to better string manipulation capabilities in JAVA. The K-means implementation involved generating K random seeds (assumed to represent centroids) and then Cosine Similarity of all the documents in the entire document set is computed with the centroids. A document is assigned to a cluster for which it has highest cosine similarity value (can be interpreted as lowest distance). After initial assignments of documents in clusters are done, the centroids of the clusters are recomputed by using feature vectors of the documents assigned to it. These recomputed centroids now become new K centroids and the process of *Cosine Similarity* computation and assignments is repeated until the algorithm converges. We have used a fixed number of iterations or no change in RSS value (whichever is earlier) as the convergence (terminating) condition.

The heuristic version of K-means that we have implemented uses precomputed centroids as initial K seeds. We first select 10 documents each for each of the K initial points to be selected. Every Kth point is taken as the computed centroid of the 10 random documents selected for that seed. This is done for all the K initial points to be

Table 2 Summary of the clustering results for two datasets **Reu-01**

Feature	SWRS[a]	Stemming	RSSK[b]	RSSHK[c]	PurityK[d]	PurityHK[e]
tf	No	No	185904.190	184997.068	0.246	0.234
tf	No	Yes	221097.820	221073.323	0.246	0.247
tf	Yes	No	192014.380	191828.020	0.233	0.244
tf	Yes	Yes	212180.330	211532.554	0.247	0.244
tf.idf	No	No	962.546	951.077	0.337	0.365
tf.idf	No	Yes	945.039	942.378	0.499	0.535
tf.idf	Yes	No	950.139	958.318	0.334	0.270
tf.idf	Yes	Yes	951.644	943.216	0.462	0.397
boolean	No	No	55652.980	55569.106	0.245	0.247
boolean	No	Yes	55256.887	55354.850	0.251	0.243
boolean	Yes	No	56619.953	56769.938	0.251	0.219
boolean	Yes	Yes	52349.23	52464.353	0.242	0.230

[a] Stop words removal; [b] RSS(K-means); [c] RSS (Heuristic K-means); [d] Purity (K-means); [e] Purity (Heuristic K-means)

Table 3 Summary of the clustering results for two datasets **Reu-02**

Feature	SWRS	Stemming	RSSK	RSSHK	PurityK	PurityHK
tf	No	No	206562.250	206031.978	0.217	0.213
tf	No	Yes	242818.500	238654.083	0.212	0.223
tf	Yes	No	212044.580	212427.641	0.219	0.219
tf	Yes	Yes	314737.300	309218.167	0.210	0.227
tf.idf	No	No	1014.815	1000.483	0.242	0.321
tf.idf	No	Yes	994.168	992.657	0.281	0.345
tf.idf	Yes	No	991.810	997.801	0.389	0.400
tf.idf	Yes	Yes	993.102	988.963	0.342	0.280
boolean	No	No	60954.975	60558.726	0.205	0.204
boolean	No	Yes	61062.982	61012.031	0.205	0.206
boolean	Yes	No	62383.274	62032.518	0.205	0.216
boolean	Yes	Yes	57428.170	57661.446	0.206	0.208

Table 4 Summary of the results of the algorithms on two datasets

Algorithm	Dataset	Purity (Best Value)
K-means	Reu-01	0.499
K-means	Reu-02	0.389
K-means	Cla-01	0.250
K-means	Cla-02	0.295
Heuristic K-means	Reu-01	0.535
Heuristic K-means	Reu-02	0.400
Heuristic K-means	Cla-01	0.476
Heuristic K-means	Cla-02	0.513
Fuzzy C-means (0.6)	Reu-01	0.682
Fuzzy C-means (0.6)	Reu-02	0.504
Fuzzy C-means (0.6)	Cla-01	0.496
Fuzzy C-means (0.6)	Cla-02	0.586

selected. The rest of the algorithm is then same as that of regular K-means clustering. The initial seeds selected in this case are assumed to represent better cluster centers than regular K-means. The Fuzzy C-means (FCM) algorithm has been implemented in MATLAB with feature vector computation done in JAVA. In FCM we chose C initial points to represent initial cluster centers. Then Cosine Similarity of all the documents in the entire document set is computed with these centroids. However, unlike hard clustering here we compute a degree of membership of every document for all the C clusters. Rest of the process of re-computation of centroids and resultant virtual assignments is similar to K-means, except that it uses a weighted average kind of centroid computation. In this case a cluster may belong with different degree of memberships to more than one cluster. To evaluate the quality of clustering result we have computed both internal and external performance evaluation criteria, namely RSS and $Purity$.

4 Results

Our experimental work on K-means, Heuristic K-means and Fuzzy C-means obtained interesting results. While some past performance evaluation experiments [16, 17] have been performed on performance evaluation of clustering algorithms, the main aim of these experiments have been to compare performance of flat and hierarchical clustering algorithms. This experimental work in contrast aims to evaluate performance of different flat-clustering algorithms, more appropriately hard and fuzzy flat-clustering algorithms. We have attempted do an extensive evaluation of three important flat clustering algorithms (K-means, Heuristic K-means and Fuzzy C-means) with a number of variations of representation schemes (*tf, tf.idf and Boolean*) and feature selections (*with or without stop words and with or without stemming*). The preliminary results obtained present some commonly speculated and few relatively less known findings. Both the representation scheme used and the feature selection schemes affect the quality of clustering result. Moreover, the magnitude of their effect varies with different algorithms. Tables 2, 3 and 4 present a part of the clustering result obtained. FCM results are for hard partition threshold of 0.6.

5 Conclusion

We have done experimental evaluation of performance of three flat-clustering algorithms namely, K-means, Heuristic K-means and Fuzzy C-means. The experimental setup involved selecting different representation schemes (*tf, tf.idf and Boolean*) and employing different feature selection schemes (*with or without stop words, and with or without stemming*). The experimental results demonstrate that out of the three representations: *tf, tf.idf and Boolean; tf.idf* obtains best clustering results with all the clustering methods across different data sets. The tf representation produces results worse than tf.idf but marginally better than Boolean representation. Although Boolean presence representation produces relatively good results for classification task but for clustering it may not be an appropriate choice. Similarly, the choice of feature selection scheme employed: with or without stop words, and with or without stemming; also determines the clustering quality. The stemming alone produces best clustering results. Stop word removal though improves the quality in some cases, but overall its effect is relatively not that significant. Surprisingly, doing stemming and stop word removal together does not produce results any better than stemming alone. Out of the three flat clustering methods employed, heuristic K-means produce better results than regular K-means. However, the qualities of results in different runs in both cases vary. To get the best values, they need to be executed multiple times. This is primarily due to the initial selection of seeds being done stochastically. FCM proves to be a better and more stable method and our preliminary results obtained using the FCM implementation validates the same. In general, out of the three flat-clustering algorithms, K-means is a good choice if computational efficiency is a major concern with the possibility of improving the results by switching to Heuristic K-means. FCM is probably a better flat-clustering algorithm, though a hard clustering still needs to be computed from its results for practical uses.

References

1. Jain, A.K., Murty, M.N., Flynn, P.J.: Data clustering: a review. ACM Comput. Surv. **31**(31), 477–488 (1995)
2. Jain, A.K.: 50 years beyond K-means. In: 19th International Conference on Pattern Recognition, Tampa, FL (2008)

3. Jain, A.K., Dubes, R.C.: Algorithms for Clustering Data. Prentice-Hall Advanced Reference Series, Prentice Hall, NJ (1988)
4. Manning, C.D., Raghvan, P., Schutze, H.: Introduction to Information Retrieval, Cambridge University Press, Cambridge (2008)
5. Duda, R.O., Hart, P.E., Stork, D.G.: Pattern Classification, 2th edn. Wiley-Interscience, New York (2000)
6. Valente De Oliviera, J., Pedrycz, W.: Advances in Fuzzy Clustering and its Applications, pp. 3–30. Wiley, Hoboken (2007)
7. Bezdek, J.C.: Pattern Recognition with Fuzzy Objective Function Algorithms. Plenum Press, New York (1981)
8. Alag, S.: Collective Intelligence in Action. Manning, New York (2009)
9. Rand, W.M.: Objective criteria for evaluation of clustering methods. J. Am. Stat. Assoc. **36**(31), 846–850 (1971)
10. Bjorner, L., Aone, C,: Fast and Effective Text Mining Using Liner time Document Clustering. In: Knowledge and Data Discovery' 1999, California. (1999)
11. Van Rijsbergen, C.J.: Information Retrieval. 2nd edn, Butterworths, London (1979)
12. Lewis, D., Reuters-21578. http://www.research.att.com/lewis/reuters21578.html. Accessed Jan 2011
13. Classic 4 (7095). ftp://ftp.cs.cornell.edu/pub/smart. Accessed Jan 2011
14. Newsgroups. http://www.ai.mit.edu/prople/jrennie/20-newsgroups. Accessed Jan 2011
15. Porter, M.F.: An Algorithm for Suffix Stripping. Program **14**(3), 130–137 (1980)
16. Steinbach, M., Karypis, G., Kumar, V.: A comparison of document clustering techniques. In: KDD Workshop on Text Mining (2002)
17. Xiong, H., Wu, J., Chen, J.: K-means clustering versus validation measures—a data distribution perspective. In: KDD'06, ACM Press (2006)

Enhancing System Usability with a Natural Language Interface

Erqiang Zhou and Bing Wu

Abstract A Natural Language Interface (NLI) is studied and developed for retrieving the recorded domain information within a prototype system, which manages complex information within the paradigm of recording live interactive records of information so that the recorded information can be queried and replayed. After introducing the background of the prototype system, concrete querying examples are presented so as to demonstrate the *effectiveness* and *efficiency* of using the developed NLI. This is done by comparing the natural language queries and the corresponding SQL statements, which are automatically gained within the prototype system. This paper shows that an NLI can greatly reduce the efforts for writing or understanding queries of information systems, thus can enhance the system usability.

1 Introduction

A broad definition of *usability* given by the International Standards Organization (ISO) is "the extent to which a product can be used by specified users to achieve specified goals with effectiveness, efficiency, and satisfaction in a specified context of use" [1]. Tullis and Albert [2] explained the three aspects of usability as:

- *effectiveness*: being able to complete a task;
- *efficiency*: the amount of effort required to complete the task;
- *satisfaction*: the degree to which the user was happy with his or her experience while performing the task.

E. Zhou (✉) · B. Wu
School of Computing, Dublin Institute of Technology,
Kevin St., Dublin 8, Ireland
e-mail: erqiang.zhou@dit.ie

B. Wu
e-mail: bing.wu@dit.ie

M. Kudělka et al. (eds.), *Proceedings of the Third International Conference on Intelligent Human Computer Interaction (IHCI 2011), Prague, Czech Republic, August, 2011*, Advances in Intelligent Systems and Computing 179, DOI: 10.1007/978-3-642-31603-6_7, © Springer-Verlag Berlin Heidelberg 2013

For an information system, a usable interface is a desirable component which can facilitate users to achieve their goals of adding, retrieving and updating domain information. Natural language has been treated as an option for fulfilling such an interface for a long time. Earlier research of Natural Language Interfaces (NLIs) can be traced back to the *Baseball* question/answer system [3] and the *LUNAR* system [4].

Although many efforts have been devoted into NLIs, the application of NLIs is still far from mature. One of the main barriers of this fact is that to enable systems (computers) to "understand" an arbitrary natural language sentence still remains an open problem. However, this open problem does not necessarily mean certain natural language sentences or phrases can not be processed and used to facilitate the interactions between humans and computers.

In this paper, we demonstrate how natural language can be used to enhance system usability when comparing to traditional methods. In detail, this paper discusses a prototype system where an NLI is developed to enhance the system usability when retrieving domain information. The support of natural language query aims at a *natural* way to extract certain piece of information units from information systems. The word '*natural*' here means users neither need to receive systematic training nor need to know the background structure of the stored information. Instead of introducing the approach for developing such an NLI, this paper demonstrates what the prototype system can do by presenting concrete examples so as to show the *effectiveness* and *efficiency* of the prototype system.

The rest of the paper is organised into four sections. Section 2 presents the related work. Section 3 introduces the background knowledge of the prototype system. Section 4 analyses concrete examples supported by the prototype system. Finally, Sect. 5 concludes this paper.

2 Related Work

As the front component of a system, an user interface directly reflects the functionalities that the system provides, thus the design of the interface has a huge influence on the usability of the system [5]. Accordingly, developing a usable interface becomes a critical factor for a good interaction [6].

Recently, the research field of NLIs covers a wide range of applications, such as (i) the NaLIX system [7] which targets to an XML database; (ii) an NLI for video databases [8]; (iii) a system which searches a corpus of surveillance video for clips that match spatial language queries [9] and; (iv) NLIs for structured information (ontologies) [10, 11] and for spatial databases [12]. While most of researchers focus on a single querying sentence, there are researchers for creating *conversation-based* NLI [13].

As to the usability of NLIs, Popescu et al. [14] pointed out that "people are unwilling to trade reliable and predictable user interfaces for intelligent but unreliable ones", therefore NLIs systems have to reduce the misunderstanding of their queries to a very low rate if at all. It should also be noted that the usability of a system depends on the intended use of the system. For example, the software for a museum kiosk should require minimum training, while the software for a bank cashier would require training but should be highly efficient [15]. Accordingly, a usable NLI not only needs to be able to complete certain tasks in a reliable manner, but also needs to be suitable to its application domain.

Although the research of NLIs has a long history and covers a wide range of applications, the support of complex sentences, such as sentences with nested relative clauses and (or) complex time phrases, has not been found, to the best of our knowledge. This paper demonstrates an NLI of a prototype system which accepts such kind of complex sentences.

3 Background of the Prototype System

This section briefly introduces the background of the prototype system so that readers can have a general idea of the whole system, thereafter can have a better understanding on the NLI.

3.1 The COINS Modelling Approach

The COINS modelling approach [16] manages complex information within the paradigm of recording live interactive records of information so that the recorded information can be queried and replayed.

The COINS model consists of three components: (1) a structure model employing Entity Relationship (ER) approach; (2) a rule mechanism which deals with logic constraints among attributes, entities and relationships and; (3) a recorder component which supports a mechanism of recoding and replaying the history of evolving information. Three kinds of recorders are provided in the COINS model: value recorders, rule recorders, and scene recorders, targeting at different type of information to be recorded respectively.

In the structure model, related entities are connected by relationships. At any given time unit, these connected entities and their attributes form a cluster of complex information within which atomic information units hold some logic connections with each other. Such a cluster of complex information is defined as an *information scene*, which is a snapshot of linking-relations among certain entities and relationships at a given time unit. When such entities and relationships are of concrete values/contents for an information scene, it yields an instance of the information scene.

3.2 The COINS Language

While the COINS modelling approach targets on the management of complex information in conceptual level, the counterparts in logic level and physical level are all necessary for systems that intend to employ the COINS approach. Accordingly, a specification language named COINS language has been proposed. The COINS language not only incorporates the COINS modelling approach, but also aims at being able to express the modelled domain knowledge in logic level so as to split the semantics of domain knowledge from programming languages.

The Extensible Markup Language Schema (XML Schema) is selected as the meta-language for the COINS language. The approach is to formulate certain rules for mapping a COINS model to the corresponding XML document, say a COINS document.

3.3 The Replaying Mechanism

Generally an information/database system responses to queries with static information. The ability of such an information/database system to responses to queries on past history with dynamic and evolving information has not been fully developed. Based on the recording mechanism of the COINS modelling approach, a replaying mechanism providing such an ability can be developed.

A simple way for replaying recorded instances of information scenes is to present them based on time order. That is, the changing history of the recorded information will be shown with the time that changes occur. A recorded instance of information scene can also be taken as a reference for displaying records of other scenes. In other words, the representation of the replaying history will focus on evolution of one scene, therefore its changing history will be shown sequentially, while other scenes whose time-stamps matches the time-stamp of the focused information scene can also be displayed accordingly. A more complex way to replay recorded information is to display designated scenes within a time interval or more generally, under given conditions. Such an approach would integrate the query mechanism and the visualisation techniques into the replaying process so as to facilitate the examination or inspection of the archived information.

3.4 Scenario of Information Flow

From a system's perspective, the proposed model and approach focus on different stages of developing and using the system. Figure 1 illustrates a scenario of information flow that an information system can adopt when incorporating the proposed model and approach.

For a real world application domain, the COINS modelling approach is used to capture the structure and semantics of information involved in the domain. There are two ways to represent the captured domain knowledge: (1) the diagram notation and; (2) the XML document. While the diagram notation is easy for humans to understand, the XML document is easy for computers to process.

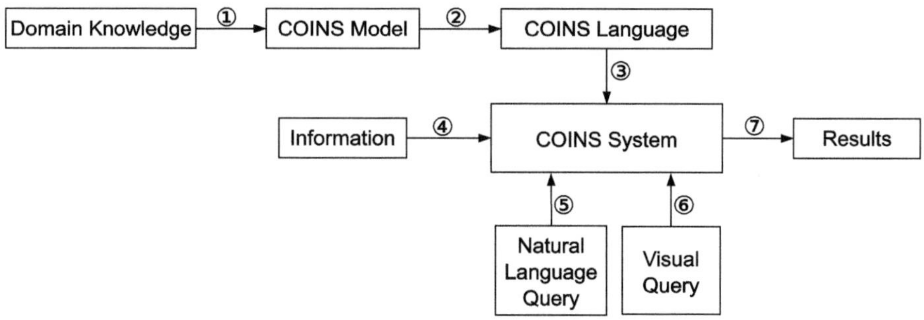

Fig. 1 Scenario of information flow

After implementing and integrating the COINS model into an information system, the defined information will be automatically recorded when users interact with the system. Ignoring the domain dependent functions of the system, one common shared requirement for the system is to retrieve the recorded (stored) information. Two query methods, natural language query and visual query, can be incorporated for fulfilling the requirement. Accordingly, the expected query results can be gained by using the two query methods.

3.5 The Prototype System

The puzzle game Sudoku is employed as the testbed for illustrating the proposed model and approach. A classical Sudoku puzzle is a Sudoku skeleton being filled with numbers for its cells in such a way that each of its 81 cells can either be blank, or contain a number within the range of 1 to 9. The key constraint for such Sudoku puzzles is that the number of a cell in (1) each row, (2) each column and, (3) each square must be unique within its corresponding row, column and square, respectively. The purpose the Sudoku game is to solve a given Sudoku puzzle by filling all its blank cells under such a constraint. A valid Sudoku puzzle must have and only have one solution.

A prototype system, which adopts the COINS modelling approach, is developed to demonstrate the recording/replaying mechanism as well as the ability to query recorded information with natural language. After users sign in the system, users can play the classical Sudoku game and the system will in turn automatically record the solving process as well as other related information. In addition to playing Sudoku, users can query the recorded information with the developed NLI and the visual query method.

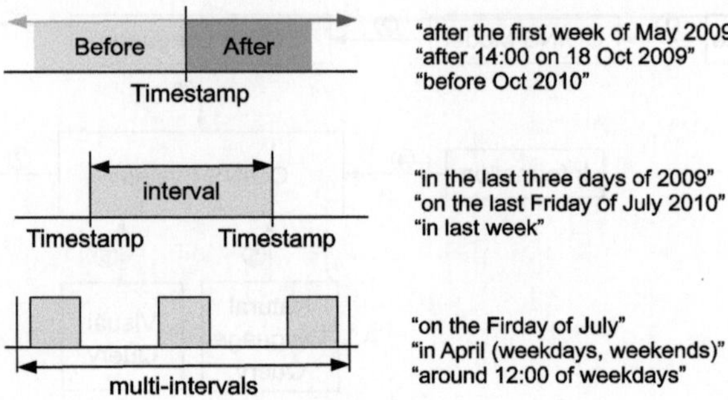

Fig. 2 Examples of three kinds of time phrases

4 Examples of Natural Language Query

This section shows example queries supported by the prototype system and the corresponding SQL statements, which are automatically translated by the prototype system.

4.1 Queries with Complex Time Phrases

As to the meaning of a time phrase, three categories can be classified: (1) an interval identified by a timestamp and a preposition "before" or "after"; (2) an interval determined whether by two timestamps or by a timestamp with another time interval; (3) multi-intervals which are given by a "day of week" or a "month of year". These three kinds of phrases are illustrated in Fig. 2 with examples.

The first two queries we consider are about the first category: an interval identified by a timestamp and a preposition "before" or "after".

> **Query 1:** list the advanced puzzle which Tom play before 14:00 on the 3rd Friday in Oct in 2009

```
SELECT user.name AS "user name",   play.city AS "play city",
   play.time AS "play time",           puzzle.name AS "puzzle name",
   puzzle.strvalue AS "puzzle strvalue", puzzle.runtime AS "puzzle runtime",
   puzzle.id AS "puzzle id"
FROM puzzle, play, user
WHERE TRUE
   AND puzzle.runtime >= 4           AND puzzle.runtime <= 6
   AND puzzle.id = play.puzzleid     AND user.id = play.userid
   AND user.name='Tom'
   AND play.time < timestamp '2009-10-16 14:00:01'
LIMIT 333
```

Query 2: list the advanced puzzle which Tom play after the first week in the 5th month in 2009

```
SELECT user.name AS "user name",   play.city AS "play city",
   play.time AS "play time",            puzzle.name AS "puzzle name",
   puzzle.strvalue AS "puzzle strvalue", puzzle.runtime AS "puzzle runtime",
   puzzle.id AS "puzzle id"
FROM puzzle, play, user
WHERE TRUE
   AND puzzle.runtime >= 4            AND puzzle.runtime <= 6
   AND puzzle.id = play.puzzleid      AND user.id = play.userid
   AND user.name='Tom'
   AND play.time > timestamp '2009-05-07 23:59:59'
LIMIT 333
```

Query 3:list the advanced puzzle which Tom play in the second three weeks in 2010

```
SELECT user.name AS "user name",   play.city AS "play city",
   play.time AS "play time",            puzzle.name AS "puzzle name",
   puzzle.strvalue AS "puzzle strvalue", puzzle.runtime AS "puzzle runtime",
   puzzle.id AS "puzzle id"
FROM puzzle, play, user
WHERE TRUE
   AND puzzle.runtime >= 4            AND puzzle.runtime <= 6
   AND puzzle.id = play.puzzleid      AND user.id = play.userid
   AND user.name='Tom'
   AND "date_part" ('WEEK' , play.time) >= 4
   AND "date_part" ('WEEK' , play.time) <= 6
   AND "date_part" ('ISOYEAR' , play.time) = 2010
LIMIT 333
```

It would be easy to use the form based time picker when the exact date and time are already known. But if users only know certain attributes of the date and time, natural language would be more powerful and flexible. Considering the time phrases in *Query 1* and *Query 2*, the form based interface would be difficult to be designed for supporting these kinds of time phrases.

It should be noted that the appeared month and year in the time phrases are vital to the meaning of the whole phrases. The phrase "on the 3rd Friday in Oct in 2009" is different from "on the 3rd Friday in 2009", and these two are all different to "on the 3rd Friday in Oct", where the current year will be used. In addition, if users do not give a limit count, the default number will be used.

As an additional example for time phrases, *Query 3* shows another form of time intervals. In the phrase "in the second three weeks in 2010", week is treated as the basic time unit for reference, and the result of "the second three weeks" means "from the 4th week to the 6th week".

4.2 *Queries with Abstract Adjectives*

The next two queries demonstrate the support of "abstract" adjectives. In *Query 4*, the meaning of "new puzzles" can be defined as: "the puzzles which have not been played by the person who gives the query", therefore the attribute "new" depends upon the user and his/her past actions. Generally, an adjective is regarded as "abstract" if its meaning can be defined by certain conditions. Comparing to "abstract" adjectives, common adjectives usually describe the static property or attribute of an object, such as the word "advanced" used in *Query 3*.

In *Query 5*, the "unfinished puzzles" refers to those puzzles that have been played but not been solved. Therefore, the adjective "unfinished" in *Query 5* is also an abstract adjective. The two queries show that natural language is much simpler and easier to use than other kinds of languages, especially those formal languages like SQL.

Query 4: get 10 new puzzles

```
SELECT puzzle.strvalue AS "puzzle strvalue",    puzzle.id AS "puzzle id",
   puzzle.name AS "puzzle name"
FROM puzzle
WHERE TRUE
   AND puzzle.id NOT IN
      (SELECT play.puzzleid FROM play WHERE play.userid=0)
LIMIT 10
```

Query 5: list unfinished puzzles

```
SELECT play.city AS "play city",              play.id AS "play id",
   play.time AS "play time",                  puzzle.name AS "puzzle name",
   puzzle.strvalue AS "puzzle strvalue",      puzzle.id AS "puzzle id"
FROM puzzle, play
WHERE TRUE
   AND play.puzzleid = puzzle.id              AND play.processid NOT IN
      (SELECT solve.processid FROM play,solve
          WHERE play.processid=solve.processid)
LIMIT 333
```

4.3 *Queries About Processes*

One feature of natural language is *context sensitive*. In *Query 5*, the adjective *unfinished* is regarded as "abstract" in phrase "unfinished puzzle". However, when the word modifies the noun "process", it describes a static property or attribute of a process. In other words, an attribute *Status* can be defined for a process so as to indicate whether the process is finished or unfinished. Accordingly, an NLI needs to distinguish these situations so as to be able to correctly translate "same" words into "different" results, as shown in *Query 5* and *Query 6*.

Query 6: list unfinished processes which were produced by Tom

```
SELECT user.name AS "user name",                    process.id AS "process id",
   process.usercellnum AS "process usercellnum",
   process.status AS "process status",
   process.totalsteps AS "process totalsteps",      play.city AS "play city",
   process.blankcellnum AS "process blankcellnum",  play.time AS "play time"
FROM process, play, user
WHERE TRUE
   AND process.status='unfinished'
   AND process.id = play.processid
   AND user.id = play.userid                        AND user.name='Tom'
LIMIT 333
```

A process may refer to series of actions. In the prototype system, two actions need to be mentioned for a process: the action "play" and the action "solve", which together indicate a process of solving a Sudoku puzzle. A process generally has a duration time, therefore one kind of query is about the duration of the process. One example of this kind of queries is shown in *Query 7*. The time phrase "in 10 minutes" specifies the duration time and the word "solve" indicates the end action of the process.

It is worthy of notice that the functions of "in 10 minutes" in *Query 7* are totally different from the phrase "in Oct 2010", which is used in *Query 8*. While the former specifies the duration of a process, the later expresses a time interval in which the action "solve" happened.

Query 7: list the user who solve the difficult puzzle in ten minutes

```
SELECT user.name AS "user name",             process.usercellnum AS "process usercellnum",
   process.id AS "process id",               process.blankcellnum AS "process blankcellnum",
   solve.time AS "solve time",               process.totalsteps AS "process totalsteps",
   process.status AS "process status",       play.city AS "play city",
   solve.city AS "solve city",               play.time AS "play time",
   puzzle.id AS "puzzle id",                 puzzle.strvalue AS "puzzle strvalue",
   puzzle.runtime AS "puzzle runtime",       puzzle.name AS "puzzle name"
FROM process, puzzle, solve, user, play
WHERE TRUE
   AND puzzle.id = solve.puzzleid            AND process.status = 'finished'
   AND puzzle.runtime >= 4                   AND puzzle.runtime <= 6
   AND user.id = solve.userid                AND solve.id = process.solveid
   AND play.id = process.playid
   AND play.time - solve.time <= interval '10 minute'
LIMIT 333
```

Query 8: list the user who solve the difficult puzzle in Oct 2010

```
SELECT user.name AS "user name",     solve.time AS "solve time",
  solve.city AS "solve city",        puzzle.name AS "puzzle name",
  puzzle.strvalue AS "puzzle strvalue",  puzzle.runtime AS "puzzle runtime",
  puzzle.id AS "puzzle id"
FROM puzzle, solve, user
WHERE TRUE
  AND puzzle.id = solve.puzzleid     AND user.id = solve.userid
  AND puzzle.runtime >= 4            AND puzzle.runtime <= 6
  AND user.id = solve.userid
  AND "date_part" ('MONTH' , solve.time) = 10
  AND "date_part" ('ISOYEAR' , solve.time) = 2010
LIMIT 333
```

4.4 Queries About Action Frequency

The next three queries show the support of querying about action frequency. For *Query 9* and *Query 10*, the only difference is the position of the phrase "on Monday". However, *Query 9* emphasises that "Tom signs in from the city for more than three times", while *Query 10* emphasises that "Tom signs in on Monday for more than three times". Accordingly, the condition clauses of the corresponding SQL statements are different.

In addition to time and positions, an action frequency can also be queried for action-object pairs, as shown in *Query 11*.

Query 9: list the city where Tom sign in for more than three times on Monday

```
SELECT user.name AS "user name",   signin.time AS "signin time",   signin.city AS "signin city"
FROM signin, user WHERE TRUE
  AND user.id = signin.userid      AND user.name='Tom'
  AND signin.city IN (SELECT signin.city FROM signin
        GROUP BY signin.city HAVING count(signin.id) > 3)
  AND "date_part"('ISODOW' , signin.time) = 1
LIMIT 333
```

Query 10: list the city where Tom sign in on Monday for more than three times

```
SELECT user.name AS "user name",   signin.time AS "signin time",
  signin.city AS "signin city"
FROM signin, user WHERE TRUE
  AND user.id = signin.userid      AND user.name='Tom'
  AND to_char(signin.time, 'YYYY-MM-DD') IN
        (SELECT to_char(signin.time, 'YYYY-MM-DD') FROM signin
        GROUP BY to_char(signin.time, 'YYYY-MM-DD') HAVING count(signin.id) > 3)
  AND "date_part"('ISODOW' , signin.time) = 1
LIMIT 333
```

Query 11: list the user who play a puzzle less than 33 times

```
SELECT user.name AS "user name",    play.city AS "play city",
   play.time AS "play time",            puzzle.name AS "puzzle name",
   puzzle.strvalue AS "puzzle strvalue", puzzle.id AS "puzzle id"
FROM play, user, puzzle
WHERE TRUE
   AND puzzle.id = play.puzzleid        AND user.id = play.userid
   AND play.puzzleid IN (SELECT play.puzzleid FROM play
         GROUP BY play.puzzleid HAVING count(play.id) < 33)
LIMIT 333
```

4.5 *Queries Related to the Quantifier of Actors/Objects*

While actions have frequencies, the *quantifier* of the actors or the objects of the actions may also be given in queries. *Query 12* and *Query 13* are examples of such kind of queries. It is worth to notice that the corresponding SQL statements usually involve sub queries which need a "group by" clause, while the natural language queries are much easier to write and understand.

Query 12: list the puzzle which more than 3 users play

```
SELECT user.name AS "user name",    play.city AS "play city",
   play.time AS "play time",            puzzle.name AS "puzzle name",
   puzzle.strvalue AS "puzzle strvalue", puzzle.id AS "puzzle id"
FROM puzzle, play, user
WHERE TRUE
   AND puzzle.id = play.puzzleid        AND user.id = play.userid
   AND play.puzzleid IN (SELECT play.puzzleid FROM play
         GROUP BY play.puzzleid HAVING count(DISTINCT play.userid) > 3)
LIMIT 333
```

Query 13: list the user who played less than 6 puzzles

```
SELECT user.name AS "user name",    play.city AS "play city",
   play.time AS "play time",            puzzle.name AS "puzzle name",
   puzzle.strvalue AS "puzzle strvalue", puzzle.id AS "puzzle id"
FROM play, user, puzzle
WHERE TRUE
   AND puzzle.id = play.puzzleid        AND user.id = play.userid
   AND play.userid IN (SELECT play.userid FROM play
         GROUP BY play.userid HAVING count(DISTINCT play.puzzleid) < 6)
LIMIT 333
```

Fig. 3 Screen shot of the results of executing Query 14

4.6 Queries with Nested Relative Clauses and Complex Time Phrases

As the last example, *Query 14* shows a query which has two nested relative clauses and a "from ... to" time phrase. This example demonstrates the ability to deal with queries with complex structure. The screen shot of the results of executing this query in the prototype system is shown in Fig. 3.

> **Query 14:** list the difficult puzzle which was played from 14:00 on the last Monday in Jan 2011 to 18:00 on the 3rd Friday in May 2011 by the user who signed in from Dublin in April 2011

```
SELECT user.name AS "user name",        play.city AS "play city",
   play.time AS "play time",             puzzle.name AS "puzzle name",
   puzzle.strvalue AS "puzzle strvalue", puzzle.id AS "puzzle id"
   puzzle.runtime AS "puzzle runtime",
FROM puzzle, play, user
WHERE TRUE
   AND puzzle.runtime >= 4               AND puzzle.runtime <= 6
   AND puzzle.id = play.puzzleid         AND user.id = play.userid
   AND play.time > timestamp '2011-01-24 14:00:01'
   AND play.time < timestamp '2011-05-20 18:00:59'
   AND play.userid IN
         (SELECT signin.userid FROM public.signin WHERE TRUE
         AND public.user.id = public.signin.userid
         AND "date_part" ('MONTH' , public.signin.time) = 4
         AND "date_part" ('ISOYEAR' , public.signin.time) = 2011
         AND public.signin.city = 'Dublin' )
LIMIT 333
```

5 Conclusion

Although the research of NLIs has been engaged for years, the support of comprehensive time phrases and complex sentence structures has not been found. This paper demonstrates a prototype system with an NLI which supports complex time phrases and nested relative clauses.

Two conclusions can be drawn based upon the presented examples and the experiments we made for evaluating the prototype system: (1) natural language can be used as a query language for retrieving information from systems; (2) queries written with natural language are much easier to write and understand than formal query languages (such as SQL). Furthermore, when specifying a query, the form of using natural language is generally much shorter than the form of using a formal language.

Currently, only part of grammar rules (such as the support of relative clauses and comprehensive time phrases) are supported by the prototype system, further research needs to be done for supporting other grammar rules so that the queries can be written as natural and easy as possible.

Acknowledgments The authors would like to thank the China Scholarship Council (CSC) and Dublin Institute of Technology (DIT) for their funding and support.

References

1. ISO: Ergonomic requirements for office work with visual display terminals (vdts). ISO 9241-11 (1988)
2. Tullis, T.S., Albert, W.: Measuring the User Experience: Collecting, Analyzing, and Presenting Usability Metrics. Morgan Kaufmann, Burlington (2008)
3. Green, Jr., B.F., Wolf, A.K., Chomsky, C., Laughery, K.: Baseball: an automatic question-answerer. In: Papers Presented at the May 9–11, 1961, Western Joint IRE-AIEE-ACM Computer Conference, IRE-AIEE-ACM '61 (Western), pp. 219–224. ACM (1961)
4. Woods, W.A.: Progress in natural language understanding: an application to lunar geology. In: Proceedings of the National Computer Conference and Exposition, AFIPS '73, pp. 441–450. ACM, 4–8 June 1973
5. Jaspers, M.W., Steen, T., van den Bos, C., Geenen, M.: The think aloud method: a guide to user interface design. Int. J. Med. Inform. **73**(11–12), 781–795 (2004)
6. Seneler, C.O., Basoglu, N., Daim, T.: Interface feature prioritization for web services: case of online flight reservations. Comput. Human Behav. **25**(4), 862–877 (2009)
7. Li, Y., Yang, H., Jagadish, H.: Constructing a generic natural language interface for an xml database. In: Advances in Database Technology—EDBT 2006, pp. 737–754. Springer Berlin/Heidelberg, Munich, Germany (2006)
8. Erozel, G., Cicekli, N.K., Cicekli, I.: Natural language querying for video databases. Inf. Sci. **178**(12), 2534–2552 (2008)
9. Tellex, S., Roy, D.: Towards surveillance video search by natural language query. In: Proceeding of the ACM International Conference on Image and Video Retrieval (CIVR '09), pp. 38:1–38:8. ACM, Santorini, Fira, Greece (2009)
10. Tablan, V., Damljanovic, D., Bontcheva, K.: A natural language query interface to structured information. In: Proceedings of the 5th European Semantic Web Conference on The Semantic

Web: Research and Applications (ESWC'08), pp. 361–375. Springer-Verlag, Tenerife, Canary Islands, Spain (2008)
11. Wang, C., Xiong, M., Zhou, Q., Yu, Y.: Panto: A portable natural language interface to ontologies. In: Proceedings of the 4th European conference on The Semantic Web: Research and Applications, ESWC '07, pp. 473–487. Springer-Verlag, Innsbruck, Austria (2007)
12. Mador-Haim, S., Winter, Y., Braun, A.: Controlled language for geographical information system queries. In: Proceedings of Fifth International Workshop on Inference in Computational Semantics, pp. 57–66. Buxton, England (2006)
13. Owda, M., Bandar, Z., Crockett, K.: Conversation-based natural language interface to relational databases. In: WI-IATW '07: Proceedings of the 2007 IEEE/WIC/ACM International Conferences on Web Intelligence and Intelligent Agent Technology—Workshops, pp. 363–367. IEEE Computer Society, Washington, DC, USA (2007)
14. Popescu, A.M., Etzioni, O., Kautz, H.: Towards a theory of natural language interfaces to databases. In: Proceedings of the 8th International Conference on Intelligent User Interfaces, pp. 149–157. ACM (2003)
15. Ferré, X., Juristo, N., Windl, H., Constantine, L.: Usability basics for software developers. IEEE Softw. **18**, 22–29 (2001)
16. Zhou, E., Wu, B., Ding, Y., Wu, J.: Coins: a modelling approach for recording & replaying complex information. In: Proceedings of The IADIS International Conference Information Systems, pp. 37–44. Avila, Spain (2011)

Building a Standard Amazigh Corpus

Siham Boulaknadel and Fadoua Ataa Allah

Abstract Natural language processing is showing more interest in the Amazigh language in recent years. Suitable resources for Amazighe are becoming a vital necessity for the progress of this research. Corpora are an important resource but Amazighe lacks sufficient resources in this field, therefore we have been conducted to build an Amazighe corpus. In this paper, we present preliminary results experiment with a corpus for Standard Amazighe. We selected samples of published data from different Amazighe varieties. The selection was driven mainly by the amount of data available. We still demonstrate the completeness and representativeness of this corpus using metrics and show its suitability for language engineering experiments.

1 Introduction

Corpora of a variety of types have been developed for a different research and teaching purposes. English, being the main international language, has received the greatest attention among the research community [1]. Amazighe is a language which is spoken in Morocco, Algeria, Tunisia, Libya, and Siwa and rivalled English in number of mother-tongue speakers. However, little attention has been devoted to Amazighe. Although there has been some effort in Europe but the progress in this field is still limited. Generally speaking, there is widespread ignorance of Amazighe in western universities, due not only to historical and cultural separation but also to the complexity of the Amazighe language structure and its unique script. Furthermore, progress has been obstructed by lack of efficient corpus analysis tools specifically tokenisers,

---

S. Boulaknadel (✉) · F. Ataa Allah
Institut Royal de la Culture Amazighe, Avenue Allal El Fassi Madinat Al Irfane, Rabat, Morocco
e-mail: boulaknadel@ircam.ma

F. A. Allah
e-mail: ataaallah@ircam.ma

M. Kudělka et al. (eds.), *Proceedings of the Third International Conference on Intelligent Human Computer Interaction (IHCI 2011), Prague, Czech Republic, August, 2011*, Advances in Intelligent Systems and Computing 179, DOI: 10.1007/978-3-642-31603-6_8, © Springer-Verlag Berlin Heidelberg 2013

taggers, morphological analyzers and optical character readers, which are required for developing and exploiting an Amazighe corpus. In this paper we are interested in contributing a publicly available Amazighe corpus of written text to the natural language processing community. To our knowledge the corpus presented in this paper is the first one to deal with Amazighe. The rest of the paper is structured as follows: in Sect. 2 we discuss the utility to develop a corpus for Amazighe language. Then, in Sect. 3 we describe Amazighe language characteristics. In Sect. 4 we present the corpus peculiarities. Sects. 5 and 6 give an assessment and an analysis of our corpus. Finally, in Sect. 7 we draw some conclusions and describe the work to be done in the future.

2 Why an Amazighe Corpora

Since Amazighe texts are becoming available on the web, and some linguists advocate use of Web data for research, one may ask the question: do we really need a corpus? The use of a computer corpus is not only to collect a big file of different texts and store it on the computer, but rather to arrange the texts and put them in a certain format so that they can be used by search tools and the results of the search can be displayed in a way that is significant and helpful to the linguist, teacher and learner particularly at the higher level. For example, teachers and learners can discover the use of a word in different types of texts to see how commonly this word is used, how many meanings it has, what syntactic environment it occurs in, whether the word has the same frequency of occurrence in all types of texts. Teachers can identify the most frequent words and select them as a basis for their material. There is also the study of analyzing syntactic structures. The importance of corpus to linguistic studies is appreciated. Corpora enable the linguist to make objective statements and to contextualize their analyses of language. However, interest in corpus-based empirical research is on the increase in Amazighe linguistics. Due to these reasons we aimed to develop an Amazighe corpus that would help in the study of Amazighe language.

3 Peculiarities of the Amazighe Language

3.1 Historical Background

The Amazighe language, known as Berber or Tamazight, stretches over an area that is so vast in Africa, from the Canary Islands to the Siwa Oasis in the North, and from the Mediterranean coast to Niger, Mali and Burkina-Fasso in the South. Historically, the Amazighe language has been autochthonous and was exclusively reserved for familial and informal domains [2]. The language is characterized by the proliferation of varieties due to historical, geographical and sociolinguistic factors. Their most outstanding feature is the great distance that exists between the countries where Amazigh is spoken, which hinders the processes of language exchange and planning among them. This situation causes the existence of greater differences and more heterogeneity among the varieties of Amazighe. In Morocco, we may distinguish three big regional varieties, depending on the area and the communities: Tarifiyt in the Northern, Tamazight in the Center and Tashlhiyt in the South-west and the High Atlas of the country. According to official statistic data compiled in 1994, the Amazigh language is spoken by some 30 10 Million inhabitants. As regards the Amazighe varieties, there are 2,5 million Tachelhit-speakers in the south of Morocco-known as Sus-, 3 million Tamazight-speakers in the Atlas Mountains, and about 1,7 million Tarifit-speakers in the Riff [3]. Politically, the Amazighe language has enjoyed its status at different levels, depending on the country where this language exists. In Morocco, the status of Amazighe has achieved the most advanced level, especially by the foundation of the Royal Institut of Amazighe Culture (IRCAM) in the Dahir on October 17th 2001, which stipulates that the Amazighe culture is a "national matter" and, this being the case, is a concern of all the citizens.

3.2 Amazighe Characteristics

As regards the origin of the Amazighe language, several efforts have been made to find out its exact source and its subsequent expansion, but in the end the most widely accepted theory among specialists is the one that suggests an Afro-Asiatic origin, a family in which languages like old Egyptian, Cushitic, or Chadic are found [4].

Amazighe Graphical System

Like any language passes throw oral to a written mode, the Amazighe language has been in need of a graphic system. In this aspect, there were in Morocco three competing graphic systems for Amazighe. These competing systems are those of Arabic, widely used for religion and rural poetry writing; Latin supported by the International Phonetic Alphabet (IPA), used particularly by berberists since early works of missionaries; and Tifinaghe, the ancestral writing system which has been preserved by Touareg. In Morocco, the choice ultimately fell on Tifinaghe for technical, historical and symbolic reasons. Since the Royal declaration on February, 11th 2003, Tifinaghe has become an official graphic system for writing Amazighe, particularly in schools. Thus, the IRCAM has developed an alphabet system called Tifinaghe-IRCAM. This alphabet is based on a graphic system towards phonological tendency. This system does not retain all the phonetic realizations produced, but only those that are functional (For more details see [5]). The Tifinaghe-IRCAM alphabet contains:

- 27 consonants;
- 2 semi-consonants;
- 4 vowels.

Directionality

Historically, in ancient inscriptions Amazighe language was written horizontally from left to right, from right to left; vertically upwards, downwards or in boustrophedon. However, the orientation most often adopted in Amazighe language script is horizontal and from left to right, which is also adopted in IRCAM-Tifinaghe writing.

Tifinaghe Encoding

Over the last several years, the encoding of scripts in Unicode/ISO 106461 for rendering complex scripts has advanced rapidly. However, encoding the Tifinaghe script was not a straight prospect as one might hope—partly owing to the many variants of Tifinaghe used in different parts of the amazighe world. The encoding is composed of four Tifinaghe character subsets: the basic set of IRCAM, the extended IRCAM set, other Neo-Tifinaghe letters in use, and modern Touareg letters. The two first subsets constitute the sets of characters chosen by IRCAM. While, the first is used to arrange the orthography of different Moroccan Amazighe dialects, the second subset is used for historical and scientific use. The letters are classified in accordance with the order specified by IRCAM. Other Neo-Tifinaghe and Touareg letters are interspersed according to their pronunciation. Thus, the UTC accepts the 55 Tifinaghe characters for encoding in the range U+2D30..U+2D65, U+2D6F, with Tifinaghe block at U+2D30..U+2D7F [6].

4 Building the Corpus

A corpus is not only a collection of texts. Rather, a corpus seeks to represent a part of a language or even a language. Therefore, the suitable design for a corpus depends upon what it is meant to represent. The representativeness of the corpus determines the kinds of research questions that can be addressed. In this context, we planned to constitute a corpus which will contain 1 million words in future for representing and supporting various linguistic studies in Amazighe language. In developing the collection of texts for our corpus, a number of decisions related to corpus design such as representativeness, diversity, balance and size were taken into consideration:

- Some sources of texts were determined.
- A range of different genres of writing would be gathered for inclusion in the corpus.
- Each genre would be separated into text samples.
- Collecting the texts to express all regional variations of the Amazighe Language.

4.1 Pre-processing Stage

There are a number of universal considerations to stand in mind when starting the process of computerizing the written corpus. The pre-processing stage is an important

Table 1 Corpus description

Corpus description	
Token number	76,437
Distinct token	14,014

stage for the compiled texts to be organized to assist in analysis and search processes on the corpus. The collected text needs some pre-processing stages to be prepared.

1. HTML to TXT conversion
   As the texts to be integrated in a corpus will be edited with a word processing program, it could be tempting to save digital text in a file format used by a wordprocessing program. However, it is from the onset to save text in TXT format, as this is the standard format used for text integrated in corpora, and to employ a simple text editor to work with texts rather than a standard word-processing program.
2. Removing HTML codes
   Some hidden characters that were found when HTML to TXT conversion occurred need to be striped because such hidden characters may cause problems in both search and analysis stages. So removing HTML codes is an important stage that helps in cleaning up any of these hidden characters from the TXT files.
3. Coding file names
   When developing a corpus, it is matter-of-fact to save individual text in separate files stored in directories that reflect the hierarchical structure of the corpus. Arranging a corpus into a serie of directories and subdirectories makes working with the corpus much easier. Our corpus consists of a main directory that containing all the written texts. This directory is separated into a serie of subdirectories containing the main types of writing that are collected. Each text obtains an additional file extension, coding files name with meta information in turn to specify the directory and subdirectory that the text file fits in. It is very essential to rename the file according to some parameters to help search processes and to avoid overlapping among file names in case they are located in the same folder. This coding process of text file names will enable us in the future to customize the search in any part of the corpus. This work is still in progress, the current status of our corpus is presented in Table 1.

5 Corpus Assessments

Using available tools we first experimented by applying some statistical and probability tests, such as Zipf's law and the Mandelbrot formula. These tests are practical for describing the frequency distribution of the words in the corpus. Also they are well-known tests for gauging data sparseness and providing evidence of any imbalance of the dataset. According to Zipf's law, if we count up how often each word occurs in a corpus and then list these words in the order of their frequency of

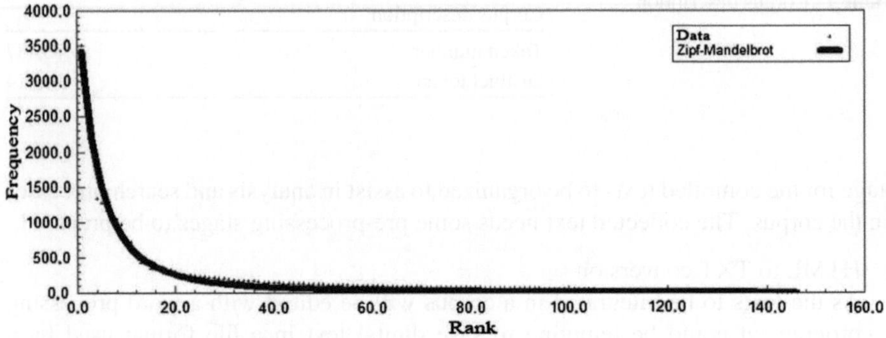

Fig. 1 Word frequency versus rank

occurrence, then the relationship between the frequency of a given word f and its rank r in the list will be a constant k such that:

$$f.r = k \tag{1}$$

So in our corpus, we started with one file and increasingly adding more files to a corpus and checking the behavior of the relation between the rank and the frequency. An improved theory of the Zipf's law is the Mandelbrot distribution. Mandelbrot annotes that "although Zipf's formula gives the general shape of the curves, it is very bad in reflecting the details" [7]. So to reach a closer fit to the empirical distribution of words, Mandelbrot derived the following formula for relation between the frequency and the rank:

$$f = P(r + R)^{-B} \tag{2}$$

where P, B, and R are parameters of the text that collectively measure the richness of the text's use of words. The common factor is that there is still a hyperbolic relation between the rank and the frequency as in the original equation of Zipf's law, (see Fig. 1).

6 Analyses

We start assessing the data collected after the preprocessing to have the data in a usable format that the tools could work on. First, we began with one dataset and checked the contribution of every document to the corpus construction, by checking the number of distinct words added by every document and interpolating in a function that simulates the behavior of this process. Table 2 and Fig. 2 show details about this. From the function simulating the corpus construction, we conclude that the speed

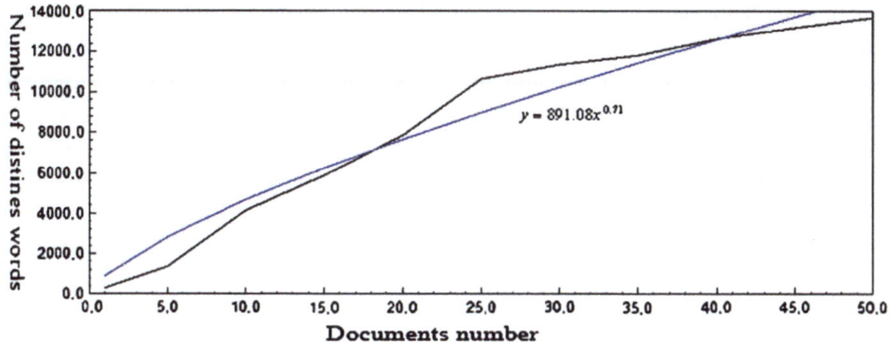

Fig. 2 Number of distinct words versus number of documents

Table 2 Contribution of the document to the corpus

Number of files	Number of words	Number of distinct words
10	16,424	4,093
20	42,553	7,854
30	60,956	11,350
40	68,169	12,640
50	76,437	13,682

and the amount of data added after a certain size of the corpus will not contribute significantly to the vocabulary richness of the corpus.

For the analysis purpose, we used the function got from the interpolation of Table 2

$$y = k.x^{-a} \tag{3}$$

For this function the acceleration will be the second derivative of the function(3),

$$y' = -k.a.x^{-a-1} \tag{4}$$

$$y'' = k.a.(a+1).x^{-a-2} \tag{5}$$

Figure 2 shows a representation of the Eq. 5 for the values of a = 0.71 and k = 891.08

7 Conclusions and Future Work

It is very essential to know that building a significant corpus for the purpose of studying the language is important, due to the issue that this kind of research could not depend on individual cognitive perception. In this view, corpus-based approaches are very helpful. This can be very clear in lexical studies, grammar, semantics, NLP and can be a boost for improving different researches in Information Retrieval, Machine Translation in general, with an authentic quality and adequate quantity. Although the corpus was tested for completeness and representativeness, expanding the collection for more texts and including other types of literature (i.e. official transcripts, stories, etc) will improve the quality and give more confidence for the results induced from it.

References

1. McEnery, T., Wilson, A.: Corpus linguistics. Edinburgh University Press, Edinburgh (1996)
2. Boukous, A.: Société, langues et cultures au Maroc: Enjeux symboliques. Najah El Jadida, Casablanca, Maroc (1995)
3. Moustaoui, A.: The Amazigh language within Morocco's language policy, Dossier 14, University of Autònoma de Madrid (2003)
4. Ouakrim, O.: Fonética y fonología del Bereber. Survey, University of Autònoma de Barcelona (1995)
5. Ameur, M., Bouhjar, A., Boukhris, F., Boukouss, A., Boumalk, A., Elmedlaoui, M., Iazzi, E.M., Souifi, H.: Initiation à la langue amazighe. IRCAM, Rabat, Morocco (2004)
6. Andries, P.: Unicode 5.0 en pratique, Codage des caractères et internationalisation des logiciels et des documents. Dunod, France (2008)
7. Manning, C., Schütze, H.: Foundations of Statistical Natural Language Processing, MIT Press, Cambridge (1999)

Simple Stemming Rules for Arabic Language

Hussein Soori, Jan Platoš and Václav Snášel

Abstract Processing of Arabic language is eminent for the fact that currently the number of computer and Internet users in the Arab word is growing tremendously. The problem of stemming is very important in information retrieval, knowledge mining and language processing. Arabic has very complex morphology and stemming rules that must deal with many specific properties of Arabic. This paper describes very simple rules for stemming of Arabic words. Two of these rules are universal, i.e. they are applicable to any word category, and one rule for each of the four categories: nouns, verbs, adverbs and adjectives. The rules were more successful in case of adverbs. As for nouns, verbs and adjectives, some errors occurred especially in case of suffix processing.

1 Introduction

The importance of conducting research in the area of Arabic language processing is called for two main reasons: the rapidly growing number of computer and internet users in the Arab world and the fact that the Arabic language is the sixth most used language in the world today. It is also one of the six languages used in the United Nations (www.un.org). Another important factor is, after the Latin alphabet (Latin alphabet), Arabic alphabet is the second-most widely used alphabet around the world. Arabic script has been used and adapted to such diverse languages as the

H. Soori (✉) · J. Platoš · V. Snášel
Department of Computer Science, FEECS, VSB-Technical University of Ostrava,
17. listopadu 15, Poruba, 70800Ostrava, Czech Republic
e-mail: sen.soori@vsb.cz

J. Platoš
e-mail: jan.platos@vsb.cz

V. Snášel
e-mail: václav.snášel@vsb.cz

M. Kudělka et al. (eds.), *Proceedings of the Third International Conference on Intelligent Human Computer Interaction (IHCI 2011), Prague, Czech Republic, August, 2011*, Advances in Intelligent Systems and Computing 179, DOI: 10.1007/978-3-642-31603-6_9,

Slavic tongues (also known as Slavic languages), Spanish, Persian, Urdu, Turkish, Hebrew, Amazigh (Berber), Swahili, Malay (Jawi in Malaysi and Indonesia), Hausa, and Mandinka (in West Africa), Swahili (in East Africa), Sudanese, and some other languages [1].

1.1 Challenges to Be Considered When Working with Arabic Texts

One of challenges faced by researchers in this area has to do with the special nature of Arabic script in Arabic. Arabic is written horizontally from right to left. The shape of each letter depends on its position in a word-initial, medial, and final. There is a fourth form of the letter when written alone. The letters alif, waw, and ya (standing for glottal stop, w, and y, respectively) are used to represent the long vowels a, u, and i. This is very much different from Roman alphabet which is naturally not linked. Other orthographic challenges can be the persistent and widespread variation in the spelling of letters such as hamza (ء) and ta' marbuTa (ة), as well as, the increasing lack of differentiation between word-final ya (ي) and alif maqSura (ى). Another problem that may appear is that typists often neglect to insert a space after words that end with a non-connector letter such as ز، ر، و [2]. In addition to that, Arabic has eight short vowels and diacritics (see Fig. 1). Typists normally ignore putting them in a text, but in case of texts where they exist, they are pre-normalized -in value- to avoid any mismatching with the dictionary or corpus in light stemming. As a result, the letters in the decompressed text, appear without these special diacritics.

Diacritization has always been a problem for researches. According to Habash [3], since diacritical problems in Arabic occur so infrequently, they are removed from the text by most researchers. Other text recognition studies in Arabic include, Andrew Gillies et. al. [4], John Trenkle et al. [5] and Maamouri et al. [6]. The problem of syllabification was studied in [7].

This paper is organized as follows: Sect. 2 describes Arabic morphology, Sect. 3 describes the rules for stemming, Sect. 4 illustrates the experimental results, and the last section restates the main points and sums up the contents of this paper.

2 Arabic Morphology

Arabic is considered as one of the highly inflectional languages with complex morphology. Unlike most other languages, it is written from right to left. It consists of 28 main letters. As mentioned earlier, the shape of each letter depends on its position in a word-initial, medial, and final, as well as, a fourth form of the letter when written alone (separate). One example of this can be given for the letter (ع) as depicted in Table 1. The full arabic alphabet is depicted in Table 2.

Fig. 1 Short vowels and diacritics marks

(◌َ , ◌ِ , ◌ُ , ◌ْ , ◌ً , ◌ٍ , ◌ٌ , ◌ّ)

Table 1 Shape of a character in dependence on its position in a word

Initial	Medial	Final	Separate
عـ	ـعـ	ـع	ع

Table 2 Arabic alphabet(s)

خ	ح	ج	ث	ت	ب	ا
kha	haa	jiim	thaa	taa	baa	alif
ص	ش	س	ز	ر	ذ	د
saad	shiin	siin	zaay	raa	thaal	daal
ق	ف	غ	ع	ظ	ط	ض
qaaf	faa	ghayn	ayn	thaa	taa	daad
ي	و	ه	ن	م	ل	ك
yaa	waaw	ha	nuun	miim	laam	kaaf

Table 3 Application of diacritics changes word meaning

Word	Meaning	Part of speech
رجُلٌ	Man	Noun (subject)
رجُل	Man	Noun (subject)
رِجْل	Foot	Noun
رَجِل	To go an foot (rather then e.g., ride a bike)	Verb

In addition to the 28 letters, in some cases we may encounter a dual letters such as (لا) which is a combination of two letters (ل + ا). These dual letters are considered as one letter.

Other than letters, other factor determine the word identity and in many instances can change the meaning and part of speech. This factor is the eight short vowels and diacritics (see Fig. 1). An example for (رجل) is given in the following table where we can see the total change in word category and meaning as a result of adding the diactricals which resulted in producing three different words in meaning and three different parts of speech for the same three letter (رجل) as depicted in Table 3.

Never the less, it is always advised that these vowels and diacritics are often normalized before processing in most light stemming or morphological approaches [3]. Mainly the reasons for not including them in the word processing is the claim that they do occur so infrequently, and that in Modern Standard Arabic (MSA), people tend not to use them and, as a result of that, the meaning is left for the native speaker's intuition, or, in some cases, can be determined from the context. This problem is still waiting for a challenging attempt where the processor is ready to process words with or without diacritics, without needing to normalize words.

Another morphological feature in Arabic is that, unlike Roman letters which are separated naturally, Arabic has an agglutinated nature (as mentioned above) where letters are linked to each other in some cases, while unlinked in some other cases, depending on position of the letter in the root, stem and word level. For example, in English the pronoun (he) in (he plays) is separated from the following verb (plays), while in Arabic the pronoun is represented by the letter (يـ) which is linked to the root verb لعب to form يلعب (he plays). The same is true when it comes to different kinds of Affixes.

Arabic has four types of affixes. Prefixes: these are letters (normally one) that change the tense of the verb from past to present, such as the letter (يـ) in case of the verb لعب and يلعب above. Suffixes: these represent the inflectional terminations (endings) of verbs, as well as, the female and dual/plural markers for the nouns. Postfixes: these are the pronouns attached at the end of the word. Antefixes: these are prepositions agglutinated to the beginning of words.

One of the problems that always occur during word processing is when another word, of the same or different category as (يلعب) above starts with the letter (يـ), as in يمن (Yemen) or يئس (felt desperate), while ya (يـ) in these two last words is not an affix, but part of the roots of both words. Most processors would remove the ya (يـ) resulting in undesirable output. This calls for a deeper stemming that requires involving morphology, as well as, syntax, but this is beyond the limit of this paper.

3 Simple Stemming Rules

Our rules are divided into five categories. The first two categories are focused on any word type. The first necessary step is to normalize the words using the following rules: convert all different types of alif (أ) and (إ) to bare alif (ا) and change alif maqSura (ى) to ya (ي).

After normalization, remove all possible prefixes (preposition, definite articles and conjunctions) using following process:

- Remove baa (ب) from all beginning of every word;
- Remove waaw (و) from all beginning of every word, if the remaining is three or more characters.

Then remove the following linked characters from the beginning of a word, if the remaining number of characters is three or more characters.

- Remove: alif (ا) and laam (ل);
- Remove: alif (ا), waaw (و) and laam (ل);
- Remove: alif (ا), baa (ب) and laam (ل);
- Remove: alif (ا), kaaf (ك) and laam (ل);
- Remove: alif (ا), faa (ف) and laam (ل).

The next rule is designed for suffixes. The rule removes linked characters if the remaining is two or more characters:

- Remove: ha (ه) and alif (ا);
- Remove: alif (ا) and noon (ن);
- Remove: alif (ا) and taa (ت);
- Remove: waaw (و) and noon (ن);
- Remove: ya (ي) and noon (ن);
- Remove: ya (ي) and ha (ه);
- Remove: ya (ي) and ta' marbuTa (ة).

The third rule is designed for Adjectives and Adverbs, more precisely for changing adjectives and adverbs from feminine into muscular form:

- Remove ta' marbuTa (ة) if found as the last character in a word;
- Remove ha (ه) if found as the last character in a word.

The fourth rule is designed for nouns; it is to change nouns from plural into singular form. In this rule, following linked characters are removed if the remaining is longer than two characters. If the word has only three characters, remove the following characters if the additional conditions are fulfilled.

- Remove alif (ا) from the beginning of every word;
- Remove alif (ا) if it is found as the character before the last one;
- Remove waaw (و) if it is found as the character before the last one;
- Remove alif (ا) from the beginning of every word, only if the character before the last is also alif (ا).

If the word has four characters, remove the following characters if the additional conditions are fulfilled.

- Remove alif (ا) from the character before the last one
- Remove alif (ا) if it is found as the third character

The fifth rule is designed for verbs. It converts inflectional forms of the verb into stem. The characters are removed from the end of the word, if the remaining is at least 2 characters.

- Remove: taa (ت)
- Remove: yaa (ي)
- Remove: alif (ا)
- Remove: alif (ا) if found as the second character in a word, only if alif (ا) is found as the last character in a word.
- Remove: waaw (و)
- Remove: miim (م)
- Remove: noon (ن) and alif (ا)
- Remove: taa (ت) and alif (ا)
- Remove: taa (ت) and miim (م)
- Remove: taa (ت) and ya (ي)

Table 4 List of words for each word type

a) Nouns
تمساح حصان سمك أرنب حوت خد خدود إصبع إصبعان إصابع قلب أسنان مكتبة مكتب دجاجة دجاج لحم قميص زوجة زوج مهندس لغات لغة قلم الصين بلد مكتبة خليج صحراء ضباب ثلج عاصفة سكرتيرة سكرتير الإجهاض المحامي الخريف الكرة
b) Verbs
سمح إعترفت سأل عض كسر اتصل اعترف ركز تعامل رسم تزوج تغلب على زرع كتب رن استمتع توسل
c) Adverb
قادر نشيط غاضب مستيقظ سيء أحسن أفضل يغلي مكسور
d) Adjectives
كبيرة كبير صغيرة صغير جميلة جميل أزرق زرقاء أبكم بكماء كبير كبيران أزرق أزرقان زرقاوان

- Remove: miim (مـ) and alif (ا)
- Remove: miim (مـ), taa (ت) and alif (ا)
- Remove: noon (ن), taa (ت) and alif (ا)
- Remove alif (ا) if it is the second in a word. After that, Remove taa (ت) at the end of the word
- Remove: taa (ت) and alif (ا).

4 Experiments

The suggested rules must be tested against real data. For this purpose, we used some news articles, from the BBC Arabic and Al Jazeera Arabic news portals, of which we selected some of the most frequent nouns, verbs, adverbs and adjectives in those articles. These words were divided into group files according to words' part of speech. Each group consists of several words, selected precisely to cover all type of words. The lists of words used in the experiments are shown in Table 4. All source data were encoded using codepage 1256 Arabic.

In our experiments, we use the rules described in the previous section. There are three common rules used for each word type and one specific rule per group. Accordingly, we process each word by four rules and therefore, we have five versions

Table 5 Results for nouns

Original	Step 1	Step 2	Step 3	Final
تمساح	تمساح	تمساح	تمساح	تمساح
حوت	**حوت**	**حوت**	**حوت**	**حت**
خد	خد	خد	خد	خد
خدود	خدود	خدود	خدود	خد
إصبع	اصبع	اصبع	اصبع	اصبع
إصبعان	اصبعان	اصبعان	اصبع	اصبع
اصابع	**اصابع**	**اصابع**	**اصابع**	**اصابع**
أسنان	اسنان	اسنان	اسن	سن
دجاجة	دجاجة	دجاجة	دجاجة	دجاجة
دجاج	**دجاج**	**دجاج**	**دجاج**	**دجج**
قميص	قميص	قميص	قميص	قميص
زوجة	زوجة	زوجة	زوجة	زوجة
لغات	**لغات**	**لغات**	**لغات**	**لغت**
الصين	الصين	صين	صين	صين
بلد	**بلد**	**لد**	**لد**	**لد**
صحراء	صحراء	صحراء	صحراء	صحراء
ضباب	**ضباب**	**ضباب**	**ضباب**	**ضبب**
سكرتيرة	سكرتيرة	سكرتيرة	سكرتيرة	سكرتيرة
سكرتير	سكرتير	سكرتير	سكرتير	سكرتير
الإجهاض	الاجهاض	اجهاض	اجهاض	اجهاض
الكرة	الكرة	كرة	كرة	كرة

of a word during its processing. The first version is a original word. The second version has the replaced alif and alif maqSura. The third version shows the word after it has already processed prepositions, definite articles and conjunctions. The fourth version contains the processed suffixes. The fifth version uses the rule specific for each word type. The errors are highlighted using bold font.

Our results for the nouns list are depicted in Table 5. The results for the noun category produced some undesirable output especially in case of plural. Some other letters were deleted by the processor as they were taken as suffix prepositions.

The verbs list results are depicted in Table 6. The verb category result produced good results in case of two, three, four and five-letter stems. However, some letters were removed by the processor from the three letter verbs from the middle and the end. Modification of rules is to be considered in the future.

Our results for adverbs are depicted in Table 7. The results for the adverb category produced very good results in both muscular and feminine cases.

The results for adjectives are depicted in Table 8. Three undesirable results were produced in the adjective category of which two are in case of singular feminine and

Table 6 Results for verbs

Original	Step 1	Step 2	Step 3	Final
سمح	سمح	سمح	سمح	سمح
إعترفت	اعترفت	اعترفت	اعترفت	اعترف
سأل	**سال**	**سال**	**سال**	**سل**
عضت	عض	عض	عض	عض
اتصل	اتصل	اتصل	اتصل	اتصل
إعترفا	اعترف	اعترف	اعترف	اعترف
تعامل	تعامل	تعامل	تعامل	تعامل
تزوج	تزوج	تزوج	تزوج	زوج
علي	**علي**	**علي**	**علي**	**عل**
زرعت	زرع	زرع	زرع	زرع
رن	رن	رن	رن	رن
استمتع	استمتع	استمتع	استمتع	استمتع
توسل	توسل	توسل	توسل	توسل

Table 7 Results for adverbs

Original	Step 1	Step 2	Step 3	Final
قادر	قادر	قادر	قادر	قادر
نشيط	نشيط	نشيط	نشيط	نشيط
مستيقظ	مستيقظ	مستيقظ	مستيقظ	مستيقظ
مستيقظة	مستيقظ	مستيقظ	مستيقظ	مستيقظ
سيء	سيء	سيء	سيء	سيء
أحسن	احسن	احسن	احسن	احسن
أفضل	افضل	افضل	افضل	افضل
يغلي	يغلي	يغلي	يغلي	غلي
مكسور	مكسور	مكسور	مكسور	مكسور

one in case of dual feminine. Morphological rules have to be enhanced in the future to overcome this problem.

The summarization of the experiments is depicted in Table 9. As may be seen, the best results were achieved for adverbs. Two errors were produced for verbs and adjectives and six errors showed up in case of nouns.

Table 8 Results for adjectives

Original	Step 1	Step 2	Step 3	Final
كبيرة	كبيرة	كبيرة	كبيرة	كبير
كبيران	كبيران	كبيران	كبير	كبير
كبير	كبير	كبير	كبير	كبير
صغيرة	صغيرة	صغيرة	صغيرة	صغير
جميلة	جميلة	جميلة	جميلة	جميل
جميل	جميل	جميل	جميل	جميل
أبكم	ابكم	ابكم	ابكم	ابكم
بكماء	بكماء	كماء	كماء	كماء
أزرق	ازرق	ازرق	ازرق	ازرق
زرقاء	زرقاء	زرقاء	زرقاء	زرقاء
أزرقان	ازرقان	ازرقان	ازرق	ازرق
زرقاوان	زرقاوان	زرقاوان	زرقاو	زرقاو

Table 9 Summery of the experiments results

Word type	Errors	Word count
Nouns	6	21
Verbs	2	13
Adverbs	0	9
Adjectives	3	12

5 Conclusion

In this paper we defined very simple rules for stemming Arabic words. We defined two universal rules, one rule for adjectives and adverbs, one rule for nouns and one rule for verbs. As may be seen from the experiments results, the rules were more successful in case of adverbs. As for nouns, verbs and adjectives, errors were produced. Most errors were occurred in case of suffix processing. In the future, these rules must be improved and enhanced to cover more inflectional cases and be tested against wider vocabulary and bigger number of words.

Acknowledgments This work was supported by the Grant Agency of the Czech Republic under grant no. P202/11/P142.

References

1. Encyclopaedia Britannica Online. Alphabet. http://www.britannica.com/EBchecked/topic/17212/alphabet (2011)
2. Buckwalter, T.: In: Ide, N., Veronis, J., Soudi, A., Bosch, A.v.d., Neumann, G. (eds.) Arabic Computational Morphology, Text, Speech and Language Technology, vol. 38, pp. 23–41. Springer, The Netherlands (2007). http://dx.doi.org/10.1007/978-1-4020-6046-53

3. Habash, N.Y.: Synthesis lectures on human language technologies. **3**(1), 1 (2010).10.2200/S00277ED1V01Y201008HLT010. http://www.morganclaypool.com/doi/abs/10.2200/S00277ED1V01Y201008HLT010
4. Gillies, A., Erl, E., Trenkle, J., Schlosser, S.: In: Proceedings of the Symposium on Document Image Understanding Technology (1999)
5. Trenkle, J., Gilles, A., Eriandson, E., Schlosser, S., Cavin, S.: In: Symposium on Document Image Understanding Technology, pp. 159–168 (2001)
6. Maamouri, M., Bies, A., Kulick, S.: In: Proceedings of the British Computer Society Arabic NLP/MT Conference (2006)
7. Soori H, Platos J, Snášel V, Abdulla, H.: In: Snášel, V., Platos, J., El-Qawasmeh, E. (eds.) Digital Information Processing and Communications, Communications in Computer and Information Science, vol. 188, pp. 97–105. Springer, Berlin (2011). http://dx.doi.org/10.1007/978-3-642-22389-19

Part III
Applications

Fuzzy Logic Ranking for Personalized Geographic Information Retrieval

Garnett Wilson, Rodolphe Devillers and Orland Hoeber

Abstract This work describes a novel fuzzy logic system designed to meet the real world demand of providing intelligent ranking to large repositories of documents previously encoded with non-fuzzy (crisp) metadata. The fuzzy logic prototype was tested in practice to complement the GeoConnections Discovery Portal, which is a web portal for specialized search and retrieval of Canadian geographic data resources via an associated web service. Users of the portal are able to query the system and then filter their search results by selecting topic categories, spatial and temporal extents, and resource types. The authors present a fuzzy logic information retrieval system that utilizes document metadata, and compare it to an unranked listing, standard term frequency-inverse document frequency (TF-IDF) ranking, and a TF-IDF/fuzzy hybrid system. Results indicate that the fuzzy logic system provided the overall highest precision among the top ranked documents for searches by an expert user, and that these results were robust with respect to the number of results returned by a number of different query types.

G. Wilson (✉) · O. Hoeber
Department of Computer Science, Memorial University of Newfoundland,
St. John's, NL, Canada
e-mail: gwilson@mun.ca

O. Hoeber
e-mail: hoeber@mun.ca

R. Devillers
Department of Geography, Memorial University of Newfoundland,
St. John's, NL, Canada
e-mail: rdeville@mun.ca

M. Kudělka et al. (eds.), *Proceedings of the Third International Conference on Intelligent Human Computer Interaction (IHCI 2011), Prague, Czech Republic, August, 2011*, Advances in Intelligent Systems and Computing 179, DOI: 10.1007/978-3-642-31603-6_10,

1 Introduction

Many fuzzy logic systems that aim to improve the ranking of the search results require existing specialized fuzzy metadata [3, 8] or the creation of supporting fuzzy ontologies [5, 7]. However, large repositories, such as the ones accessed in this work, are often encoded with crisp metadata. The fuzzy logic system we present in this work is designed to take advantage of XML encodings in large online repositories of existing documents that do not contain fuzzy data. Design choices for the system were guided by an actual end user need to intelligently rank these documents using existing metadata, resulting in the design of a novel fuzzy logic solution that uses existing non-fuzzy (crisp) ontologies.

The GeoConnections Discovery Portal[1] provides search capabilities for geospatial data resources contained within multiple repositories. However, the search facilities simply provide a filtering mechanism, without any ranking capabilities. The web portal has a connected web service that can be used to return XML documents that contain the supplied search terms. As a research project in collaboration with GeoConnections, we created a system that implements an intelligent ranking of the documents returned by the web service. To this end, we chose to provide a personalized ranking based on per session user profiles because a typical end user of the portal may search for information on many unrelated topics. The performance of the fuzzy logic retrieval system is compared to unranked documents returned by the web service, the standard information retrieval ranking of term frequency-inverse document frequency (TF-IDF) [2], and a hybrid TF-IDF/Fuzzy system.

2 Related Work

The traditional approach to ranking search results has been to order documents that match the search query based on measures such as term frequency-inverse document frequency (TF-IDF) [2]. Such an approach gives ranking preference to query terms that appear frequently within the search results, but are rare with respect to their use across all of the documents in the collection. Although such ranking methods can be quite effective, there are opportunities for improvement. One such improvement that we explore within this research is to personalize the ranking of the search results based on additional information and preferences regarding the searcher's needs [9].

Castellano et al. [3] (and related work in Mencar et al. [8]) have described modelling of user preferences with fuzzy profiles. In particular, their work focuses on a system where individual users have multiple (often diverging) sets of interests in a user profile. An adaptive algorithm is designed to learn these multiple profiles of users and thus personalize their search results. For the practical purposes of a user researching a particular area of interest in one setting (for an individual professional contract or student project, for example) it made little sense in our implementation

[1] http://geodiscover.cgdi.ca

to keep likely unrelated user profiles from past searches. For instance, if a user is doing a search regarding road infrastructure, the search on caribou habitat that they performed last week would inhibit effective ranking. Our work uses fuzzy logic foundations behind the modelling of user profiles (as in [3]). However, the method of ranking in the fuzzy system used in this work is significantly different and makes use of existing non-fuzzy metadata in documents, rather than requiring the *a priori* encoding of fuzzy data within all metadata as in [3, 8].

There is a large body of work dedicated to geographic information retrieval (GIR), where many works focus on metadata encoding, methods to determine whether or not an arbitrary query is of a geographic nature, or ways to combine geographical information (such as location) with other geographical information (such as a city being known to be in a country) [6]. The scope of this work is different, and looks at practical decisions for a fuzzy logic ranking system that is to provide improved ranking based on metadata that happens to be geographic in nature (but can be applied to other domains). A GIR system by Leite [7] that does use geographic fuzzy relations differs significantly from our study because it uses a pre-existing ontology for Brazilian territories and climate in order to expand user queries for better search. Our work does not expand user queries for improved search, and the substantial amount of existing metadata from the geographic web services has no fuzzy metadata encoding. Furthermore, our method is a practical solution that can be easily implemented over top of existing metadata repositories and does not require additional metadata or the development of a formal ontology.

In order to explore the methods for effective ranking of documents devised in this work, we examine a standard information retrieval ranking mechanism that ranked documents based on terms in the document (TF-IDF), a fuzzy logic system that relies solely on the metadata tags in the documents stored on the online databases, and a combination of the TF-IDF and fuzzy rankings. Other researchers have attempted to combine TF-DF and fuzzy rankings. For instance, Holi et al. [5] combine a fuzzy ontology with the TF-IDF measure itself. Rubens [12] combine traditional logic ranking with TF-IDF using fuzzy logic. Rather than using fuzzy logic within the TF-IDF framework, as in [5, 12], we combine the separate results of TF-IDF and fuzzy logic by placing an appropriate weight on each method. The aim of combining the weighted measures is to allow consideration of both the textual contents and metadata for effective ranking (discussed further in Sect. 4.3).

3 Fuzzy Logic Search Portal

In order to improve the relevance of the documents provided by the GeoConnections Discovery Portal, we further filter the documents retrieved from the portal's corresponding web service. We accomplish this by creating a web interface front end for our prototype that incorporates its own web server and other back end elements. The overall system architecture is shown in Fig. 1. In particular, the public-facing web front end uses AJAX, JQuery, and sends JSON objects to a Jetty web server that

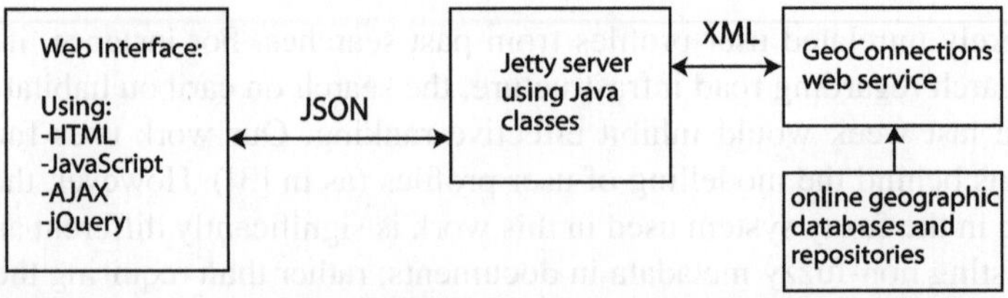

Fig. 1 Fuzzy logic retrieval prototype architecture

Fig. 2 The main page of the fuzzy logic retrieval portal

allows Java classes to be used server-side for the processing of our document ranking techniques and then returned to the web portal. The Jetty web server communicates with the web service to retrieve unranked documents encoded with XML metadata from the online geography resources.

The web interface front end is shown in Fig. 2. There is a query box in which the searcher can enter their textual query, and four buttons, each of which activates one of the four methods for ranking the search results (Unranked, TF-IDF, Fuzzy Logic, and TF-IDF/Fuzzy Logic). Titles of separate returned documents are displayed at the bottom of the page, and upon clicking the title the user can examine the corresponding document's abstract. The interface has been developed for experimental purposes, with potential future interfaces based on the findings in this research. The User Profiles section of the web interface allow users to specify, on a scale of 1 to 10, how interested they are in each of 68 topics specified in the metadata of the documents. The metadata are not custom encoded for our fuzzy logic system, but we make creative use of the large amount of existing metadata in documents of a number of geographic databases provided by the web service. In particular, there are a number of topic categories under "Categories," types of data resources under "Resource Type," and political and physiographic locations under "Locations." Examples of values for these metadata attributes used in our experiments will be provided in Sect. 5.

4 Ranking Methods

We establish a performance baseline using a standard TF-IDF ranking of documents. The standard TF-IDF is compared to a fuzzy logic implementation that takes the user specifications of interest from each of several topic categories, regions, and information resource types and returns rankings based on those user interests as encoded in the metadata. In addition, a hybrid of TF-IDF and fuzzy logic is examined. In this section, we specify the details of each ranking method implementation.

4.1 Term Frequency-Inverse Document Frequency

In information retrieval ranking based on term frequencies, each document d is considered as a vector of elements $\overrightarrow{d}$ with each dimension corresponding to a term contained in that document d. A set of search query terms is considered a special type of document vector, $\overrightarrow{q}$. In order to reduce the dimensionality of these vectors, stop word removal is used to eliminate common terms (prepositions, conjunctions, etc.) and Porter's stemming algorithm [10] strips prefixes and suffixes from terms, thereby counting terms with the same root together.

When representing the document (or query) as a vector, each dimension of the vector that corresponds to a term is assigned a weight, typically a TF-IDF weight. TF is a function of term frequency, and IDF is a function of inverse document frequency. The TF rewards a term occurring more frequently in a document, while IDF provides a lower weight to a term that appears in many documents. Let N be the total number of documents in the filtered collection based on the user-supplied query,[2] N_t be the number of those documents containing a term t, and $f_{t,d}$ be the number of occurrences of t in a document d. In this work the TF and IDF function are standards in information retrieval [2], where TF is defined as

$$TF = \begin{cases} \log(f_{t,d}) + 1 & \text{if } f_{t,d} > 0 \\ 0 & \text{otherwise} \end{cases} \tag{1}$$

and IDF is defined as

---

[2] It is not possible to know the number, or content, of all documents in all repositories served by the web portal.

$$IDF = \log(N/N_t). \tag{2}$$

TF-IDF is then the product of the term frequency and inverse document frequency,

$$TF \cdot IDF. \tag{3}$$

To measure the similarity of each document vector $\vec{d}$ to the query vector $\vec{q}$, the cosine similarity measure is used. The cosine similarity measure is actually the dot product of the document and query vectors normalized to unit length:

$$sim(\vec{d}, \vec{q}) = \frac{\vec{d}}{|\vec{d}|} \cdot \frac{\vec{q}}{|\vec{q}|}. \tag{4}$$

Since the components of the vectors are not negative, the cosine similarity measure ranges from 0 to 1. When the cosine similarity value is high, then the vectors of the document and query are similar to one another. Therefore, this measure of similarity can be used to rank the search results such that those that are most similar to the query are placed at the top of the search results list.

4.2 Fuzzy Logic

The fuzzy logic system considers user-specified values on a scale of 0 to 10 for each attribute a from the set of all possible attributes, A. The first step in the fuzzy logic implementation is to gather all the user-defined values for each attribute (ranked from 1 to 10, adjusted to a float in the interval [0, 1]) from the web site and send them to the Jetty web server for processing. More specifically, the collection of 68 metadata tags (total of possible values of all attributes) in a user profile is divided into three different attribute possibilities as indicated under "User Profile" on the web portal. These attributes are "Categories," "Resource Type," and "Locations," so $|A| = 3$ and we have $A = a_1, a_2, a_3$. Furthermore, given the possible metadata tags in the documents, the 68 possible attributes are $|Dom(a_1)| = 18$, $|Dom(a_2)| = 17$, and $|Dom(a_3)| = 33$ where $Dom(a_i)$ is the set of all possible values of attribute a_i. The unranked documents are then retrieved from the web service using the Java implementation on the Jetty web server based on the search terms. The attributes corresponding to the metadata for the returned items, which is consistent across the collection of repositories, is then analyzed by the fuzzy logic system. For the fuzzy logic system, these documents will be referred to as "items" to distinguish from the TF-IDF ranking.

We adapt the notion of component-based fuzzy user profiles similar to those applied by Castellano in [3]. Other than this component of the fuzzy logic system, we create our own methodology for forming fuzzy metadata encoding from existing crisp metadata. In addition, we use a fuzzy similarity measure more appropriate for our information retrieval-focused purposes. Each item in a set of fuzzy metadata is expressed as a couple $\langle a, \mu \rangle$ where $a \in A$ and $\mu : Dom(a) \rightarrow [0, 1]$ is a fuzzy set defined on $Dom(a)$. An item that a user is interested in can then be described as

$$I = \{\langle a, \mu \rangle \,|a \in A\} \quad (5)$$

with the constraint that each attribute does not occur more than once in a description.

User profiles represent the interests that each individual user has for the attributes of the returned items. User profiles are expressed as

$$P = \{\langle a, \mu \rangle \,|a \in A\}. \quad (6)$$

Regarding items (documents) to be compatible with user profiles, each item I is not encoded as fuzzy metadata. To convert the existing metadata labels in documents, we use a fuzzy control to create appropriate fuzzy attribute value content. In particular, the total of all values in an item I corresponding to each of the three profile attributes a_1, a_2, a_3 is totaled, and each attribute value x is divided by the total of all x values for that attribute. More formally,

$$sum_{a_i}(I) = \sum_{x \in Dom(a_i)} \mu_I(x) \quad (7)$$

if μ_I is a fuzzy set of items. Each $x \in a_i$ is then replaced with x' where

$$x' = x/sum_{a_i}(I). \quad (8)$$

This production of fuzzy data from the existing metadata labeling means that if the article has more metadata tags labeling it under one of the three attributes of a user profile, each metadata tag has less value. This makes intuitive sense: If an article is labeled exclusively as one value under one of the user profile's attributes, and the user places importance on that value, then that item ought to be of higher relevance to the user than another article where it is labeled with several values under that same attribute. As a practical example of this, suppose one item in a search for "birds" is labeled as only "Biology and Ecology" under the Category attribute, and another item is labeled with all of "Atmosphere and Climate," "Biology and Ecology," and "Business and Economic." There will then be more emphasis on the former item if the user is very interested in "Biology and Ecology" and has little or no interest in atmospheric conditions effecting birds or associated economic issues.

The user profile P is a fuzzy set μ_P and the set of items returned from the web service prior to ranking is a fuzzy set of items, μ_I. In the fuzzy logic system, we instantiate all possible attribute values in each item I and user profile P, so each vector is of length 68 (all possible attribute values, the $Dom(A)$). If a user does not classify an attribute value as being of interest, it has a value of 0. Similarly, if the attribute value is not present in an item, it is given a value of 0 for that attribute. To calculate the compatibility degree between the fuzzy set of items μ_I and the user profile μ_P of attributes, we use the dot product of μ_P and μ_I:

$$K(\mu_I, \mu_P) = \mu_I \cdot \mu_P \tag{9}$$

The dot product is one of the standard functions used in engineering-based fuzzy systems, ranging from 0 (complete dissimilarity) to 1 (identical), and it is suitable for pairwise comparison of vector-based data points [11] (individual items and the user profile). This dot product operation is essentially the same as used in the TF-IDF method in Eq. 4 (only the vectors are based on entirely different information). Additionally, the vector returned by the dot product operation of the matrix of multiple items and the user profile vector is a list of the compatibilities (in the range 0 to 1) of each item to the user profile. These compatibilities are then used to rank the items as displayed to the end user on the web portal.

4.3 TF-IDF/Fuzzy Logic Hybrid

TF-IDF, considered in isolation, will provide ranking of documents based on the terms contained within them in relation to the user's search terms. Fuzzy logic, considered in isolation, will provide ranking of document (items) based on the user's ranking of particular topics contained in their metadata. Thus, to combine the information contained in both the text and metadata of a document (item in fuzzy logic), we use a hybrid ranking similarity measure

$$sim'(\vec{d}, \vec{q}, \mu_I, \mu_P) = \alpha \cdot sim(\vec{d}, \vec{q}) + \beta \cdot K(\mu_I, \mu_P) \tag{10}$$

where $sim(\vec{d}, \vec{q})$ and $K(\mu_I, \mu_P)$ are as defined in Eqs. 4 and 9, respectively. The coefficients α and β are used to put appropriate weights on the similarity and compatibility measures, respectively, are in the interval [0, 1] and add up to 1.0.

It is convenient to combine the similarity measure of the TF-IDF ranking and the compatibility measure of fuzzy logic into one measure of similarity since both measures are in the interval [0, 1] and there is a measure for each document (item). Due to the significant difference in the dimensionality of the document vectors used in calculating the TF-IDF ranking compared to the fuzzy vectors used in calculating the fuzzy ranking, we observe that typically the TF-IDF similarity measures are 2-3 orders of magnitude smaller than the compatibility measures between the fuzzy vectors and each item. In this work we therefore use weightings of $\alpha = 0.99$ and $\beta = 0.01$, which we found to provide a suitable balance between the two document (item) sets returned by TF-IDF and fuzzy measures.

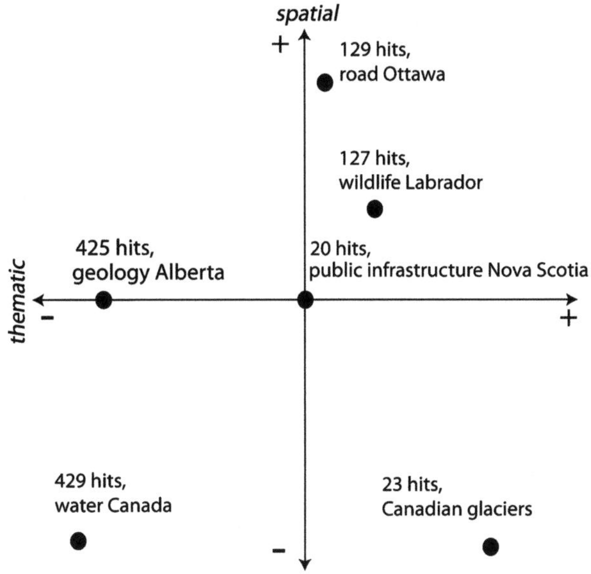

Fig. 3 Specificity of spatial and thematic component of expert searches

5 Results

We assess the ranking of the documents returned from the web service (which is known to be unranked, and is designated "Unranked"), TF-IDF that is our baseline measure from standard information retrieval ("TF-IDF"), fuzzy logic ranking ("Fuzzy"), and the combination of TF-IDF and fuzzy logic as in Eq. 10 ("TF-IDF/Fuzzy"). An expert user formulated six searches designed to test the effectiveness of the ranking mechanisms across a spectrum of specificity with respect to both spatial and thematic components of the search query. The searches chosen by the expert are plotted along the thematic and spatial spectra in Fig. 3. The number of documents returned by the web service for each search query is also provided.

For each of these separate searches, we calculate the effectiveness of the rankings using precision. Recall is commonly used as an effectiveness measure with precision, giving the proportion of relevant documents in the result set. We do not use recall here because it is not possible to know the number of relevant documents in the large online databases served by the GeoConnections-associated web services. Precision measures the proportion of the result set that is relevant and is defined as

$$precision@k = \frac{|\mathrm{Rel} \cap \mathrm{Res}[1..k]|}{|\mathrm{Res}[1..k]|} \tag{11}$$

for the top k documents returned by a search query [1]. The expert user judged relevance of a document based on whether or not the document contained useful

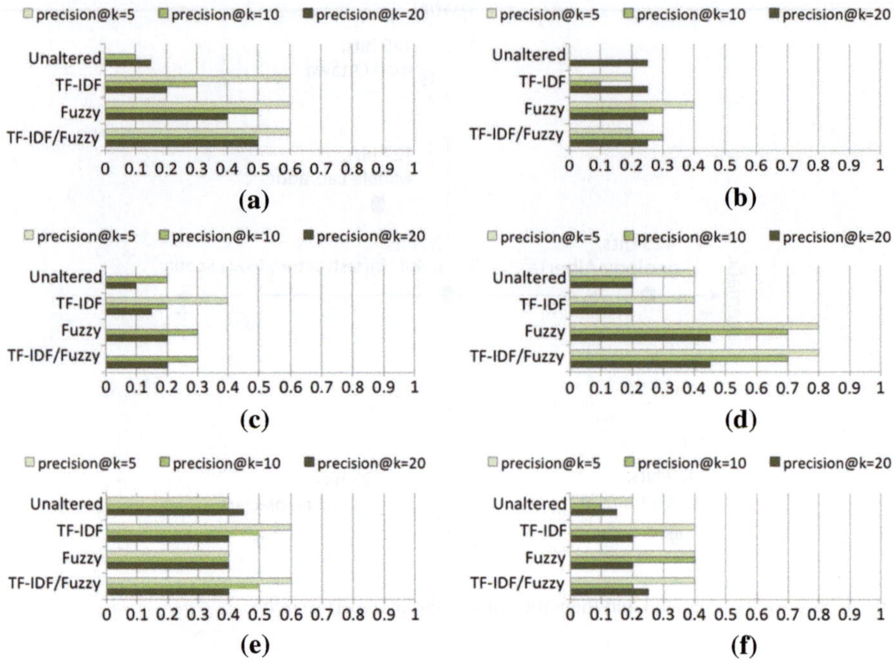

Fig. 4 Precision for top k documents returned per search query. **a** Geology Alberta, **b** public infrastructure Nova Scotia, **c** road Ottawa, **d** wildlife Labrador, **e** Canadian glaciers, **f** water Canada

information given his search based on both title (and abstract if necessary). The precision results for $k = 5, 10, 20$ are shown in Fig. 4. The fuzzy logic user profile attribute values chosen by the expert are provided in Table 1.

The precision results in Fig. 4 show that, as expected, all ranking methods outperformed the unranked search results. The fuzzy logic system performed as well as (or better than) the traditional TF-IDF ranking in all searches for the top $k = 20$ documents, and all but one instance ("Canadian glaciers") for $k = 10$. This result clearly indicates the value of geographic metadata over document text for effective ranking of the results from the online repositories. Results at $k = 5$ provided no discernible trend when comparing TF-IDF and fuzzy solutions. As k increases, the precision decreases for almost all fuzzy searches, indicating that it provides an effective ranking mechanism with relevant documents high in the list and non-relevant documents lower in the list. Considering results at $k = 10$ and $k = 20$, as the spatial specificity increased from Canada (Searches e and f) to province ("Alberta" and "Nova Scotia") or region in a province ("Labrador"), the fuzzy logic implementation provided the most benefit. However, for the most spatially specific search (Search c, involving the city "Ottawa") the benefit of the fuzzy search was moderate for $k = 10$ and $k = 20$. In Search d, where the fuzzy logic search outperformed TF-IDF considerably, the expert user provided a detailed fuzzy profile (Table 1) and a large

Table 1 User profile attribute values set by expert

Search 1: geology Alberta
<Categories, {Geology & Geophysical = 1.0} >, <Locations, {Alberta = 1.0} >
Search 2: public infrastructure Nova Scotia
<Categories, {Public Health & Disease = 0.5} >,
<Categories, {Public Safety & Security = 0.3} >,
<Categories, {Transportation Networks = 0.8} >,
<Categories, {Utilities & Communication = 1.0} >,
<Location, {Nova Scotia = 1.0} >, <Location, {Atlantic Canada = 0.5} >
Search 3: road Ottawa
<Categories, {Transportation Networks = 1.0} >, <Locations, {Ontario = 1.0} >
Search 4: wildlife Labrador
<Categories, {Biology & Ecology = 0.5} >,
<Categories, {Environment & Conservation = 0.7} >,
<Location, {Newfoundland = 1.0} >, <Location, {Atlantic Canada = 0.5} >
Search 5: Canadian glaciers
<Categories, {Atmosphere & Climate = 0.5} >,
<Categories, {Environment & Conservation = 0.5} >,
<Categories, {Inland Water Resources = 0.5} >,
<Locations, {Territorial North = 0.7} >, <Locations, {Innuition = 0.5} >,
<Categories, {Alberta = 0.7} >, <Location, {British Columbia = 1.0} >,
<Location, {Newfoundland = 0.5} >, <Location, {Northwest Territories = 0.5} >, <Location, {Nunavut = 0.5} >, <Location, {Yukon = 0.5} >,
Search 6: water Canada
<Categories, {Inland Water = 1.0} >

number of documents were returned by the query (a total of 127 documents, see Fig. 3).

The attempt to balance term-based and fuzzy searches (TF-IDF/Fuzzy) show that only in two of six searches at $k = 20$ did an improved ranking result compared to either TF-IDF or fuzzy. Both of these searches (Searches a and f) where the TF-IDF/fuzzy logic hybrid worked best were the most thematically non-specific (see Fig. 3). Examining the specificity of the searches in Fig. 3 and precision in Fig. 4, it is not readily apparent that the raw number of hits has a direct relationship to the precision metric in general, or the performance of one ranking mechanism over another. Result independence with respect to specificity indicates that the ranking mechanisms are robust with respect to the size of the document set.

The computational complexity of TF-IDF and fuzzy systems are identical, and are dominated only by the step of vector-matrix multiplication. In each algorithm, the representation of the vector space can be created by repeated passes through the set of returned documents, using $O(n)$ time. The following step is to find the dot product of the search values and the document values, which is done in $O(n^2)$ time. Following evaluations, sorting will be at worst $O(n^2)$. The TF-IDF/fuzzy logic hybrid is thus also dominated by $O(n^2)$. All algorithms are of reasonable time complexity for document ranking [4] and run in practical time.

6 Conclusions

In this work, we provide intelligent ranking of documents returned by a web portal that accesses multiple online Canadian-focused geographic data resources. To provide the ranking desired by an end user (the expert), we produced our own web portal prototype designed around a novel fuzzy logic ranking system for documents returned by the existing GeoConnections Discovery Portal. The precision of the top search results of five different ranking systems was examined: unranked, TF-IDF (an information retrieval baseline), fuzzy logic, and a TF-IDF and fuzzy logic hybrid solution. The expert user identified relevant documents using searches he devised to test the capabilities of the system across both a spatial and thematic specificity spectra. The precision results indicated that the fuzzy logic system outperformed TF-IDF in almost all searches, so the geographic metadata in the returned set of items proved more valuable for effective ranking than the text content of the documents in this study. The fuzzy logic system generally provided increased effectiveness in ranking documents with spatial specificity at a moderate to high level. The fuzzy ranking system examined is also robust with respect to the number of documents in the result set. Future work may examine the possibility of improving the intelligent fuzzy logic search presented here by adding initial filtering on the GeoDiscovery Portal that uses spatial queries created by bounding box rectangles of quantitative latitude and longitudes and temporal queries.

Acknowledgments The authors would like to acknowledge programming assistance provided by Derek Leblanc, the financial assistance of a Canadian GEOIDE Network grant held by the second author, and the assistance of the GeoConnections initiative in accessing their web services.

References

1. Bütcher, S., Clarke, C., Cormack, G.: Information Retrieval: Implementing and Evaluating Search Engines. MIT Press, Cambridge (2010)
2. Baeza-Yates, R., Ribeiro-Neto, B.: Modern Information Retrieval: The Concepts and Terminology Behind Search. 2nd edn. Addison-Wesley, Harlow (2011)
3. Castellano, G., Dell'Agnello, D., Fanelli, A.M., Mencar, C., Torsello, M.A.: A competitive learning strategy for adapting fuzzy user profiles. In: 10th International Conference on Intelligent Systems Design and Applications, ISDA 2010, pp. 959–964, Nov 29–Dec 1 (2010)
4. Han, E.-H., Karypis, G.: Centroid-based document classification: analysis and experimental results. In: Proceedings of the 4th European Conference on Principles of Data Mining and Knowledge Discovery, PKDD '00, pp. 424–431, Springer-Verlag, London (2000)
5. Holi, M., Hyvönen, E., Lindgren, P.: Integrating tf-idf weighting with fuzzy view-based search. In: Proceedings of the ECAI Workshop on Text-Based, Information Retrieval (TIR-06) (2006)
6. Jones, C.B., Purves, R.S.: Geographical information retrieval. Int. J. Geogr. Inf. Sci. **22**(3), 219–228 (2008)
7. Leite, M.A.A., Ricarte, I.L.: Document retrieval using fuzzy related geographic ontologies. In: Proceeding of the 2nd International Workshop on Geographic Information Retrieval, GIR '08, pp. 47–54, ACM, New York (2008)
8. Mencar, C., Torsello, M., Dell'Agnello, D., Castellano, G., Castiello, C.: Modeling user preferences through adaptive fuzzy profiles. In: 9th International Conference on Intelligent Systems Design and Applications, ISDA 2009, pp. 1031–1036, Nov 30–Dec 2 (2009)
9. Micarelli, A., Gasparetti, F., Sciarrone, F., Gauch, S.: Personalized search on the world wide web. In: Brusilovsky, P., Kobsa, A., Nejdl, W. (eds.) The Adaptive Web: Methods and Strategies of Web Personalization, pp. 195–230. Springer-Verlag, Berlin (2007)
10. Porter, M.: An algorithm for suffix stripping. Program **14**(3), 130–137 (1980)
11. Ross, T.: Fuzzy Logic with Engineering Applications. Wiley, West Sussex (2004)
12. Rubens, N.O.: The application of fuzzy logic to the construction of the ranking function of information retrieval systems. Comput. Model. New Technol. **10**(1), 20–27 (2006)

Virtually Cloning Real Human with Motion Style

Manoj kumar Rajagopal, Patrick Horain and Catherine Pelachaud

Abstract Our goal is to capture style from real human motion so it can be rendered with a virtual agent that represents this human user. We used expressivity parameters to describe motion style. As a first contribution, we propose an approach to estimate a subset of expressivity parameters defined in the literature (namely spatial extent and temporal extent) from captured motion trajectories. Second, we capture the expressivity of real users and then output it to the Greta engine that animates a virtual agent representing the user. We experimentally demonstrate that expressivity can be another clue for identifiable virtual clones of real humans.

1 Body Motion Style and Expressivity

Non-verbal behavior, including body language, is an important communication channel. It conveys user's mental and affective states. Non-verbal behavior reflects the cognitive and emotional states [15]. In face-to-face communication, body language can convey up to 55 % of the information on feelings and intentions [21]. Hand gesture is part of the cognitive process in communication [1].

Gallaher [12] defined "style", or expressive movement, in reference to the way behavior is performed. She found style has four dimensions namely expressiveness, animation, expansiveness and coordination. Human motion style and human expressivity are interlinked to each other [7].

M. K. Rajagopal (✉) · P. Horain
Institut Telecom, Telecom SudParis,
9 rue Charles Fourier, 91000Evry, France
e-mail: manojkumar.rg@gmail.com, Patrick.Horain@Telecom-Sudparis.eu

C. Pelachaud
CNRS Telecom ParisTech,
37-39 rue Dareau, 75014Paris, France
e-mail: catherine.pelachaud@telecom-paristech.fr

M. Kudělka et al. (eds.), *Proceedings of the Third International Conference on Intelligent Human Computer Interaction (IHCI 2011), Prague, Czech Republic, August, 2011*, Advances in Intelligent Systems and Computing 179, DOI: 10.1007/978-3-642-31603-6_11, © Springer-Verlag Berlin Heidelberg 2013

1.1 Motion Style

To our knowledge, Tenanbaum et al. [27] were the first who separated motion style and content. They used a bilinear model to identify the font from hand written characters.

Elgammal et al. [10] proposed to separate style and content through non-linear dimensionality reduction.

Hsu et al. [16] proposed to transform the style of human motion while preserving its original content, using a linear time invariant model.

Wang et al. [31] used multi-factor Gaussian process models for style and content separation in human motion. They used a non-linear basis function to generalize multi linear models and a Gaussian process latent variable model to marginalize over the weights.

Human identification based on gait analysis has been widely studied [5, 9, 18]. Gait, which can be regarded as a walking style, does convey information on the identity of a person.

Noot and Ruttkay [23] developed a markup language called GESTYLE to specify the gesturing style of an Embodied Conversational Agent (ECA). They defined style for virtual human in terms of style dictionaries. Style dictionary contains typical for an individual, professional or cultural group or people of certain age, sex or personality (e.g. to accommodate meanings depending on his culture, specific meanings belonging to his profession, personal habits, etc.).

Since all these approaches work frame-by-frame, they consider neither motion speed nor acceleration.

1.2 Expressivity

Some studies [2, 22] decode the emotional states to movement qualities and investigate from expressive body movements. Drosopoulos et al. [8] analyzed and classified expressive gesture from image processing (body movements and facial expressions) for multimodal emotion recognition. Camurri et al. [3, 4] classified expressive gesture in human body movement during music and dance performances. Wallbott et al. [30] classified the bodily behavior using five criterions: slow or fast, small or expensive, weak or energetic, small or large movement activity and unpleasant or pleasant. Behavior expressivity has been correlated to energy in communication, to the relation with characteristics of gestures and/or personality/emotion. For Wallbott [29], it is related to the mental state (e.g. emotion). Behaviors encode not only content information (information communicated through gesture shape) but also expressive information (the how it is communicated, the manner of execution).

Following Wallbott's approach, Hartmann et al. [14] focused on expressivity (behavior expressivity) based on surface realization of movement from hand motion. They have defined a set of parameters that describe expressivity for a virtual agent based on the wrists 3D trajectories. They integrated expressivity control in the animation engine of the Greta ECA [25], a three-dimensional virtual agent that is able to communicate with real humans using a rich palette of verbal and nonverbal behaviors.

Expressivity is also conveyed by face expressions. Vasilescu and Terzopoulos [28] proposed a multi-linear analysis of tensor faces based on a tensor extension of the conventional matrix Singular Value Decomposition (SVD). Mancini et al. [20] map acoustic cues and emotion to expressivity parameters they use to control the expressive virtual head of the Greta ECA. Pelachaud and Poggi [24] aim at combining the Greta facial expressions in a complex and subtle way, just like humans do, by assessing and managing the multimodal communicative behavior of a person when different communicative functions have to be displayed at the same time.

1.3 Our Approach

We aim at animating a virtual agent representing a real human with the expressivity learnt from that person. In this work, we propose an approach to estimate two of the Hartmann's et al. [14] expressivity parameters from 3D motion data. We validate this approach against synthesis animations with controlled expressivity generated using Greta. Then we estimate expressivity from motion captured from real users, and we feed the user's expressivity to the extended Greta ECA animation engine. Finally, we discuss a possible application for a user virtual representative that would clone gesturing expressivity, in complement to visual appearance and voice, turning expressivity into another key clue for user identity in virtual inhabited worlds.

1.4 Organizaion of the Paper

In this paper, Sect. 2 defines Hartmann et al. [14] spatial and temporal extent parameters which is input for Greta ECA. Also in this section we have given the inside view of expressivity parameters. In Sect. 3, we explain our method of determining the spatial and temporal parameters from real human motion. In Sect. 4, we validate our spatial and temporal extent estimation against synthesized motion. In Sect. 5, we explain our experiments conducted with real human data. In the process of experiments we animated virtual agent with estimated spatial and temporal extent as input. Also in this section we have given user reviews for animated virtual agent. In Sect. 6 we have given the conclusion and future work.

2 Expressivity Parameters

Hartmann et al. [14] animated the Greta ECA [25] with a set of six expressivity features: overall activation, spatial extent, temporal extent, fluidity, power and repetitiveness. Each of the parameters is float-valued and defined over the interval $[-1, 1]$, where zero point corresponds to the action without expressivity control. We briefly introduce those two parameters we are using, namely spatial extent and temporal extent.

2.1 Spatial Extent

Spatial extent describes the space used by a person for making gestures. Hartmann et al. [14] define spatial extent as a parameter controlling the centers of the McNeill's [19] sectors where hand movement occurs. Spatial extent value +1.0 (resp. 1.0) means the wrist moves further (resp. closer) to the coordinate origin, which is set at the sacroiliac vertebra.

They replace the wrist positions $p = (p_x, p_y, p_z)^T$ in the neutral spatial extent trajectory with positions $p' = (p'_x, p'_y, p'_z)^T$:

$$\begin{bmatrix} p'_x = (1 + SPC \cdot Agent_x)p_x \\ p'_y = (1 + SPC \cdot Agent_y)p_y \\ p'_z = (1 + SPC \cdot Agent_z)p_z \end{bmatrix} \tag{1}$$

where $\cdot_x$, $\cdot_y$ and $\cdot_z$ refer to the lateral, vertical and frontal directions, Agent are scaling factors in the respective directions used in the Greta animation engine (resp. 1.3, 0.6 and 0.25 resp. in the x, y and z directions and positive SPC, and resp. 0.7, 0.25 and 0.25 resp. in the x, y and z directions and negative SPC), and SPC is the spatial extent in the range [1.0, +1.0].

2.2 Temporal Extent

Temporal extent (TMP) describes how fast a gesture is performed. According to Hartmann et al. [14] time for performing a gesture is segmented into three intervals, namely preparation, stroke and retraction. A stroke is that part of an expressive gesture that conveys meaning. It is preceded by a stroke preparation period when the wrist moves from the initial position, and followed by retraction where the wrist returns to the rest position after completing the stroke.

Temporal extent is related to the stroke time interval. Hartmann et al. [14] derive the time taken for each segment from a simplification of Fitt's law [11]:

$$T = a + b\log_2(\left\| x_{(n)} - x_{(n+1)} \right\| + 1) \tag{2}$$

where T is the duration of the stroke, a is a time offset, b is a velocity coefficient, x_n is the wrist positions at stroke start, x_{n+1} is the wrist positions at the stroke end. The velocity coefficient is defined by Hartmann et al. [14] as:

$$b = (1 + 0.2 \cdot TMP)/10 \tag{3}$$

where TMP is the temporal extent where -1.0 value corresponds to lower speed and $+1.0$ to higher speeds.

3 Estimating Expressivity from a Motion Trajectory

We estimate the spatial and temporal extents from 3D motion trajectories described as sequences of poses p for the two wrists. We use the Greta animation engine to generate example motions with controlled expressivity, from which we tune parameters for estimation.

3.1 Estimating Spatial Extent

The spatial extent defines a scale of coordinates to the wrists positions $p'_.$ with respect to an origin. Following the convention of the Greta animation engine, we set that origin at the sacroiliac vertebra, which stands approximately between the rest positions of the two wrists.

Taking the logarithm, Eq. 1 turns into an additive expression, where the term SPC is to be estimated:

$$Log(p'.) = Log(1 + SPC \cdot Agent.) + Log(p.) \tag{4}$$

Taking the mean over input samples, we get:

$$Log(1 + SPC \cdot Agent.) = \overline{Log(p'.)} - \overline{Log(p.)}$$

where $\overline{Log(p.)}$ can be learnt from a set of motion trajectories generated with null spatial extent. Finally, we estimate the spatial extent SPC of our input trajectory as the mean of the directional spatial extents:

$$SPC. = \frac{e^{\left[\overline{Log(p'.)} - \overline{Log(p.)}\right]} - 1}{Agent.} \tag{5}$$

where . stands for either x, y or z and $\left[\overline{-}\right]$ is the mean of $[-]$ and $\overline{Log(p.)}$ is learnt from a trajectory generated with null spatial extent. Finally, we estimate the spatial extent SPC of our input trajectory as the mean of the directional spatial extents:

$$SPC = \frac{1}{3}(SPC_x + SPC_y + SPC_z) \tag{6}$$

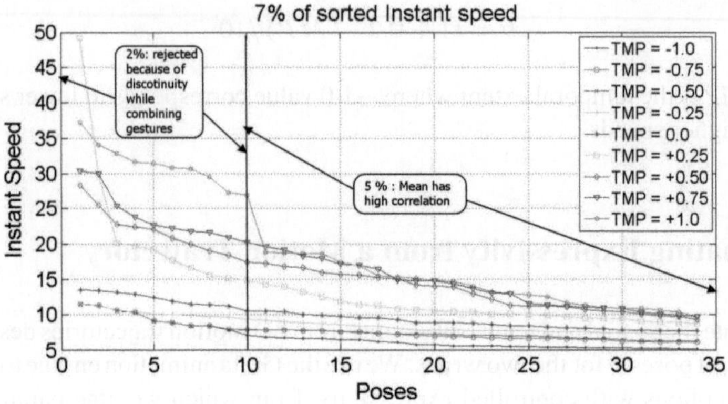

Fig. 1 Highest instant speeds from example synthesized trajectories

3.2 Estimating Temporal Extent

As already mentioned, the higher the temporal extent TMP, the higher the speed is. From this monotonous relation, we derive a heuristic to estimate TMP from the observed instant speeds along the trajectory, i.e. the distance between poses at successive time intervals.

In the work by Hartman's et al. [14] expressivity parameters affect synthesized gestures during stroke time only. Rather than segmenting strokes, which is a difficult problem [26] we estimate expressivity parameter from whole motion trajectories. From synthesis trajectories, we find that high TMP values give high speed only during the strokes, while the preparation and retraction exhibit low speeds. Therefore, TMP can be estimated by considering only some quantile of the highest speeds.

Based on this, we sort the instant speeds of example trajectories in descending order, and we set the quantile limit at the sample with highest correlation with the TMP values that were originally input to the animation engine for generating the example trajectories. From the learnt trajectories we considered, we found that the mean speed up to 7 % upper quantile determines the temporal extent. The correlation for mean speed up to 7 % upper quantile is 93 %. After rejecting 2 % upper quantile, the correlation improves to 97 %. This improvement is because there may be a chance for discontinuity while combining gestures to form a motion, this decreases the correlation value. So we reject initial 2 % of upper quantile for the mean calculation and mean speed with the remaining 5 % of upper quantile we achieve best mapping to temporal extent in the range [1.0, +1.0] through linear regression (Fig. 1). Using the obtained linear regression from learnt trajectories, we estimated temporal extent for other motion trajectories. The experimental results show that our estimation method is robust.

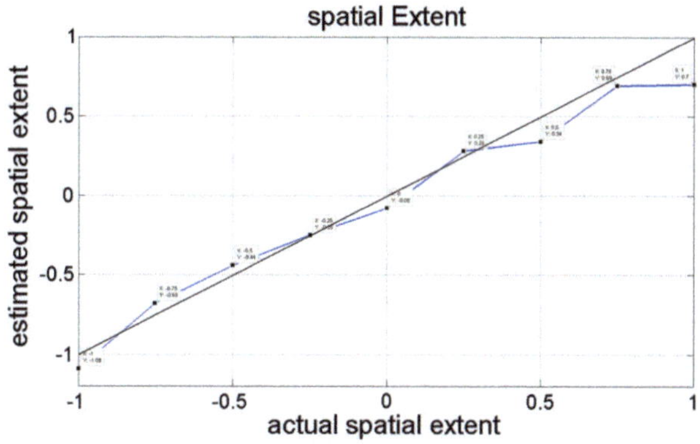

Fig. 2 Estimated spatial extent versus input spatial extent for synthesized gestures

4 Validating Expressivity Estimation

We evaluate the above approach for estimating the previous two expressivity parameters against a corpus of ten communicative gestures with controlled expressivity that are generated using the Greta animation engine [25]. Each of the ten gestures has been rendered with nine different values of spatial extent (−1.0, −0.75, −0.5, −0.25, 0.0, +0.25, +0.5, +0.75 and +1.0). We calculate the wrist positions of the synthesized motions through forward kinematics [6]. From the wrist positions, spatial extent is calculated as described in Sect. 3.1. The estimated spatial extent and actual spatial values are plotted as shown in Fig. 2.

From Fig. 2 we see that estimated spatial extent not exactly lying with actual spatial extent due to error in the estimation process. The error in the estimation process is calculated in terms of absolute mean error. The absolute mean error for spatial extent estimation is 0.14. In the range [−1.0, +1.0] there are 20 samples of size 0.1. It is understood that our spatial extent estimation process produces the result with 7 % of error.

The estimation of TMP is also validated against synthesized motion with varying controlled expressivity. Instant speeds of the wrists are calculated using forward kinematics and are sorted in increasing order. We map instant speed to temporal extent values using the linear regression found at Sect. 3.2. The estimated and actual temporal extent values are plotted in Fig. 3.

Similar to the spatial extent, the error measure for TMP is also done by absolute mean error. Absolute mean error for TMP estimation is 0.15. (i.e.,) our TMP estimation method causes 7.5 % error in determining the TMP from 3D motion data.

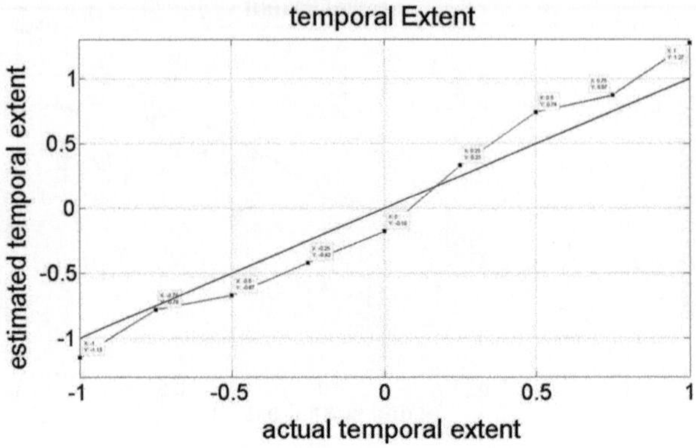

Fig. 3 Estimated temporal extent versus input temporal extent for synthesized gestures

5 Experiments

We are interested in reproducing real user's expressivity in an ECA animation. Starting from input 3D motion data, we estimate expressivity parameters that we feed to the Greta ECA and then compare the synthesized animation against user's motion.

5.1 Estimating Expressivity

Our input is 3D human motion data that can be captured e.g. using a consumer system such as newly available Kinect [17]. Alternatively, we used motion data captured by computer vision from two video lectures (hereafter named V1 and V2 , see Figs. 4 and 5) using software developed by Gómez Jáuregui et al. [13] that works by registering a 3D articulated model of the human body onto videos and outputs the upper body joints angles. 3D wrist positions are obtained from upper body joint angles through forward kinematics [6].

The spatial extent is estimated as explained in Sect. 3.1 for both videos of V1 and V2. It yields the spatial extent for V1 as $+0.8$ and the spatial extent for V2 as $+0.6$. By seeing the V1 and V2 snap shots, we can say V1 has more spatial extent than V2. Experiment results also confirm this.

The estimated TMP is -1.0 for V1 and -0.7 for V2. This shows the user in V2 has more temporal extent than user in V1. The estimated TMP cannot be compared from the snap shots of V1 and V2. The TMP estimation for V1 and V2 is analyzed in Sect. 5.2.

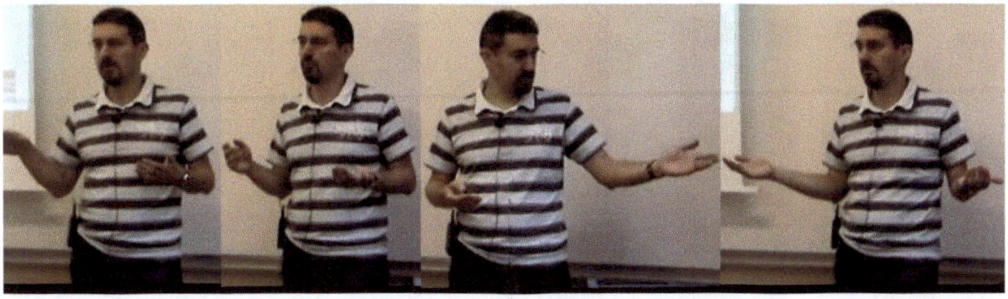

Fig. 4 Set of poses of real human motion in video sequence V1

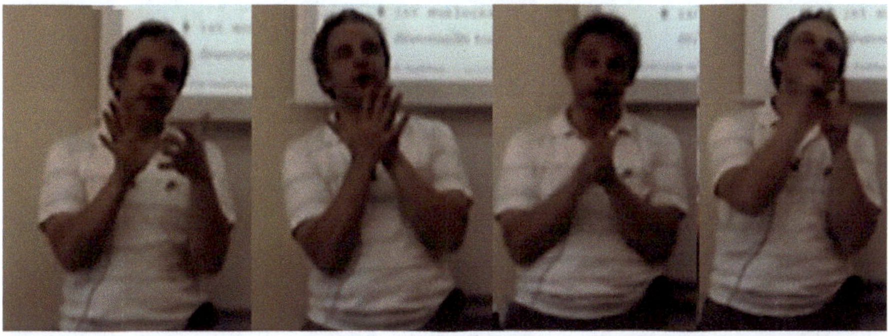

Fig. 5 Set of poses of real human motion in video sequence V2

5.2 Animating Virtual Agent

We animate a virtual agent with new gesture based on spatial and temporal extent value which we estimated based on Sects. 3.1 and 3.2. From this estimation we synthesize new gesture with Greta having the spatial extent of V1 and of V2 (resp. Figs. 6 and 7).

Using estimated temporal extent for the users in videos V1 and V2, motion trajectories are synthesized. Variation of temporal extent for synthesized trajectories is explained by Y position variation of wrist with respect to chest. From the synthesized motion, the variation of Y position of the wrist with respect to chest is shown in Fig. 8. In Fig. 8, the higher TMP attains peak earlier and finishes the gesture earlier than lower TMP motion. As we know the gesture which is done with higher speed will end earlier, this is reflected in the Fig. 8. Also this shows that user in V2 has higher temporal extent than user in V1 comparatively.

5.3 User Reviews on Animation

We validated our results with user reviews. We have chosen some of our friends who do not work on expressivity parameters or animation field as our users to validate

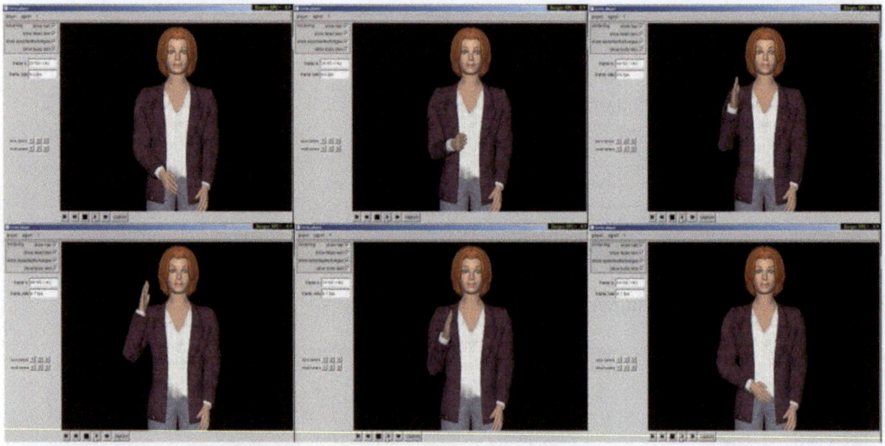

Fig. 6 Animated gesture using Greta for the spatial extent of V1

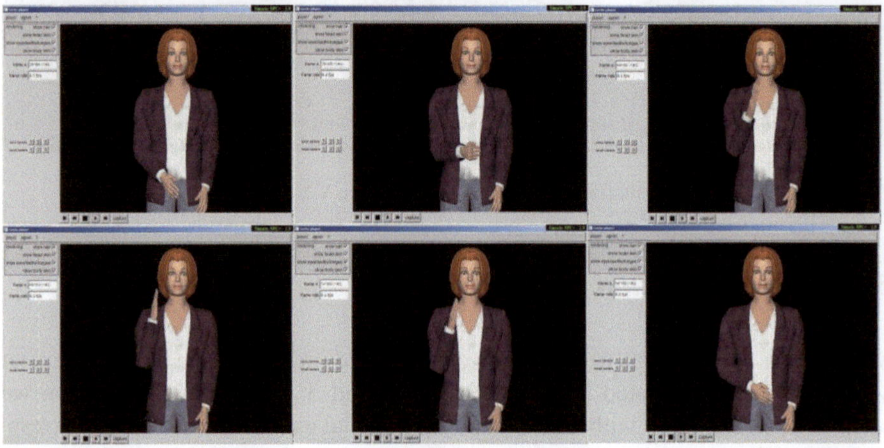

Fig. 7 Animated gesture using Greta for the spatial extent of V2

our results. We believe their reviews will be from the lay man view on animated virtual agent. So far we have seventeen users reviewed our results. We have conducted three tests. In the first test, the animated video contains only spatial extent parameter variation in the generated motion. In the second test, the synthesized motion has variation based on temporal extent parameter. In the third test, we generated synthesized video having variation of both spatial and temporal extent variation. Users are asked to identify whether they can recognize the synthesized motion similar to real human motion. First test result reveals that out of seventeen users, thirteen users correctly identified the synthesized motion with real human motion. In case of second test, fourteen users rightly recognized the animated video to the real human video. In the third test, fourteen users could recognize the synthesized motion matching with

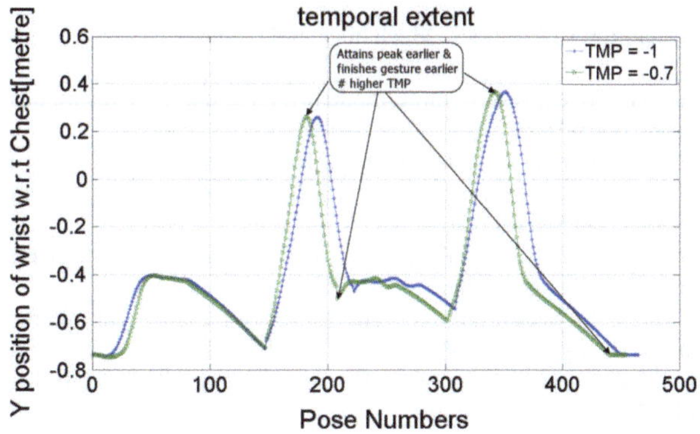

Fig. 8 Wrist trajectories synthesized using Greta animation engine with temporal extent from V1 and V2

real human motion. These results show that, our result has more than 75 to 83 % consistency among defining the style of a human.

6 Conclusion and Future Work

By estimating the spatial and temporal extents from motion by a real user, we can then animate the virtual agent to render the expressivity captured from a real user. This animation can be played virtually when the user is not available to control the avatar. Rendering the expressivity parameters allows generating personalized animations, so that the viewer can have the feeling of interacting with an expressive virtual human. In the future, more expressivity parameters such as power and fluidity can be considered for fine control of the virtual character.

References

1. Bavelas, J.B., Chovil, N.: Visible acts of meaning:an integrated message model of language in face to face, dialogue. **19**, 163–494 (2000)
2. Boone, R.T., Cunningham, J.G.: Children's decoding of emotion in expressive body movement: the development of cue attunement. Develop. Psychol. **34**, 1007–1016 (1998)
3. Camurri, A., Castellano, G., Ricchetti, M., Volpe, G.: Subject interfaces: measuring bodily activation during an emotional experience of music. Gesture Hum.-Comput. Interact. Stimulat. **3881**, 268–279 (2006)
4. Camurri, A., Lagerlöf, I., Volpe, G.: Recognizing emotion from dance movement: Comparison of spectator recognition and automated, techniques. Int. J. Hum.-Comput. Stud. **59**, 213–225 (2003)
5. Chellappa, R., Roy-Chowdhury, A.K., Kale, A.: Human identification using gait and face. In: Proc. of IEEE Conference on Computer Vision and Pattern Recognition (CVPR), Minneapolis, 2007. pp. 1–2

6. Craig, J.J.: Introduction to Robotics: Mechanics and Control, 3rd edn. Prentice hall, London (1986)
7. Davis, J.W., Gao, H.: An expressive three-mode principal components model of human action style. **21**, 1001–1016 (2003)
8. Drosopoulos, A., Mpalomenos, T., Ioannou, S., Karpouzis, K., Kollias, S.: Emotionally-rich man-machine interaction based on gesture, analysis. Human Computer Interaction International, vol. 4, pp. 1372–1376 (2003)
9. Ekinci, M.: A new approach for human identification using gait recognition. In: Proc. of ICCSA, vol. 3, pp. 1216–1225, Glasgow (2006)
10. Elgammal, A., Lee, C.S.: Separating style and content on a nonlinear manifold. In: Computer Vision and, Pattern Recognition (CVPR), 2004. pp. 478–485
11. Fitts, P.M.: The information capacity of the human motor system in controlling the amplitude. J. Exp. Psychol. **47**, 381–391 (1954)
12. Gallaher, P.E.: Individual differences in nonverbal behavior: Dimensions of style. **63**, 133–145 (1992)
13. Gómez Jáuregui, D.A., Horain, P., Rajagopal, M.K., Karri, S.S.K.: Real-time particle filtering with heuristics for 3d motion capture by monocular vision. In: Proc. of IEEE Conference on Multimedia Signal Processing. Saint-Malo, 2010. pp. 139–144
14. Hartmann, B., Mancini, M., Pelachaud, C.: Implementing expressive gesture synthesis for embodied conversational agents. In: Gesture Workshop, LNAI, pp. 188–199. Springer, Heidelberg (2005)
15. Hassin, R., James, U., Bargh, J. (eds.): The New Unconcious. Oxford University Press, Oxford (2005)
16. Hsu, E., Pulli, K., Popović, J.: Style translation for human motion. ACM T. Graphic. **24**, 1082–1089 (2005)
17. Kinect: Microsoft corporation (2010), http://www.xbox.com/en-GB/kinect
18. Lee, L., Grimson, E.: Gait analysis for recognition and classification. In: Fifth IEEE International Conference on Automatic Face Gesture Recognition, Washington, 2002. pp. 734–742
19. McNeill, D.: Hand and Mind what gestures reveal about thought. The university press of chicago press, Chicago (1992)
20. Mancini, M., Bresin, R., Pelachaud, C.: A virtual-agent head driven by musical performance. IEEE Trans. Audio Speech Lang. Process. **15**, 1883–1841 (2007)
21. Mehrabian, A., Wiener, M.: Decoding of inconsistent, communications. J. Pers. Soc. Psychol. **6**, 109–114 (1967)
22. Meijer, M.: The contribution of general features of body movement to the attribution of emotions. J. Nonverbal Behav. **13**, 247–268 (1989)
23. Noot, H., Ruttkay, Z.: The gestyle language. Int. J. Hum. Comput. Stud. **62**, 211–229 (2005). (Special issue: Subtle expressivity for characters and robots)
24. Pelachaud, C., Poggi, I.: Subtleties of facial expressions in embodied agents. J. Vis. Comput. Anim. **13**, 301–312 (2002)
25. Pelachaud, C.: Greta, http://perso.telecom-paristech.fr/~pelachau/Greta/
26. Quek, F., McNeill, D., Bryll, R., Kirbas, C., Arslan, H.: Gesture, speech, and gaze cues for discourse segmentation. Comput. Vis. Pattern Recognit. (CVPR). **2**, 247–254 (2000)
27. Tenanbaum, J.B., Freeman, W.T.: Separating style and content with bilinear models. Neural Comput. **12**, 1247–1283 (2000)
28. Vasilescu, M.O., Terzopoulos, D.: Multilinear analysis of image ensembles: Tensorfaces. In: Proceeding of the European Conference on Computer Vision (ECCV'02), Copengan, May 2002. pp. 447–460
29. Wallbott, H.G.: Bodily expression of emotion. Eur. J. Soc. Psychol. **28**, 879–896 (1998)
30. Wallbott, H.G., Scherer, K.R.: Cues and channels in emotion recognition. J. Pers. Soc. Psychol. **51**, 690–699 (1986). (American Psychological Association)
31. Wang, J.M., Fleet, D.J., Hertzmann, A.: Multifactor gaussian process models for style-content seperation. In: ICML, Corvallis, 2007. pp. 975–982

WordVis: JavaScript and Animation to Visualize the WordNet Relational Dictionary

Steven Vercruysse and Martin Kuiper

Abstract Today's data needs flexible visualization, and the advent of new technologies defined under HTML5 has opened up new possibilities for rich web-applications. We used JavaScript and the new HTML5 'canvas' element, to create a software module *GraphVis* for online visualization and exploration of large relationship networks. As a first test-case we applied the module to Princeton's WordNet English thesaurus, and made the result available as the web-application *WordVis*. WordVis is an exceptionally agile, online synonym-dictionary explorer. It shows a customizable graph of interlinked words and meanings, which are arranged by real-time animation, and behave like repelling magnets that are also rearrangeable by the user. *WordVis* shows the merits of using JavaScript: the visualizer can be fully integrated inside a web page, and that enables communication with other HTML elements in the page. It demonstrates the latest developments in web technology, and its agility may serve as an inspiration for other web-applications. As an intuitive, online synonym-dictionary explorer, it helps to 'find the right word' while writing English texts. It offers an online, free and more powerful browsing experience than any existing word-network visualizer.

1 Introduction

JavaScript is the prominent scripting language on the internet. JavaScript code is embedded in a web page's HTML code or sent along with it. It enables programmatic and user-guided manipulation of webpage components (DOM elements), and

S. Vercruysse (✉) · K. Martin
NTNU, Norwegian University of Science and Technology, Systems Biology group
Høgskoleringen 5, 7491Trondheim, Norway
e-mail: vercruys@nt.ntnu.co

K. Martin
e-mail: kuiper@nt.ntnu.no

M. Kudělka et al. (eds.), *Proceedings of the Third International Conference on Intelligent Human Computer Interaction (IHCI 2011), Prague, Czech Republic, August, 2011*, Advances in Intelligent Systems and Computing 179, DOI: 10.1007/978-3-642-31603-6_12,

so supports rich and responsive user interfaces (e.g. it drives the client-side of Gmail and Facebook). Until recently, complex animation was not well supported with JavaScript. To move objects required multiple updates of their DOM-tree entry, a painfully slow process which is only advisable for small numbers of elements. To create smooth larger animations, programmers had to resort to third-party plugins like Java-applets or Flash-objects, which means that an isolated piece of code is running inside the webpage. This makes it hard or even impossible to let this embedded container and the other elements in the same webpage interact: event-signals do not travel across the container's boundaries, and each part of the webpage can not be made aware of (or react to) changes in the other part. This weakens the potential for rich web-applications where all components would be integrated.

To alleviate these problems, the WHATWG group defined the specification for a new HTML-element '<canvas>' as part of the new HTML5 standard [1]. A canvas element reserves a rectangular area on the page (a bitmap), onto which JavaScript code can directly draw pixels, lines, circles, text, etc. Modern browsers like Firefox, Chrome, Opera, Safari have supported canvas for a long time, and recently also Internet Explorer (IE)'s version 9 supports it. In the still widely used IE 8, canvas will also work (although slower) if the webpage sends along the 'excanvas' module [2], which translates basic functions to IE 8s older graphics method. The canvas element has opened up opportunities for interactive and animated visualizations, integrated inside web pages.

2 Results

We developed the software module *GraphVis* based on JavaScript and canvas, for browsing through a large graph, by displaying parts of it via an animated, interactive, force-based layout algorithm. Although we designed this as a general visualizer for exploring large graphs (such as molecular interaction networks in biomedical data), we here demonstrate it using the publicly available dictionary database WordNet [3–5]. As such, we created a visually browsable English synonym dictionary, and made this freely usable online as the web-application WordVis: http://wordvis.com. WordVis lets users look up words in the English dictionary, inspect synonyms grouped by meaning, and browse them all in an animated, interactive web of words and meanings (see Fig. 1).

The module GraphVis can display a subset of a large graph that is for instance stored in an online database. After bringing up an initial subgraph, a user can browse the graph by bringing up new adjacent parts, or removing other parts. Also filters can be applied to the data. During the addition or removal of nodes and edges (connections), GraphVis visualizes the changes by transforming the old subgraph into a new one, via real-time animated re-layout. Animation and fluid movements enable the user to keep track of moving nodes and how the graph rearranges during re-layout. It also increases interactivity: users can drag around (manually reposition) nodes,

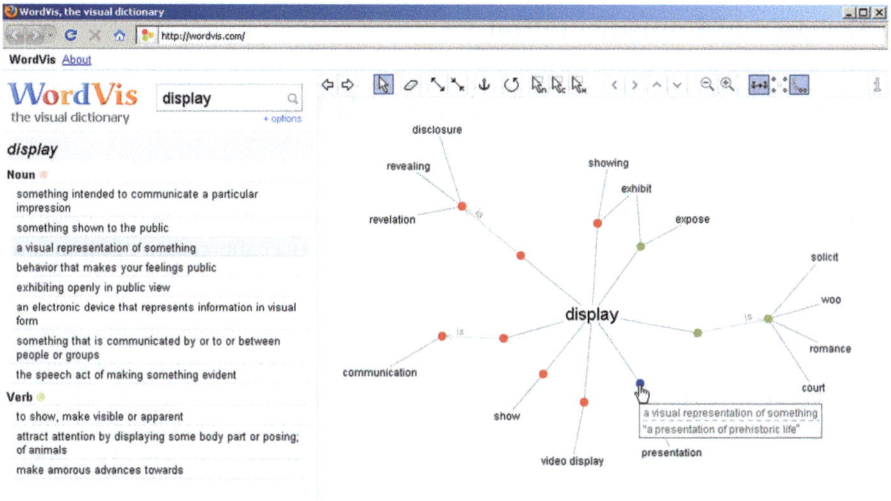

Fig. 1 Screenshot of a web-browser that shows the WordVis main page A search term was entered (*shown next to the logo*). The visualizer (at the *right*) centers this word in a web of related meanings (*balls: red for nouns, green for verbs*) and synonyms (*words: attached to a particular meaning*). The results panel (at the *left*) shows full definitions linked to visible meanings (*ball-nodes*). Each ball or text node can be clicked, which will bring up a new part of the WordNet-network. The network is arranged in real-time by a JavaScript-based, force-based layout algorithm, which responds interactively to rearrangements by the user. The toolbar shows operations to edit the graph or change the view

while surrounding nodes continuously slide towards new optimal positions. All this results in a clear and streamlined browsing experience.

WordVis is an application of our module GraphVis. It takes both WordNet's words and meanings (also called 'synsets' or 'concepts') as graph nodes, and represents them as string labels and balls resp. (ball color indicates: noun, verb, ...). Word-to-meaning relationships are represented as plain, straight lines. Semantic and other relationships between meanings ('is a', 'has part, 'see also', ...) are lines tagged with the relationship's abbreviation (or full name when mouse-hovered). WordVis also provides filters to exclude certain relationship types.

WordVis shares similarities with TouchGraph [6], ThinkMap Visual Thesaurus [7, 8], VisuWords [9], Visual Browser [10], etc. But in addition to using the modern technologies JavaScript and canvas, and its high user-friendliness, it also surpasses existing visualizers with its extensive array of graph-editing functions, which empower a detailed customisation during the exploration of information webs (see design details).

3 Design and Implementation

3.1 GraphVis: The Browser/Visualizer Module

Here we give an insight into key aspects of GraphVis' design.

Force-based layout algorithm. Nodes of the (sub)graph currently in the visualizer are modeled as objects that per animation frame undergo a capped sum of up to four different physical forces:

1. Nodes repel each other 'electrically', whereby a maximum interaction-distance limit supports computational efficiency. A node is modeled either as a charged point (think of an electron), a line-segment, or a rectangle. GraphVis automatically treats nodes representing a single line of text as linear charges, and multiline text-nodes as rectangular charges. A line-segment charge interacts like a point-charge placed on the segment where it is closest to the interactor. For example, for a point charge located left of a segment charge, the horizontal component of the distance is calculated as to the segment's left side. This prevents that text labels overlap: they look and feel like magnetic rods that repel each other over their whole length.
2. Edges are modeled as springs that attract their two nodes the further they are away than the spring's preferred length, and repel when closer. (GraphVis also supports 'edges' and spring-like connections between multiple nodes, a feature not used in WordVis).
3. 'Hidden springs' act like normal springs, but are not associated with actual edges. They can be temporarily placed around a node that 'has focus' (see below), and invisibly connect this node with indirect neighbouring nodes, limited to some fixed path-length. In WordVis for instance this path-length is simply 2, and so they link a focused node with nodes not directly connected, but reachable by following two connected edges). Their initial spring length is proportional to the average sum of lengths of 'visible' springs that they span. Like this they arrange nodes more clearly into 'levels' (concentric rings) around a focused node.
4. Any edge can be given any preferred orientation, enforced by an 'orienting force'. This can e.g. cause 'is-a' relationships to become organized as a top-down hierarchy. An edge's orienting force is implemented efficiently, by connecting one of its two nodes via a spring-force to a precalculated target offset (the ideal rotated position) relative to the edge's moving center, and applying a mirrored force to the other node.

The layout algorithm skips the intermediate step of keeping a speed variable, and updates positions directly based on the sum of forces. This makes the animation less 'slowly-wobbly': nodes don't bounce in and out until they come to a rest; instead they move and stop directly at their optimal position. However, this makes densely connected nodes (undergoing many opposing spring-forces) tend to jump continually between two nearby positions (one jump per animation step). This effect was mitigated by dampening the spring force around such nodes, putting them at the average of their last two force-layout-calculated positions, and eliminating their electric charge (otherwise their weak springs could not prevent them from drifting away by global electrical repulsion).

In order to spare processor power, the layout algorithm stops when only small forces remain, or after a set time-interval. Certain user interactions (manual re-layout, graph modifications) always reactivate the animation.

Manual layout. Users can move around nodes and edges (i.e. its nodes together) to modify the graph's layout at all times, by dragging them with the mouse. Force-based animation immediately responds to these changes, and continues to steer all other nodes towards new optimal positions. When a node or edge is hovered or closely approached by the mouse, its movement it frozen so it can easily be grabbed.

User interaction. Next we will describe several other ways of manual interaction on the visible graph. All are enabled via GraphVis' toolbar, or the mouse plus modifier keys (Shift, Ctrl, Alt), or keyboard shortcuts, or several of these. This is detailed by tooltips when hovering toolbar buttons. A webpage can run several instances of GraphVis; then keyboard events are transmitted only to the 'currently active' instance: the one that received the last mouse click.

Expanding a node. Clicking on a node launches an 'expand' command that looks up and then shows its 'environment'. A node's 'environment' is defined as the connected subgraph that should be shown around it, following this 'expand' command. Its particular implementation is up to the application using the GraphVis module. It could e.g. include neighbours up to path length X, or only certain types of neighbours (see below for WordVis in particular). It can also depend on current filter settings; when these change, GraphVis can update all its currently visible 'expansion environments'.

Expansion track. Several nodes can be expanded consecutively, and are kept on a list: the 'expansion track'. Then in order to keep the visualization clean, this track has an optional maximum length whereby older expanded nodes are automatically un-expanded: their environment is removed, except for nodes/edges still in some other expanded node's environment. WordVis simply uses a track-length of 1, showing only the last expanded node-environment. By toggling off one of the rightmost buttons on GraphVis's toolbar, all expanded environments will be kept visible.

Removing nodes/edges. Via the eraser tool (accessible via the toolbar, among others), any node or edge can be removed. Sweeping this tool (mouse-down + drag) removes all nodes and edges on its path. Removing a node also removes its connected edges. 'Orphaned' nodes that are hereby no longer connected to an edge are removed automatically (unless disabled via the toolbar).

Fade-in/out. For viewing pleasure, added/removed objects are smoothly faded in/out. This process piggy-backs on the animation loop, and blocks it from stopping until fading is complete. When a node/edge is fading out and queued for removal, it stops interacting in the layout algorithm.

Edge length. To clarify the layout of complex graphs, one can manually adjust the length of any edge, or in fact, the preferred length of the edge's associated spring.

Automatic edge length optimization. Nodes that are densely connected (i.e. have many incoming edges) need more space around them to lay out connected nodes without overlap. Therefore GraphVis automatically increases edge length around dense nodes. A small fixed length is added to all connected edge-springs after a new edge connects to a node, or subtracted when any is removed. (Note: this addition doesn't apply to hidden-springs: these are rebuilt after graph structure changes).

Anchoring nodes. An emphasis or 'focus' can be put on any number of nodes, by 'anchoring' them. While anchored, a node doesn't undergo force-based layout and stays at a fixed position on the screen. It is drawn larger (and connected springs are lengthened), and hidden springs are added that connect it with some levels of indirect neighbours. As described above, this makes it a center of attention by organising its surroundings clearer into levels. GraphVis can automatically anchor a newly expanded node and remove all previous anchors. (Default setting in WordVis; can be toggled off in the toolbar).

Grouping. Multiple objects can be selected via Shift+dragging, which marks them with colored background. A command on one selected object is launched for all selected objects (anchor, remove, edge length, ...). Selected objects can be dragged simultaneously, either without or with undergoing force-based-relayout (Ctrl+Shift+drag).

Viewport changes: panning, zooming, rotating. Panning (shifting all) happens by dragging the graph's background, or via toolbar buttons or keyboard arrows. GraphVis can automatically shift to center a newly expanded node. During zooming in/out, edge lengths increase/decrease linearly with zoom level, while shape and font sizes change more slowly (inverse power law) to keep them longer recognizable or readable at high zoom levels. While rotating, node labels stay horizontal.

Undo/redo. The undo/redo history stores graph structure changes (add, remove, anchor, edge length), but not manual layout drags, automatic layout, or viewport changes. Automatic center-shifts onto expanded nodes are stored as well. The positions of removed nodes are stored too, for fair re-placement upon undo/redo actions. Certain undoable events are grouped into a single undoable action: group commands, fast successive commands, or e.g. eraser sweep events.

Initial positioning. Newly added nodes are placed around the expanded node, so as to reduce the force-based layout movements that remain necessary. Nodes on a path-length of 1 are placed on a circle around the expanded node (except if it already had 1 incoming edge, in which case they are put on a half-circle that faces away). For path-length 2 and up, nodes are placed on ever narrower and further, outwards oriented circle segments. Preferred edge orientation overrides this method.

Visual styles. GraphVis works with an extensible system of visual styles that determine how a node/edge should look, and each node/edge is linked to one style. This made it easy to define particular WordVis styles. Nodes can be text or any 2D-shape. Edges can have any color, weight, or dashes. Edges can carry several decorations (e.g. arrows or text) at their start, middle or end; then when several relations exist between the same two nodes, all symbols or labels are put on the same edge (always drawn as a straight line). Text and e.g. asymmetric arrow-heads on edges are rotated or mirrored resp. (relative to the edge's direction), to prevent being shown upside down. For clarity, edges (and arrowheads) are only drawn up to rectangular or circular nodes' borders, not their center.

Context menus. These are customizable based on object type or contents, e.g. 'Google search' for text nodes.

Event notification. Essential for integration with external JavaScript modules, is that these can be notified of events happening inside the visualizer. Modules registered as 'GraphVis-listeners' can be made aware of, and then respond to, events like mouse-hovering, expand, remove, undo, etc.

3.2 WordVis: GraphVis Applied to WordNet

Now we describe more aspects of WordVis: the web-application that embeds one GraphVis module and makes WordNet browsable as a large graph of words and meanings.

Autocompleting search-box. This helps to find a word that exists in WordNet, and upon Enter, shows its environment as the initial graph. Entering a comma-separated list of words will bring up several focused nodes together. Auto-suggestions are queried from the server, ranked by commonness, and insensitive to case or punctuation.

Node environments, as defined for WordVis: The environment of a text-node (word) includes all its connected meanings, represented by ball-nodes. Each of these meanings can in turn be linked to zero or more other words (synonyms), which are also included. If a meaning-node is not linked to any other word, then WordVis makes its meaning clear in the graph in another way: it looks up other meaning-nodes that are linked e.g. as parent in a 'is-a' relationship, or via 'see-also', etc.; it includes these meaning-nodes, and it includes their synonyms. This gives a general idea of what the original, synonym-less ball-node represents. The environment for a meaning-node (ball) consists of all its directly linked text-nodes, as well as all its linked (via any relation) meaning-nodes plus their connected text-nodes.

JavaScript code sends node-environment XMLHttpRequests to the WordVis server, based on a text string or a meaning-ID. PHP-code on the server then calculates this environment, based on queries on a MySQL-database version of WordNet. It then sends back simple, tab-delimited data, from which the JavaScript code builds a node-and-edge datastructure.

Filters. The 'options' link on the site unfolds a checklist of word and relation types that should be included (all, by default) or excluded in the visualizer. The excluded types are sent along with each node-environment request, and the server sends back a filtered environment. After changes in these filter-settings, GraphVis is notified to refresh the environments of all nodes on its expansion track.

Results panel. Lists the definitions of all meaning-nodes currently in the visualizer. The panel is an example of integration between the JavaScript/canvas-based visualizer, and other HTML elements in the page: hovering over either a definition or its corresponding meaning-node will highlight both of them; and a click on a definition will send a node-expansion command to the visualizer. Also, the results panel will update after it receives a 'graph changed' notification from the visualizer.

Precalculation. Calculating a node-environment in WordVis requires evaluating several levels of graph connections, what makes that database-query time may take up to 1–2 s on a standard webserver. Therefore we pre-calculated the (only not-filtered) environments for WordNet's over 305,000 nodes. When the server now receives a (not-filtered) node-environment request, it needs only one query to find the answer. This dramatically increased WordVis' response speed.

Page load speed optimizations. (1) When a JavaScript module (.js file) is included in the 'head section' of an HTML-page, page-loading is blocked at that point until the entire .js file is loaded. Instead we included a small piece of code in the main page, that loads the GraphVis module after the main page is loaded (it dynamically adds a new '<script>' tag for the module). Then the search page loads fast and users can enter a word while the visualizer's code is still loading in the background, in parallel with other extra files (this can be verified e.g. in the Net panel of Firefox's 'Firebug' extension). (2) The Javascript code was minified by combining functionality from [11] and [12] (short variable and function names, comments stripped, and compressed), which reduced it to 15 % of its original size. (3) Toolbar images are not loaded as dozens of separate files, but as a single combined image from which each <img>tag shows a portion (via its CSS style).

4 Conclusion

We presented WordVis, an application of our JavaScript and animation based network visualizer. We showed that JavaScript web-technology is mature enough to support smooth animation for a rich user experience. The JavaScript-based visualizer could be fully integrated inside a web page, enabling direct interaction with other HTML elements. WordVis is a free, versatile, online synonym-dictionary explorer, that helps to 'find the right word' while writing English texts.

5 Availability

The WordVis web-application is accessible at http://wordvis.com. WordVis was tested to work in recent versions of five major browsers: Mozilla Firefox 3.6 up to 5, Internet Explorer 8 and 9, Google Chrome 10, Opera 11, and Safari 5.

References

1. WHATWG: http://www.whatwg.org/specs/web-apps/current-work/multipage/the-canvas-element.html
2. Excanvas: http://code.google.com/p/explorercanvas/

3. Miller, G.A., Beckwith, R., Fellbaum, C., Gross, D., Miller, K.: WordNet: An on-line lexical database. Int. J. Lexicography **3**(4), 235–312 (1990)
4. Miller, G.A.: WordNet: A lexical database for english. Comm. ACM **38**(11), 39–41 (1995)
5. WordNetSQLbuilder: http://sourceforge.net/projects/wnsqlbuilder/files/wnsqldatabase/
6. TouchGraph: http://sourceforge.net/projects/touchgraph/
7. ThinkMap: http://www.thinkmap.com
8. Visual Thesaurus: http://www.visualthesaurus.com
9. VisuWords: http://www.visuwords.com
10. Visual Browser: http://nlp.fi.muni.cz/projekty/vizualni_lexikon/
11. Javascript obfuscator: http://javascript-source.com
12. Jasob 4: http://www.jasob.com

A Reduce Identical Event Transmission Algorithm for Wireless Sensor Networks

Hong-Chi Shih, Shu-Chuan Chu, John Roddick, Jiun-Huie Ho, Bin-Yih Liao and Jeng-Shyang Pan

Abstract This paper proposed a Reduce Identical Event Transmission Algorithm (RIET). The algorithm can decide that which sensor nodes could send the event to sink node when sensor nodes sense a same even. Moreover, other nodes can save power because they didn't send the same event. In our simulation, the RIET algorithm can enhance sensor nodes' life time about 12.9 times and saving power consumption about 52.43 % than tradition algorithms.

1 Introduction

Recent advances in micro processing, wireless and battery technology, and new smart sensors have enhanced data processing, wireless communication [1, 2, 6] and detecting capability. In general, sensor node is a small and cheap device. Hence, the wireless sensor network is usually constituted by hundred or thousand nodes. In the wireless sensor network, sensor nodes might detect a same event because they were deployed in a large-scale randomly. The sensor nodes will consume lots of power if they send the same event to sink node independently in the tradition algorithm.

This paper proposed a Reduce Identical Event Transmission Algorithm (RIET). The algorithm can decide that which sensor nodes could send the event to sink node when sensor nodes sense a same even. Hence, the RIET algorithm can solve the problem of sending a same event repetitively and reduce sensor nodes' power consummation when the sensor nodes' detect area are overlapping and sense a same

H.-C. Shih (✉) · B.-Y. Liao · J.-S. Pan
National Kaohsiung University of Applied Sciences, Kaohsiung, Taiwan
e-mail: hqshi@bit.kuas.edu.tw

S.-C. Chu · J. Roddick
Flinders University of South Australia, Adelaide, SA, Australia

J.-H. Ho
Cheng Shiu University, Kaohsiung, Taiwan

M. Kudělka et al. (eds.), *Proceedings of the Third International Conference on Intelligent Human Computer Interaction (IHCI 2011), Prague, Czech Republic, August, 2011*, Advances in Intelligent Systems and Computing 179, DOI: 10.1007/978-3-642-31603-6_13,

event. Moreover, the algorithm will avoid the transmission delay when sensor nodes commute each other.

2 Related Work

This paper proposed the RIET algorithm which is based on Grade Diffusion algorithm (GD) [6] to reduce the power consummation when sensor nodes sense a same event. We introduce some tradition routing algorithms and the GD algorithm as below.

2.1 Ad-hoc On-Demand Distance Vector

The Ad-hoc On-Demand Distance Vector (AODV) was proposed by C. E. Perkins and E. Royer in 1999 [4], and it applies for sensor nodes are fixed or can be moved. The AODV algorithm employs three processes: route request, route replying and route maintenance, to build and maintain the routes required by active nodes for data transmission. The AODV algorithm creates the routing tables for sensor nodes which were on the transmission path when the even appeared, but the routing table just a part of the wireless sensor network. After, the routing table is rebuilt again and consumes lots of power when other nodes have to transfer event.

Please always cancel any superfluous definitions that are not actually used in your text. If you do not, these may conflict with the definitions of the macro package, causing changes in the structure of the text and leading to numerous mistakes in the proofs.

2.2 Direct Diffusion Algorithm

C. Intanagonwiwat et al. [3] introduced the directed diffusion (DD) protocol in 2003. DD aims to reduce the data relay in order to better manage power consumption. Basically, DD is a query-driven transmission protocol. The collected data are transmitted only if they fit the query from the sink node, thereby reducing the power consumption due to data transmission. In DD, all sensor nodes are bound into a route when broadcasting the interested queries, even if the route is such that it will never be used. In addition, several circle routes, which are built simultaneously when broadcasting the queries, result in wasted power consumption and storage.

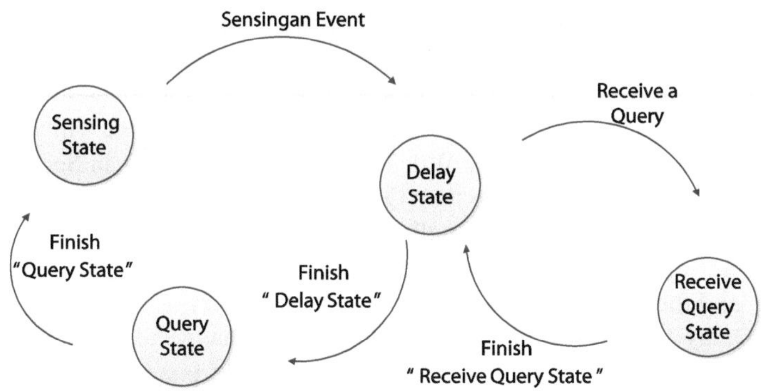

Fig. 1 RIET algorithm's final state machine

2.3 Grade Diffusion

In 2011, [6] proposed a GD algorithm which improved the LD-ACO algorithm [2] to enhance nodes' life time, raised transmission's efficiency and solve the problem of node's routing and power consumption. The GD algorithm broadcast grade creating packages from sink node to every node completely and quickly in the wireless sensor network by LD-ACO algorithm. Then, the GD algorithm proposed a routing algorithm to reduce nodes' average load, enhance nodes' life time and, reduce nodes' power consumption.

But, the GD algorithm and tradition algorithms just think about that there is just a sensor node sent an event to sink when there is an event appeared. Hence, sensor nodes will consume lots of power to send a same event if they sense a same one. So, this paper proposed a RIET which can reduce the probability of sending a same event and saving sensor nodes' power. Moreover, the RIET algorithm is based on GD algorithm.

3 Reduce Identical Event Transmission Algorithm

The RIET algorithm which is used in the wireless sensor network can reduce the probability of sending a same event, saving nodes' power, and enhance sensor nodes' lift time by sensor nodes' communication. The sensor node is not only has the ability of sensing and transferring event but also can commutation with neighbor nodes when it sensors an event. The RIET algorithm uses the finite state machine (FSM) which has "Sensing State", "Delay State", "Query State", and "Receive Query State" to avoid sensor nodes commutation with neighbor nodes at the same time in the algorithm. Hence, sensor node can query or response neighbor nodes by our algorithm and FSM, and the FSM's transition is showed on Fig. 1.

In Fig. 1, the sensor node has 4 states which are “Sensing State”, “Delay State”, “Query Start”, and “Receive Query State”. In general, sensor node is under the “Sensing State” process to wait event appeared. When sensor nodes sense an event, sensor nodes’ state will be changed to “Delay State” and each node which senses the event will get a different delay time Wd. The Wd ’s illustration is in Sect. 3.1. Then, the sensor nodes will be change to the “Query State” after the delay time. If sensor nodes’ state is “Query State”, they will commutation with neighbor nodes. Hence, the delay time can avoid the sensor nodes change their state to “Query State” at the same time, and then send event to sink node independently.

After “Delay State”, sensor node’s state will be change to “Query State” and send a query message to neighbor nodes. The message is asking neighbor nodes whether or not you sense a same event. The sensor node will be changed to “Receive Query State” when node is under the “Delay State” processing and receives a query message. If the sensor node’s delay time is short than other nodes, the sensor node’s state will be changed to “Query State” faster than others. Then other neighbor nodes will receive a query message and change their state to “Receive Query State” from “Delay State”. The illustration of “Query State” and “Receive Query State” is in the Sects. 3.2 and 3.3.

3.1 Delay State

After sensor node senses an event, the node will record an event processing state Es and the Es has three states: Un-processing (−1), Processed (+1), and In-processing (0). The “Query State” and “Receive Query State” will change the Es, and their illustrations are in the Sects. 3.2 and 3.3.

The W_d is calculated by Eq. 1, and it is a delay time mean when the sensor node can change its state from “Delay State” to “Query State”.

$$\begin{aligned} W_d &= \left(\alpha + N_l + \left(1 - \frac{E_{now}}{E_{original}}\right) + (1-\alpha) * P_r\right) * T_d \\ N_l &= \begin{cases} \left(\frac{l_v - 2}{l_v}\right), & l_v > 2 \\ 0, & otherwise. \end{cases} \end{aligned} \tag{1}$$

In Eq. 1, α is a user set value between 0 and 1, T_d means the shortest delay time, P_r is a random value between 0 and 1, l_v means sensor node’s grade value, E_{now} means sensor node’s remaining power, and $E_{original}$ means sensor node’s initial power.

In the RIET algorithm, sensor node needs to wait an $\alpha * T_d$ delay time least. And, if sensor node’s E_{now} is more than others, the sensor node’s state can change to “Query State” quickly. Moreover, the sensor node whose grade value is smaller than others will have shorter delay time W_d than others because the event might be sensed by different grade sensor nodes.

3.2 Query State

In Fig. 1, the sensor node's state will be changed to "Query State" from "Delay State" after delay time. When a sensor node is under "Query State" processing, the sensor node will check its E_s firstly. If the E_s is "Processed (+1)", the sensor node will do nothing and change its state to "Sensing State" because the event has been processed by other nodes.

On the other hand, the sensor node will query its neighbor nodes because the event hasn't been processed if sensor node's E_s is "Un-processing (−1)" or "In-processing (0)". The sensor node which sent query messages sets a minimum waiting time T_w to wait neighbor nodes' response, and the sensor node will check received information which contains E_s and W_d from neighbor nodes.

The query node check its and response node's E_s. If the response node's E_s is "Processed (+1)", the query node will do nothing and change its state to "Sensing State" because neighbor nodes had sent event to sink node.

But, if response node's E_s is "Un-processing (−1)" or " In-processing (0)", the query node will check its and response node's W_d. The query node sends the event to sink node, modify its E_s to "Processed (+1)", and change its state to "Sensing State" when its W_d is smaller than response node's. On the other hand, the query node just changes its state to "Sensing State" when its W_d is bigger than response node's or equal because neighbor nodes will send event to sink node after.

3.3 Receive Query State

The sensor node's state will be changed to "Receive Query State" from "Delay State" when it received a query message.

When the sensor node is under the "Receive Query State" processing, it will check whether or not it had sensed a same event as the query node. The sensor node will do nothing and change its state to "Delay State" if it didn't sense the same event. On the other hand, the sensor node responses its E_s and W_d to the query node, and receives the query node's E_s and W_d if the sensor node has the same event.

When the sensor node got the query node's W_d and E_s, it will check them. If the sensor node's E_s is "Processed (+1)", it will do nothing and change its state to "Delay State". Otherwise, the sensor node will check its W_d when its E_s is "Un-processing (−1)" or "In-processing (0)" because the event hasn't been sent. If the sensor node's W_d is smaller than the query node or equal, the sensor node changes its E_s to "Processing (0)" and its state to "Delay State" because the sensor node might send the event to sink node after. The query node will change its E_s to "Processed (+1)" and finish its "Query State" if its W_d is bigger than the receive node's.

The sensor node is still under the "Delay State" till the delay time W_d end when it returns from "Receive Query State" processing. If the sensor node which is under the

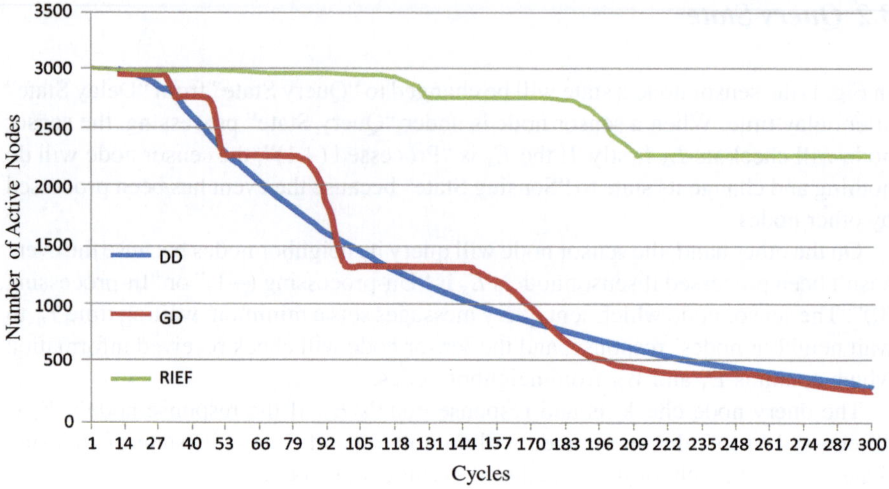

Fig. 2 Number of active nodes

"Delay State" processing and it receives a query message again, the sensor node's state will be changed to "Receive Query State" again.

Additionally, the sensor node won't into the "Qyery State" and "Receive Query State" processing at the same time because we need to avoid they change E_s at the same time.

4 Simulation

We simulate and analyze the RIET algorithm in this session. In the simulation, we build a 3-D space of 100*100*100, and the scale of the coordinate axis for each dimension is limited from 0 to 100. The radio limit of the nodes was set to 15 units. There are three non-overlay nodes in each 10*10*10 space and their Euclidean distance each other is at least 2. Thus, the space includes 3,000 nodes, and the center node in (50, 50, 50) is assigned to the sink node.

There are 300 cycles and each cycle has 300 event been appeared. Thus, there are 90,000 events in our simulation. Every sensor node's initial power is 3,600 mw and it will consume 1.6 mw to send an event. Moreover, we set the parameter α is 0.8 in Eq. 1 and the E_{th} is 10 %. In the simulation, the sensor nodes have a same event if the event's values are different less than E_{th}.

Firstly, we compare the active nodes of the DD, GD, and RIET algorithm in each cycle, and the result was showed on Fig. 2. And, the active node means that it has enough power to send event and has relay nodes in its routing table.

In Fig. 2, the DD and GD algorithm only have 303 and 176 active nodes after 300 cycles, but the RIET still has 2,240 active sensor nodes which has power can

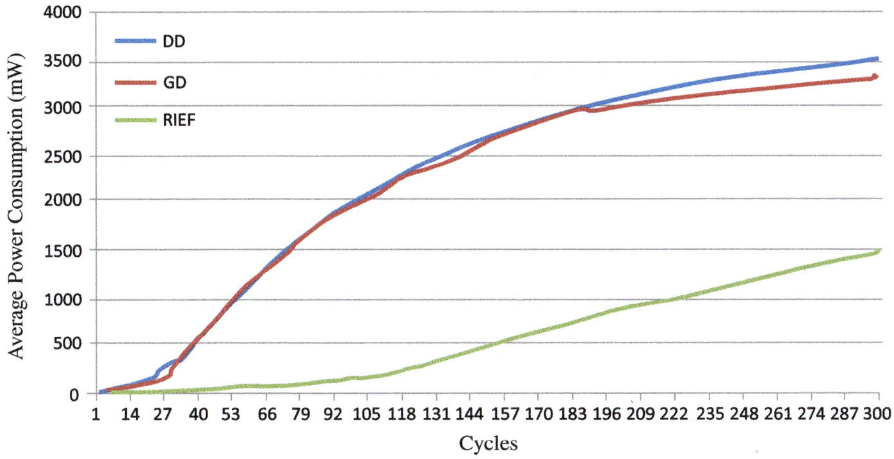

Fig. 3 Sensor nodes' average power consumption

sense and send event. The RIET algorithm rises sensor node's lift time about 7.5 times and 12.9 times than the DD and GD algorithms because the RIET algorithm can reduce the sent numbers of sensor nodes which sense a same event. But the sensor nodes will send an event to sink nodes independently if they sense the same event in the DD and GD algorithms, and they consumed lots of power. Hence, the RIET algorithm can save lots of power and rises sensor node's lift time. So, the RIET algorithm's active nodes are more than the DD and GD algorithms in Fig. 2.

Then, we compare the average power consummation of the DD, GD, and RIET algorithms, and the result was showed on Fig. 3. In Fig. 3, we can find that the RIET algorithm has least average power consummation because it can reduce the sent numbers of sensor nodes which sense a same event. Hence, the average power consummations of the DD and GD algorithms are 3497.83 and 3342.16 mw after 90,000 events appeared, but the RIET algorithm only consume 1663.94 mw. So, the RIET algorithm can reduce average power consummation about 52.43 and 50.21 %.

5 Conclusion

This paper proposed a RIET. The algorithm can reduce to send a same event repeatedly, raise sensor node's lift time, and reduce sensor node's power consummation.

In our simulation, we can find that the RIET algorithm can save sensor node's power about 52.43 and 50.21 % than tradition algorithms. Moreover, the RIET algorithm can enhance sensor node's lift time about 12.9 times.

References

1. Corson, S., Macker, J.: Mobile Ad Hoc Networking (MANET): Routing Protocol Performance Issues and Evaluation Considerations, RFC 2501, January 1999
2. Ho, J.-H., Shih, H.-C., Liao, B.-Y., Chu, S.-C.: A ladder diffusion algorithm using ant colony optimization for wireless sensor networks. In: Information Sciences, 2011
3. Intanagonwiwat, C., Govindan, R., Estrin, D., Heidemann, J., Silva, F.: Directed diffusion for wireless sensor networking. IEEE/ACM Trans. Netw. **11**, 2–16 (2003)
4. Perkins, C.E., Royer, E.: Ad Hoc on-demand distance vector routing. In: Proceedings of IEEE WMCSA, February 1999. pp. 90–100
5. Royer, E.M., Toh, C.-K.: A review of current routing protocols for Ad-Hoc mobile networks. IEEE Pers. Commun. **6**, 46–55 (1999)
6. Shih, H.-C., Ho, J.-H., Liao, B.-Y., Pan, J.-S.: Grade diffusion algorithm. In: Conference on Information Technology and Application in Outlying Islands, 2011

Business Process Improvement Framework and Representational Support

Azeem Lodhi, Veit Köppen and Gunter Saake

Abstract Business process management and improvement are vital for enterprises in competitive environments. Understanding of a process is a pre-requisite and important step for improvement. Interaction between humans, computers, and business objects provide excellent opportunities for knowledge extraction. However, the specification of a framework is required for business process improvement, which extends from data collection, analytical methods, storage, and representation of knowledge. The process models conceived for information system development are not sufficient for post execution analysis and improvement. In this paper, we specify such a framework briefly and focus on providing representational support for business process improvement. The main objective is to improve the overall improvement process by providing enriched graphical process models. Furthermore, we use a case study to explain the proposed usage and extensions of an existing modeling language for business process improvement.

1 Introduction

Business world is competitive where enterprises compete with one another to get the market share. Enterprises try to improve their existing processes in order to provide new or better services. Before making any improvements into existing operations,

A. Lodhi (✉) · V. Köppen · G. Saake
Department of Technical and Business Information Systems,
Faculty of Computer Science,
University of Magdeburg, Magdeburg, Germany
e-mail: azeem@iti.cs.uni-magdeburg.de

V. Köppen
e-mail: veit@iti.cs.uni-magdeburg.de

G. Saake
e-mail: saake@iti.cs.uni-magdeburg.de

M. Kudělka et al. (eds.), *Proceedings of the Third International Conference on Intelligent Human Computer Interaction (IHCI 2011), Prague, Czech Republic, August, 2011*, Advances in Intelligent Systems and Computing 179, DOI: 10.1007/978-3-642-31603-6_14,

first, we need to understand how processes are executed in the enterprise. We should also find out the connections at various levels like inter and intra organizational level. For example, at a lower level between resources and activities whereas on higher levels between business units like sales, manufacturing, purchasing, and other supplies. Organizational decisions like maintaining stocks, maximum processing time of orders and others are important factors for an enterprise's success.

Computers are widely used in business as it ranges from handheld devices to main servers in order to run businesses. Human interacts with business objects (physical materials), organizational resources, and computer applications in order to run daily operations of enterprises. Information systems play an important role and record the interaction between humans, business objects, and applications during the execution of processes. These interaction records are useful to make improvement in overall businesses in case of either business process structure, organizational resource structure, or human-computer interaction (HCI). Analysis of a business process helps to identify deficiencies in existing processes.

Process mining techniques use those records to extract knowledge about business processes [1]. Similarly, application usage mining (AUM) [2] attempts to improve the design and structure of applications for improved interaction. However, a specification of a framework is required which will help to improve the existing processes. This framework should encompass data collection, analysis, and representation of results. Different questions regarding the framework are important like how the data is extracted from log files and converted for analysis. What are existing approaches in this context?

Business process modeling is important for overall understanding and improvement. Most of the research in business process modeling is focused on alignment of information technology with business processes [3]. Limited research is carried out to represent process knowledge through business process models for improvement [4]. Currently, information is represented with key performance indicators (KPIs), statistics, or visualizations (e.g., pie charts or histograms) which is too abstract and does not provide process details to business analysts. Similarly, current modeling approaches do not provide details where interactions can be further improved. Therefore, a graphical representation is very important factor for evaluation and knowledge representation for business process improvement [5].

In this paper, we address these issues and specify a framework for business process improvement. Business process context, stakeholders, and their participation is discussed in the next section. We briefly explain different components of framework and steps involved in business process improvement in Sect. 3. We also mention researchers' contributions in order to solve different issues related with components of the framework. After specification of an improvement framework, we shift our focus to the main contribution which is representational support for improvement in Sect. 4. From a human-computer interaction viewpoint, we also provide directions where business process modeling can be improved with intelligent human-computer interaction support for business process management. We also propose the usage and extension of an existing modeling language for business process improvement with the help of a case study. In the end, we summarize this paper and provide an outlook in Sect. 5.

2 Business Process Context

Customers' demands are fulfilled by enterprises which do business for several reasons, like satisfying stakeholders (executives, employees, and customers), monetary gain, or increase in reputation. In order to achieve these goals, business operations are carried out in a specific way. This specific way is called business process and governed by policies (rules and conditions). Different events occur (something "happens") in processes which usually have a cause (trigger like a request from the customer) or an impact (result of an activity) [6]. In response to these events, organizational resources perform operations on inputs and transform them into defined outputs. In this way, the final output is generated in an enterprise. The fulfilment of enterprise's goal is measured by metrics that evaluates the performance of a business process and its involved elements. This context of a business process is abstractly represented in the Fig. 1.

Different stakeholders are involved in business processes at different levels. These stakeholders work on a business process at different levels. Therefore, they have different requirements and expectations from business process models. In the following, we list some of the stakeholders, levels, and their interaction with business process elements.

- Levels: Strategic level, management level, operational level
- Stakeholders: Executive, manager, employee
- Process: Business process, process, activity, task

At a strategic level, executives define goals and objectives of an enterprise. They view a business process as a black box and are concerned with overall output, profits, new acquisitions, and mergers. From a human-computer interaction (HCI) viewpoint, aggregated reports and statistics on handheld devices would be sufficient in case of remoteness. Managers, domain experts, and process managers are concerned with operations of enterprise and management of business processes. Tabletops, smart boards and other extended displays would be fruitful in analyses. Employees execute business process operations with a view of taking inputs, performing defined actions based on certain conditions using organizational resources, and then forwarding the result of those actions to other activities. We show some of the involve stakeholders at different levels in business process (BP) context meta model in Fig. 1.

Elements define at one level are propagated to the other levels as well. For example, the goal of an enterprise is propagated as a process goal at the management level and continued as an activity and task goal to the lower levels. Similarly, metrics are computed on operational and management levels as well. Therefore, an analytical improvement framework and its representational support have to accommodate the requirements of different stakeholders.

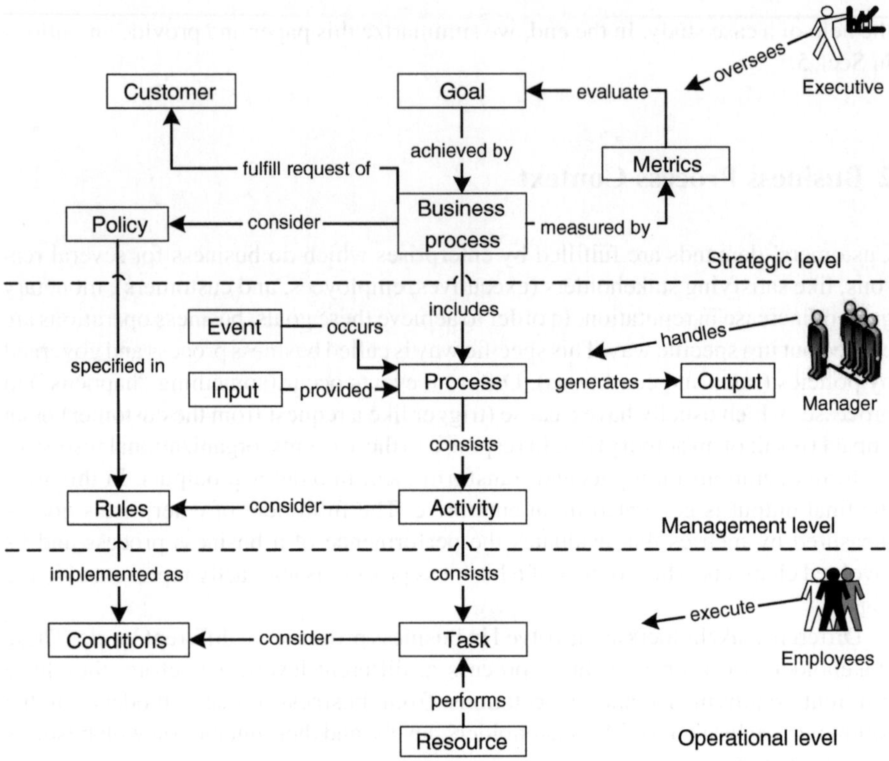

Fig. 1 Business process meta model in different layers

3 Framework for Improvement

Business process improvement is important for enterprises due to competition and changes in markets like customer demands and technological changes. Business process improvement is a challenging task because of the complexity to carry out the changes in processes. Enterprises find it difficult to decide from where to start and what changes should be carried out for improvement. Complexity in business process is due to various factors like inter-dependencies between activities, stakeholders, involved elements, their attributes, and applications. Research in business process improvement needs to address these complexity issues and provide support to decide which action should be taken for business process improvement.

In order to provide the required support for business process improvement, a complete framework for analysis is required to be specified. This framework should address issues like data collection from information systems, its computation, representation in business process models, defining steps for improvement, and mechanism to carry out the changes. An abstract picture of the overall framework is shown in the Fig. 2.

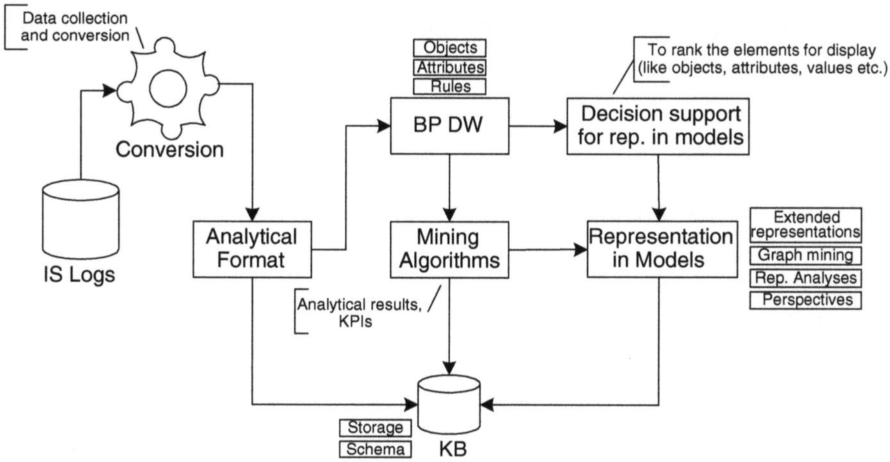

Fig. 2 Framework for business process improvement and analysis

Different techniques related with process mining [1] attempt to solve the issues related with data collection [7, 8] and its conversion [9, 10]. Similarly in [11], author proposes a data format for recording business events. This Business Process Analytics Format (BPAF) helps in correlating and aggregating business events from different systems to one location. In [12], authors store data from logs into a data warehouse for different analysis. Therefore, data is collected from log files of information system and converted into a suitable format for analysis (this data is sometimes referred as *process trace data*).

Afterwards, several techniques (like process mining algorithms [13–15] and business process intelligence [12]) are applied on process trace data to extract knowledge. Application usage mining [2] techniques analyze user's behavior with applications and attempt to improve the interaction between applications, humans and their working. Similarly, other techniques can also be applied like balance scorecards and other process metrics. A knowledge base is also required to store the gain information and knowledge from analysis techniques. Several issues are needed to be further investigated like what would be the schema design or format for storage? How the knowledge will be reflected as a company's best practices or used in enterprise operations?

The extracted knowledge and information are represented in a graphical form for intuitive understanding. For this purpose, depending on the stakeholders, certain information is filtered or provided with more details (context). Therefore, a decision support system is required which will rank the knowledge for representation with the context for stakeholders. This is also very important to reduce the cognitive load of users in graphical business process models while representing performance details and other knowledge.

Graphical representation of extracted knowledge and process trace data also provide an opportunity to apply different methods like graph mining [16]. Similarly, extended business process models help analysts to understand processes and identify deficiencies in business processes. We discuss the extended representation and benefits in the following section.

4 Representational Support

Graphical business process models are used to visualize the operations of a business process. Being graphical in nature, they are used for better understanding and analysis of business processes. They provide a medium of communication between stakeholders at different levels as described in Sect. 2. Business process modeling has increased the ability to understand business processes and to make rational decisions for organizing activities in a traceable and understandable way [17]. Business process modeling is used in different phases of a business process lifecycle, like, from identification of business processes to implementation in an information system [3].

Several gaps occur in business process lifecycle due to lack of communication, especially during the transition from one phase to another phase. Therefore, different models are proposed in literature to fulfill the requirements of stakeholders in business process lifecycle. For example, in planning phase Event-driven process chain (EPC) models [18], Flowcharts [19], and other models are used for communication. Similarly, for implementation into IT services different models are proposed, e.g., for software engineering different types of UML diagrams [20] exist. Therefore, different models and languages are devised to fulfill these gaps and improve the understanding of business processes.

Mostly, business process modeling is investigated for phases before its execution like analysis, design and implementation. Limited research is carried out to provide graphical modeling support for evaluation phase (including post execution analysis). In this section, we discuss how to provide the representational support for business process evaluation and improvement.

4.1 Evaluation and Business Objects

In evaluation or improvement of business processes, at a certain time, only specific elements (business objects) are considered in particular perspectives. For example, in organizational perspective resources of organizations are considered, like employees and machines. Similarly, in control flow perspective, sequence or collaboration of activities, and conditions are analyzed. These business objects have certain characteristics which define their state. Further characteristics of business objects become important when they get involved in business process (process executions) and interact with other objects. In different analyses, business objects are analyzed from various dimensions.

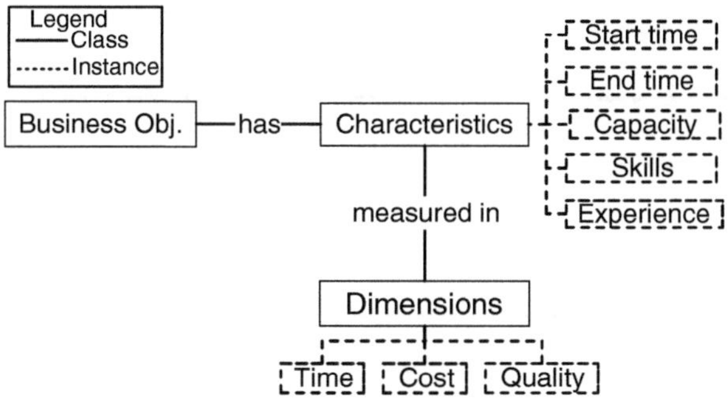

Fig. 3 Business object characteristics and dimensions

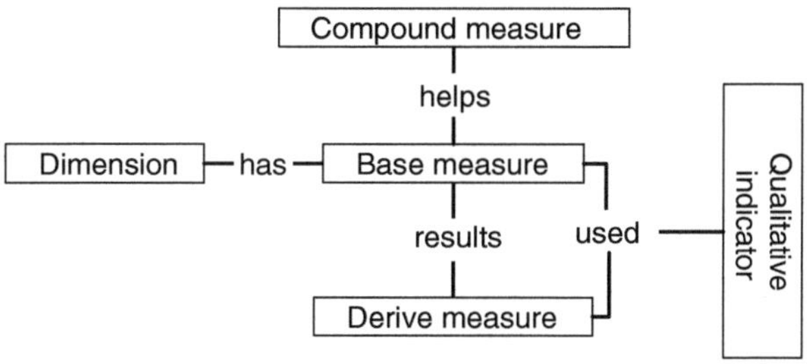

Fig. 4 Evaluation meta model with base and derive measures

Examples of these dimensions are time, cost, and quality. Different attributes of business objects and processes are analyzed under these dimensions. We show the relationship of business objects, characteristics, and dimension in the Fig. 3. We further discussed details of business objects and business processes in [5].

Evaluation of business process and business objects are carried out in different dimensions using base measures and derived measures. Base measures are main measurements from which other measurements are derived. For example, in time dimension the base measures are processing time and idle time. From base measures, different other measures like total time, minimum, maximum, and average time are derived. Compound measures are the measurements which are computed from different other measures like number of requests fulfilled in a particular time. Based on different measurements qualitative indicators are estimated like efficiency and quality as shown in the Fig. 4.

4.2 Case Study

Consider a small and simplified process given in Fig. 5 where five activities are executed to manufacture a product. In this process, first the *parts are collected* in activity A followed by activity B and activity C which are executed in parallel. These

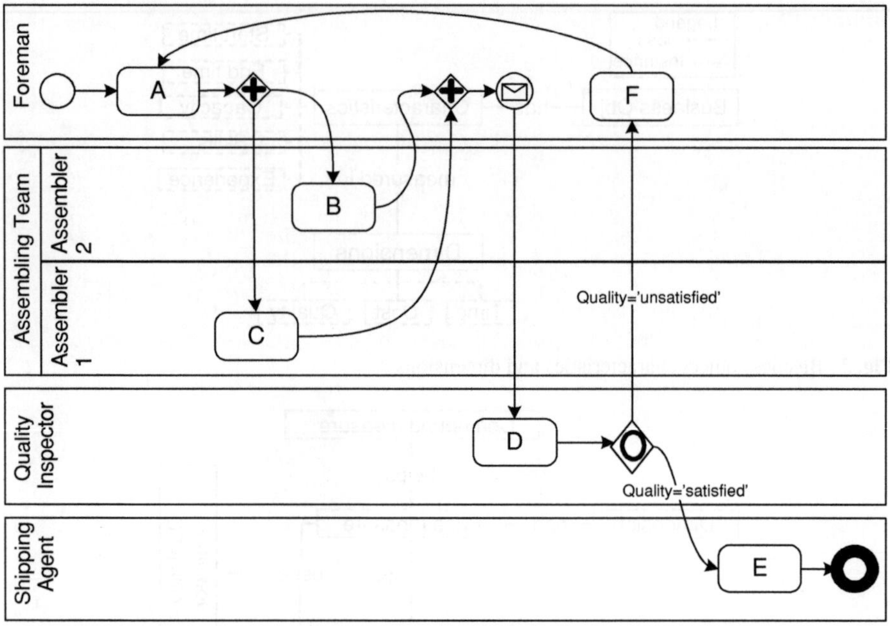

Fig. 5 Product manufacturing example using BPMN

parallel activities *assemble the specific parts* of product independently from one another. After completion, the quality of assembled parts is checked in activity D. If the assembled parts meet the quality requirements, then *product is shipped* in activity E and the process is completed. If the assembled parts do not fulfill a desired level of quality, then activity F is executed which *disassembles the parts.* Assuming that there is no loss during disassembly process, some part of the process has to be repeated again until desired level of quality is achieved. Therefore, activity B and activity C will be executed again. This process is modeled in the Fig. 5 using business process modeling notation (BPMN) [6].

BPMN is a standard defined by the Object Management Group (OMG) for modeling business processes. In BPMN specifications [6], BPMN graphical notations are divided into four basic categories. These categories are flow objects (for events, activities, and decision nodes), connecting objects (to connect flow objects), swimlanes (to represent process participants), and artifacts (for additional information).

4.3 Proposed Usage of BPMN Model for Evaluation

Availability of analytical data and records of business objects in business process execution provide means to analyze processes more carefully. Here, we discuss two

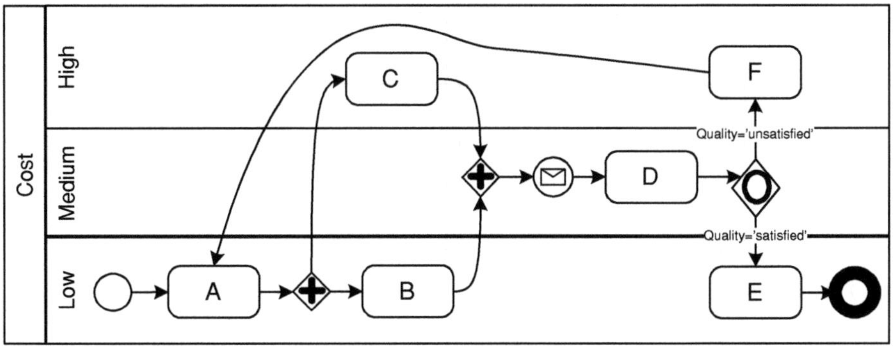

Fig. 6 Process model representation based on costs

approaches for analysis using business process models and analytical data. In the first approach, we propose a method to use process models with analytical data for analyzing activities in different dimensions. In the second approach, we extend business process models with constructs to get a better understanding of process execution and their performance. These two approaches assist one to another for detailed business process analysis in order to find out the deficiencies and for process improvement.

Swimlanes as Dimension and Attributes:

Mostly, in literature [6], Swimlanes (pool and lane) are used to represent organizational elements and their interaction during a business process. However, in this approach, first activities and involved elements are classified based on a particular dimension using analytical data. We recommend that only few classes should be defined for less cognitive loads of process models.

In our paper, we use Swimlanes (pools and lanes) to represent dimensions and different classes of these dimensions. Processes activities are aligned according to a selected dimension and its attributes (overall average values can be used for this classification) based on user defined threshold values.

We define three classes in cost dimension to arrange activities of processes and their involved elements in the Fig. 6. Similarly, different dimension attributes can also be defined as pools where lanes represent further classes of these attributes as shown in the Fig. 7. In Fig. 7, the same activity is represented into two different attributes of time dimension. This figure shows that the actual operating time would be very less but with high waiting time. For simplicity, we have not shown different other involved activities. Moreover, different dimensions and their attributes can be mixed with one to another for further business process analysis.

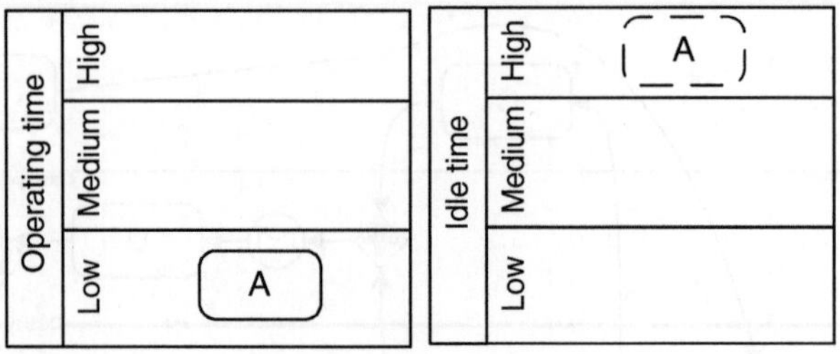

Fig. 7 Process model in time dimension and its attributes

4.4 Extending BPMN Models for Analysis

Extended notations are also used to represent the participation of the different involved elements in process execution such as organizational resources. We propose the following extensions in BPMN for business process analysis and improvement.

Performance Based Classification and Colors in Representation:

The executional data can be used to specify usage of which activities, organizational resources, and involved elements add more value to an enterprise. Different analyses provide this information in terms of cost and time. Based on this information, different colors can be assigned to activities, the process path of optimal execution, and involved elements. Similarly, the best practices process path of enterprise can also be colored.

Different colors can be used to indicate the effect of business objects like grèen for optimal cost, blue for optimal time, and red for non-optimal paths. Although, the relation of cost and time is not that simple as discussed in [21].

Colors in Process Paths:

During execution, values of operational objects (inputs) changes by involved business objects. The change in values gives us details what operations are already performed and which path is taken. During analysis, different colors can also be used to represent the instance executional history, like which path instance has taken and at which particular stage a certain decision is taken. For this, such data can also be extracted from information systems logs or operational databases.

In Fig. 8, we propose new symbols to represent different operational objects (like physical), organizational resources (if not represented in lanes/pool), and data objects with attributes. These symbols are necessary to provide more understanding of business scenarios and its corresponding process models. Similarly, the values of attributes are very important in case of business rules and conditions. Because,

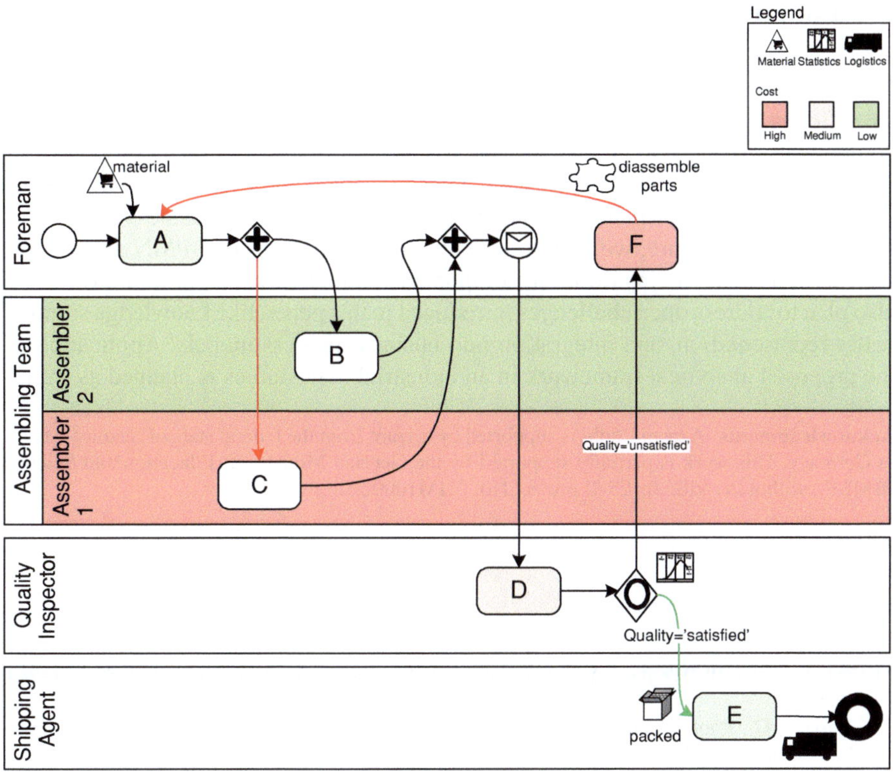

Fig. 8 Performance based extended model with involved business objects

values of attributes play a major role in deciding the route of an instance. Based on these extended notations, we represent an extended business process model of our case study in the Fig. 8. The colors red, yellow, and green represents the elements with high, medium, and low cost respectively.

Depending on the granularity of detail, a business process may consist of many processes and activities which can range up to hundred and thousand. For process improvement, process managers are interested in details of these activities with extended representations which may increase the complexity of a business process model. Apart from the decision support system as discussed in Sect. 3, tabletops and smarts boards will be very useful for providing a full picture of a business process as compared with computer screens. Management of complex business process models and their representation in tabletops need further investigation. Similarly, tangible interfaces and tabletops can also be further investigated in order to provide help for training new employees with extended representation.

5 Summary and Outlook

In this paper, we proposed a new business process improvement framework and its different components. We discussed researchers work related with different components and current challenges. We focused on representational component of improvement framework. We proposed a usage method and extension of BPMN for performance analysis which help to identify the deficiencies in a business process for improvement. In future, we plan to provide more extensions of BPMN and a complete case study involving different components of the improvement framework. We also plan to address open challenges introduced in this paper, like knowledge storage, better representation, and integration into business process models. Application of the proposed analytical framework in an industrial case studies is planned as well.

Acknowledgments Azeem Lodhi is supported by a grant from the federal state of Saxony-Anhalt in Germany. This work is partially supported by the German Ministry of Education and Science (BMBF), within the ViERforES-II project No. 01IM10002B.

References

1. van der Aalst, W., Weijters, A.: Process mining: A research agenda. Comput. Ind. **53**, 231–244 (2004)
2. Kassem, G., Rautenstrauch, C.: Application usage mining to improve enterprise workflows: ERP systems SAP R/3 as example. In: Khosrow-Pour, M. (eds.) Managing Modern Organizations Through Information Technology. Information Resources Management Association pp. 358–362. (2005)
3. Lodhi, A., Köppen, V., Saake, G.: Business process modeling: Active research areas and challenges. Technical Report 1, University of Magdeburg, Germany (2011)
4. Vergidis, K., Tiwari, A., Majeed, B.: Business process analysis and optimization: Beyond reengineering. IEEE Trans. Syst. Man Cybern. Part C Appl. Rev. **38**(1), 69–82 (2008)
5. Lodhi, A., Köppen, V., Saake, G.: Post execution analysis of business processes: Taxonomy and challenges. Technical Report 9, University of Magdeburg, Germany (2010)
6. BPMI.org, OMG: Business Process Modeling Notation Specification, Final Adopted Specification (2006)
7. Khan, A., Lodhi, A., Köppen, V., Kassem, G., Saake, G.: Applying process mining in SOA environments. In: Dan, A., Gittler, F., Toumani, F. (eds.) Service-Oriented Computing ICSOC Service Wave 2009 Workshops. Volume 6275 of Lecture Notes in Computer Science. pp. 293–302. Springer, New York (2010)
8. Ingvaldsen, J., Gulla, J.: Preprocessing support for large scale process mining of SAP transactions. Business Process Manag. Workshops **4928**, 30–41 (2008)
9. van Dongen, B., van der Aalst, W.: A meta model for process mining data. In: Missikoff, M., Nicola, A.D. (eds.) Proceedings of the Open Interop Workshop on Enterprise Modelling and Ontologies for Interoperability, Co-located with CAiSE. Volume 160 of CEUR Workshop Proceedings. Porto, Portugal (2005). CEUR-WS.org
10. Günther, C., van der Aalst, W.: A generic import framework for process event logs. In: Eder, J., Dustdar, S. (eds.) Business Process Management Workshops, Workshop on Business Process Intelligence. vol. 4103, pp. 81–92. Springer, Berlin (2006)
11. zur Muehlen, M.: Business process analytics format (BPAF) Document Number WFMC-TC-1015. Workflow Management Coalition, USA (2008)

12. Grigori, D., Casati, F., Castellanos, M., Dayal, U., Sayal, M., Shan, M.C.: Business process intelligence. Comput. Ind. **53**(3), 321–343 (2004)
13. Rozinat, A., van der Aalst, W.: Decision mining in ProM. In: International Conference on Business Process Management (BPM 2006), vol. 4102, pp. 420–425. Springer, Berlin (2006)
14. van der Aalst, W.: Business alignment: Using process mining as a tool for delta analysis and conformance testing. Requirement Eng. **10**(3), 198–211 (2005)
15. van der Aalst, W., Reijers, H.A., Song, M.: Discovering social networks from event logs. Comput. Support. Coop. Work **14**(6), 549–593 (2005)
16. Lodhi, A., Kassem, G., Köppen, V., Saake, G.: Investigation of graph mining for business processes. In: Proceedings of The International Conference on Intelligence and Information Technology ICIIT. vol. 2, pp. 293–297. IEEE Computer Society Lahore, Pakistan (2010)
17. Cumberlidge, M.: Business Process Management with JBoss jBPM: A Practical Guide for Business Analysts. Packt Publishing, Birmingham, UK (2007)
18. Scheer, A.W.: ARIS-Business Process Modeling. 2 edn. Springer, New York (1998)
19. IBM: Flowcharting techniques. Technical report, IBM Data Processing Techniques. Yorktown Heights, New York (1969)
20. Booch, G., Rumbaugh, J., Jacobson, I.: Unified Modeling Language User Guide. 2 edn. Addison-Wesley, Boston, MA (2005)
21. Vullers, M.J., Kleingeld, P., Loosschilder, M., Reijers, H.A.: Performance measures to evaluate the impact of best practices. In: Proceedings of Workshops and Doctoral Consortium of the 19th International Conference on Advanced Information Systems Engineering (BPMDS), pp. 359–368. Tapir Academic Press, Trondheim (2007)

Automatic Localization and Boundary Detection of Retina in Images Using Basic Image Processing Filters

Omar S. Soliman, Jan Platoš, Aboul Ella Hassanien and Václav Snášel

Abstract This paper proposes an automatic localization and boundary detection of retina images using basic filters to support ophthalmologists for detection and diagnoses eyes harmful diseases such as glaucoma and diabetic retinopathy accurately and diligently. The proposed system comprising three main phases including preprocessing, segmentation and detection phase. The preprocessing phase is used to enhance retinal image and to remove the noise of the retina image. The second phase is the segmentation for main parts of retinal image including optic disc, blood vessels, and fovea to extract their features. Optic disc is segmented using color intensity, the region of interest (ROI) is detected and morphological operations are applied to reduce search complexity. Also, fovea feature is extracted and the blood vessels tree is extracted from retinal image using line detection techniques. The third phase is the detection, in which identification and classifying whether the input image is left or right eye, to support ophthalmologists in identifying which eye is infected by the disease and to check it periodically. Basic image processing filters including average filter, median filter, spatial filter and morphological filter are used in all system phases. Moreover, a simple approach were used to detect left and right retinal fundus images. The proposed system is tested and evaluated using a subset of ophthalmologic images of the publically available DRIVE database.

O. S. Soliman (✉) · A. E. Hassanien
Faculty of Computers and Information, Cairo University, Cairo, Egypt
e-mail: Dr.omar.soliman@gmail.com

A. E. Hassanien
e-mail: aboitcairo@gmail.com

J. Platoš · V. Snášel
Faculty of Electrical Engendering and Computer Science,
VSB-Technical University of Ostrava, Ostrava, Czech Republic
e-mail: jan.platos@vsb.cz

V. Snášel
e-mail: václav.snášel@vsb.cz

M. Kudělka et al. (eds.), *Proceedings of the Third International Conference on Intelligent Human Computer Interaction (IHCI 2011), Prague, Czech Republic, August, 2011*, Advances in Intelligent Systems and Computing 179, DOI: 10.1007/978-3-642-31603-6_15, © Springer-Verlag Berlin Heidelberg 2013

1 Introduction

Eyes are organs that detect light, and convert it to electro-chemical impulses in neurons. The simplest photoreceptors in conscious vision connect light to movement. In higher organisms the eye is a complex optical system which collects light from the surrounding environment; regulates its intensity through a diaphragm; focuses it through an adjustable assembly of lenses to form an image; converts this image into a set of electrical signals; and transmits these signals to the brain, through complex neural pathways that connect the eye, via the optic nerve, to the visual cortex and other areas of the brain. Eye is made up of three coats, enclosing three transparent structures. The outermost layer is composed of the cornea and sclera. The middle layer consists of the choroid, ciliary body, and iris [1, 2].

The retina is the light-sensitive tissue at the back of the eye and it is necessary for good vision. It contains millions of photoreceptors that capture light rays and convert them into electrical impulses. These impulses travel along the optic nerve to the brain where they are turned into images. There are two types of photoreceptors in the retina: rods and cones. The retina contains approximately six million cones. The cones are contained in the macula, the portion of the retina responsible for central vision. They are most densely packed within the fovea, the very center portion of the macula. Cones function best in bright light and allow us to appreciate color. There are three main components of retina which are blood vessels, optic disk and fovea [3, 4].

The main goals of retinal image analysis is to support ophthalmologists to detect human retina disease as early as possible to form a successful treatment of disease. The main target of preprocessing phase is to improve images that contain an undesired distortions and to enhances some image features which is very important for further processing, as in the identification phase and segmentation phase are depended on the quality of the input image. Segmentation of blood vessels is a very important process to diagnose diabetic retinopathy and glaucoma and many other diseases. Optic disc segmentation is an essential in developing automated expert systems for eye's doctors [5]. It is also considered as key for the identification of many eye features such as identification of left and right eye. Fovea segmentation is very important as it is the one of the first parts of retina that is infected by retina diseases. Its correct segmentation allow eye specialists detect diseases as early as possible. Identification of left and right eye is used to identify whether the input retinal image is the left or right eye, and, therefore, identify infected eye [5, 6].

The aim of this paper is to develop an automatic localization and boundary detection of retina images system using basic filters to extract features that support ophthalmologists in retinal harmful diseases identification and diagnoses, such as glaucoma and diabetic retinopathy. The rest of the paper is organized as follows: Sect. 2, describes the problem background and related work. Section 3, gives a brief introduction of the basic filters used in this paper. The proposed system including preprocessed, segmentation and detection phase are introduced in Sect. 4. In Sect. 5, experimental results and analysis are presented. Where Sect. 6 is devoted to conclusions and future works.

2 Problem Background and Related Works

There are many serious retinal diseases that need to be detected as early as possible to avoid its harmful side effects on retina. We develop a retinal system that analyze digital retinal images to support ophthalmologists in detection and diagnoses of retinal diseases that are very serious for patient's and may cause blindness. If these diseases are detected early, they can be treated successfully and we can save patients from blindness. The main parts of retina that support eye specialists to detect and diagnose retina disease accurately are optic disc, fovea and blood vessels.

The problem of retina diseases detection and diagnoses is addressed as difficult problem, because it is not easy to extract accurate enhanced features of retinal image including blood vessels tree, optic disc and fovea. So, accurate extracted features retina image is an essential task in identifying and diagnosing retinal diseases. Many researches focused on this issues in the past, Zhang et al. [7] proposed an algorithm for retinal vessel extraction using matched filter with first-order derivative of Gaussian. An automated detection and characterization of retinal blood vessel tree bifurcations and crossovers in eye fundus images is proposed in [8]. Windera et al. [3] proposed an algorithms for digital image processing in diabetic retinopathy and an algorithm for identification of different stages of diabetic retinopathy using retinal optical images is introduced in [9].

Santhi et al. [5] introduced a methodology for locating optic disc center and segmenting blood vessels in retinal images for screening system of diabetic retinopathy. A segmentation approach of the optic disk in color eye fundus images using an adaptive morphological approach is proposed in [1]. Fang et al. [6] introduced an automatic segmentation of hard exudates in fundus images based on boosted soft segmentation. In [2] an automatic localization and boundary detection of optic disc using implicit active contours is presented. In [4] a novel automatic image processing algorithm for detection of hard exudates based on retinal image analysis is introduced. Garciaa et al. [10] introduce an assessment of four neural network based classifiers to automatically detect red lesions in retinal images.

3 Preliminaries: Basic Filters

This part gives a brief introduction to the basic filters used in this paper including average filter, median filter, spatial filter and morphological filters [11, 12].

3.1 Average Filter

Average filter is one of the simplest local operations over an image, which is also called as "neighborhood average method". The essential idea of a standard moving average filter is to replace the value of the center pixel $\bar{g}(x, y)$ by the average value of a predefined number of neighboring pixels $g_i(x, y)$ as shown in the following equation:

$$\bar{g}(x, y) = \frac{1}{N \times M} \sum_{i=1}^{N \times M} g_i \quad (1)$$

As may be seen, the computation of particular pixel value needs a window of size $(M \times N)$ of pixels. This is well known convolution with blurring effect.

3.2 Median Filter

A vector median filter selects one vector from set of vectors that is closest to all others. Let S be a set of vectors in F^n, where F^n is an n-dimensional linear vector space over the field F. Then the median filter $\bar{v}$ is defined as:

$$\bar{v} = arg \min_{k \neq j} \{\| v_k - v_j \|, v_k, v_j \in S\} \quad (2)$$

The Average filter can suppress isolated out-of-range noise, but the side effect is that it also blurs sudden changes (corresponding to high spatial frequencies) such as sharp edges. The median filter is an effective method that can suppress isolated noise without blurring sharp edges. Specifically, the median filter replaces a pixel by the median of all pixels in the neighborhood.

3.3 Spatial Filter

Spatial filtering is preformed directly on the pixels of an image. Its rule consists simply of moving the filter mask from point to point in an image. At each pixel (x, y) the response of the filter at that pixel is calculated using a predefined relationship. The spatial filter is defined as:
Let A and B be two images such that $B = T[A]$, where $T[.]$ represents a transformation such that:

$$B(r, c) = T[A](r, c) = f(\{A(u, v) \mid u \in \{r - s, \ldots, r, \ldots, r + s\}, \\ v \in \{c - d, \ldots, c, \ldots, c + d\}\}) \quad (3)$$

The value $B(r, c) = T[A](r, c)$ is a function of a rectangular neighborhood centered on pixel location (r, c) in A.
Thus the value of the transformed image B, at pixel location (r, c) is a function of the values of the original image A, in a $(2s + 1 \times 2d + 1)$ rectangular neighborhood centered on pixel location (r, c). There is a different neighborhood for each pixel location, but if the dimensions of the neighbor-hood are the same for each location, then transform T is sometimes called a moving window transform.

3.4 Morphological Filters

Morphology filters are especially suited to the processing of binary images and greyscale images. They are based on shapes and have two basics morphological operations: dilation and erosion. The dilation operation adds pixels to the boundaries of objects in an image, while the erosion operation deletes pixels on image boundaries. The value of each pixel in the output image is based on a comparison of the corresponding pixel in the input image with its neighbors. Different morphological operations are constructed that are sensitive to specific shapes in the input image by choosing the size and shape of the neighborhood [11]. The dilation and erosion operation are described as:

Suppose that we have two accidental set of points in discrete two dimensional space: $\{(m, n)\}$ and $\{(p, q)\}$ and lets define a function for these two sets $A(m, n)$ and $B(m, n)$, which will be called as objects A and B.

The dilation operation is defined as:

$$D(A, B)(r, s) = \max_{(j,k)\in B} (A(r - j, s - k) + B(j, k)). \tag{4}$$

And the erosion operation is defined as:

$$E(A, B)(r, s) = \min_{(j,k)\in B} (A(r + j, s + k) - B(j, k)) \tag{5}$$

where A is an image and B is a structure element used to probe the input image.

4 Automatic Localization and Boundary Detection of Retina Images System

The proposed automatic localization and boundary detection of retina image system is composed of three main phases; pre-processing, segmentation and classification phases. Figure 1 describes the structure of the proposed system.

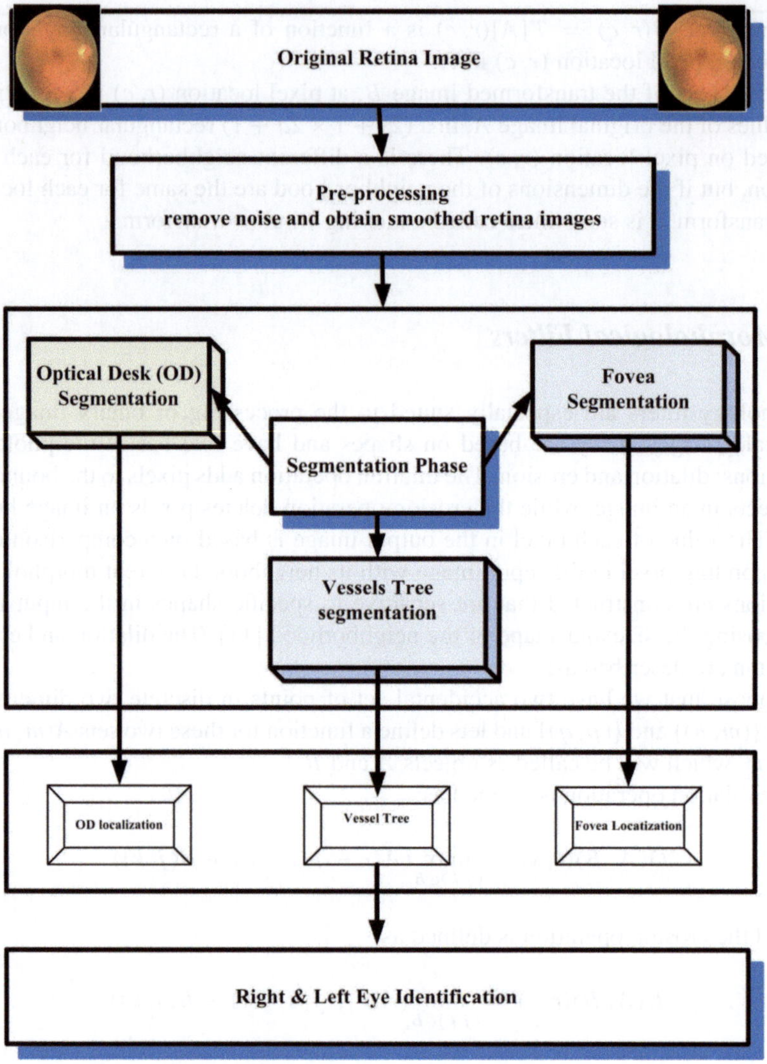

Fig. 1 Main phases of the proposed RIAS

4.1 Pre-Processing Phase

Retina image pre-processing is an essential phase. It removes noise and obtain smoothed retina images for purpose of extraction of retina image feature easily. Pre-processing does not increase image information content, but its main target is to improve images that contain an undesired distortions and it enhances some image features which are very important for further processing. The pre-processing phase is divided into three main parts: correction of non-uniform illumination, color normalization and contrast enhancement. The main steps of pre-processing algorithm are described in Algorithm 1.

Algorithm 1: Pre-processing algorithm

```
Convert the input retina color image into grayscale.
Apply average filter with mask [3 x 3] and obtain the smoothed retina image.
Apply the unsharped filter.
Apply the Histogram Equalization to enhance the contrast.
Obtain the preprocessed retina image.
```

4.2 Segmentation Phase

Retina segmentation extracts important features and regions of interest to support ophthalmologists for retinal diseases diagnoses. In this phase, three main parts of retinal image are segmented including retina blood vessels, optic disc and fovea.

Blood Vessels Segmentation

Blood vessels segmentation is done using Kirsch's templates which was originally developed as line detection algorithms. In order to enhance and remove undesired distortions of the extracted blood tree vessels, different filters and morphological operations are used. The main steps of blood vessels segmentation and enhancement summarized in Algorithm 2.

Algorithm 2: Vessel tree segmentation and enhancement algorithm

```
Get the preprocessed image.
for each pixel do
  Process spatial filtering using Kirsch's operator with different rotations
  Get the maximum result from the filter
  if maximum result > threshold then
    Set pixel as a blood vessels pixel
  end if
end for
Extracted blood vessels tree
Get the vessel tree image
Apply the median filter
Apply the wiener filter
Make thinning by morphology operation.
Apply median filter
```

4.3 *Optic Disc Segmentation*

The optic disc is the most important part in the retina, it is also the entry point for the major blood vessels that supply the retina, optic disk is characterized by the highest intensity region in the retinal image. This characteristic supports the segmentation of the optic disk. In order to segment optic disc, Optic disk segmentation and localization are implemented. Basically, the optic disk segmentation is depend on color intensity of retinal image. It is the part with highest intensity in retinal image. The segmented image is used to find the optic disk exactly and a threshold is used to detect it. The main steps of the optic disk segmentation algorithm are described in Algorithm 3.

Algorithm 3: Optic Disc segmentation algorithm

1: Get the preprocessed image.
2: Detect regions of interest using brightness and darkness degree controlling.
3: Make smoothing for region of interest to be vessel free.
4: Get binary image using thresholding with the threshold selected for each image.
5: Finding the brightest region in the binary image which is the optic disc.
6: Map the edge of the binary image on original image to detect optic disc contour approximately.

4.4 *Fovea Localization*

Fovea is a very important part in the retina as it responsible for the sharp central vision necessary for reading, watching TV and driving and many other activities. It is located at the center of macula region, fovea is the most darkest part in the retinal image. Fovea has specific geometric distance from the optic disc, which is usually used to localize fovea, also fovea can be localized using its color intensity as we mentioned it before it is the darkest part in retina image. We use the color intensity of fovea to localize it. The main steps of Fovea localization algorithm are summarized in Algorithm 4.

Algorithm 4: Fovea localization algorithm

1: Get the original image and convert it to grayscale.
2: Get image complement.
3: Apply erosion to image to make smoothing for it.
4: Convert smoothed image to binary.
5: Apply morphological operation on binary image to get fovea place correctly.
6: Obtain the white region in binary image, that will be the fovea.
7: Assign the place in original image. Repeat step 4..

4.5 Detection Phase

In this phase, the segmented optic disc in the candidate ROI area and the vessel information are extracted from the optic disc. The optic disc is divided into left and right halves. The extracted vessels from the optic disc are then summed up in its columns and projected horizontally across the x-axis. The pixel intensities in each half are compared against each other. The half which indicates lower pixel intensity denotes to the temporal sector. If the temporal sector is on the right half, it signifies a right eye, and vice versa. The main steps of optic disc detection algorithm are described in Algorithm 5.

Algorithm 5: Detection and identification algorithm

1: Localize the retinal image to get ROI.
2: Extract blood vessels in ROI.
3: Divide ROI in to two parts Left and Right.
4: Sum extracted extracted vessels up in its columns and project the left and right sides horizontally across the x-axis.
5: **if** (Right intensity < Left intensity) **then**
6: the Eye is the Right eye
7: **else**
8: the Eye is the Left eye
9: **end if**

5 Experimental Results and Discussions

The retina consist of blood vessels, optic disk and fovea, the preprocessing of retina image and segmentation for optic disk, fovea and blood vessels are done. In this experiments, we have used DRIVE database [13]. The proposed system is evaluated and tested using retina images data set. In the preprocessing phase of retina image, the retina image noisy is removed and obtaining a smoothed retina image by applying the preprocessing Algorithm 1. Figure 2 shows the original and preprocessed retina image. The preprocessed retina image is used as an input image for segmentation and detection phases. The segmented image is used to retina diseases diagnoses such as diabetic retinopathy and glaucoma.

Localization of optic disc is based on its color intensity the brightest intensity in the retinal image. It will be considered as an optic disc pixel by taking a frame around this point about $300 * 300$ to get the region of interest as shown in Fig. 5c. In order to extract ROI, firstly the original image is converted into grayscale image as in Fig. 6a, then the retinal image is smoothed and, therefore, it doesn't contain any blood vessels as shown in Fig. 6b because blood vessels are considered as a noise at the segmentation of optic disc. In the smoothed image the optic disc will be the brightest region, so then change this image to binary image. The brightest region in

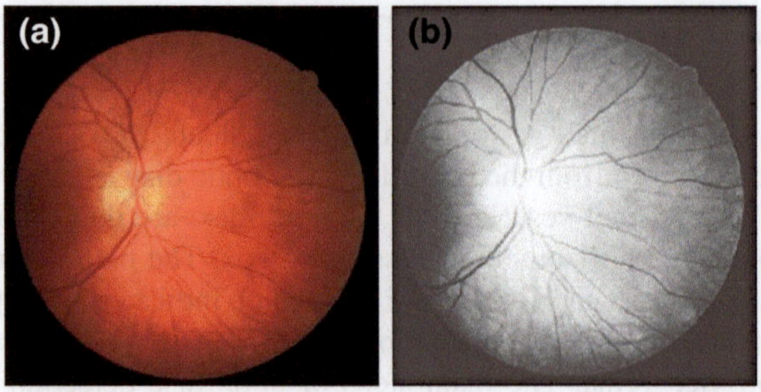

Fig. 2 Preprocessed retina image. **a** Original retina image. **b** Preprocessed retina image

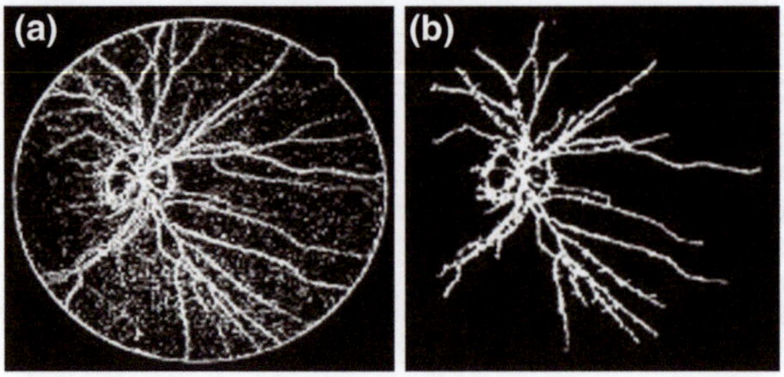

Fig. 3 Extracted and enhanced blood vessel tree. **a** Extracted vessel tree. **b** Enhanced vessel tree

the binary image will be the region of interest of optic disc so that it is extracted easily from the retinal image. Figure 6 shows stages of obtaining the center of the optic disk.

Determination of the fovea position is depicted in Fig. 7a–d.

The human retina is a thin tissue composed of neural cells that is located in the posterior portion of the eye. The retina has ten layers and it presents the surface inside the back of the eye, upon which images that have passed through the pupil are focused. The proposed system contains six analysis of retinal image. With the segmented optic disc in the candidate Region of Interest (RoI) area, we extract the vessel information from the optic disc via its green channel. It is found in our experiments that the central retinal artery and vein are most distinct in this channel. The optic disc is divided into left and right halves. The extracted vessels from the optic disc are then summed up in its columns and projected horizontally across the x-axis. The pixel intensities in each half are compared against each other. The half which indicates lower pixel intensity denotes that it is the temporal sector. If the temporal sector is on the right half, it signifies a right eye, and vice versa. If the fovea is in the right side of the optic disc, then the eye is on the right otherwise the retina is on the left eye, as shown in Fig. 8.

Figure 3 shows the segmented and enhanced blood vessels tree using "spatial filtering and thresholding" and enhanced blood vessels tree after applying morphological operations as in Algorithm 2.

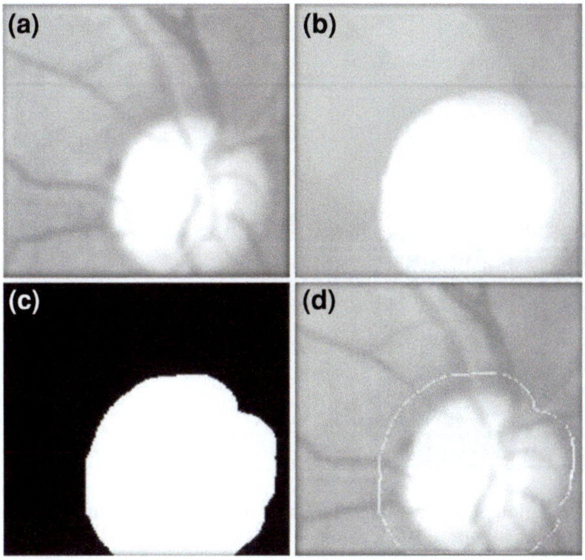

Fig. 4 Optic disc segmentation. **a** Region of interest of optic disc. **b** Smoothed region of interest without blood vessels. **c** Binary image. **d** Segmented optic disc

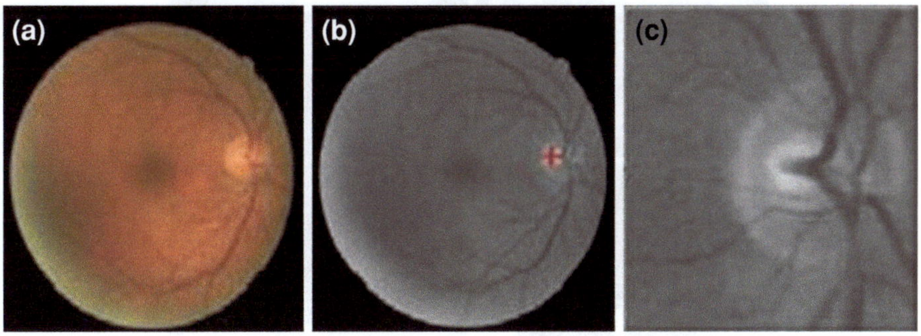

Fig. 5 Optic disk localization stages. **a** Original image. **b** Detected center of optic disk. **c** Extracted frame of ROI

Optic disc contour segmentation is usually performed after identifying the approximate center of the disc (localization of optic disc). Identifying the contour of the optic disc is a non-trivial problem. The natural variation in the characteristics of the optic disc like the differences in pigmentation and myelination of the nerve fiber layer causes significant problems in definition of the contour of the disc. Blood vessels may cross the boundary of the optic disc obscuring the rim of the disc with the edges of vessels also acting as significant noise so that we make smoothing for the gray level image. The segmentation of optic disc is preformed using Algorithm 3 and obtained segmented optic disc is shown in Fig. 4. The optic disk localization stages is shown in Fig. 5 and extracted region of interest ROI is shown in Fig. 6. The fovea is

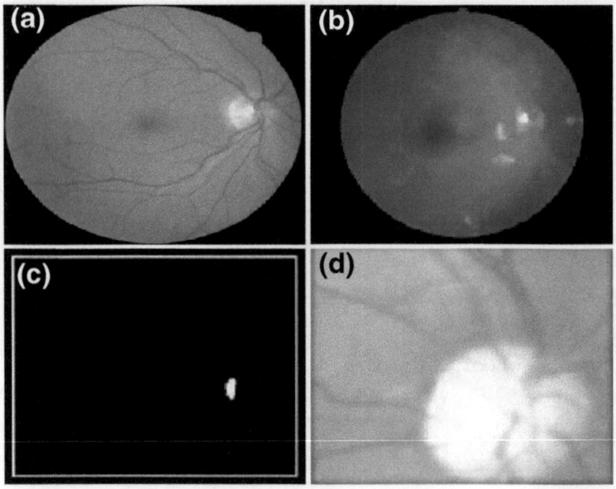

Fig. 6 Extraction of region of interest stages. **a** Show the gray level of original image. **b** Smoothed image without any blood vessels. **c** Binary image. **d** Extracted region of interest

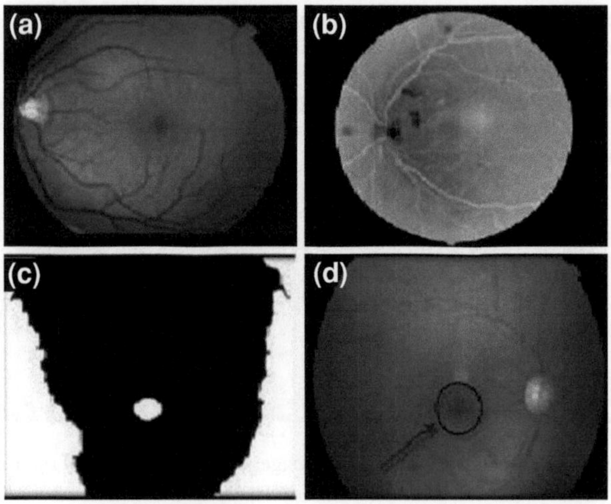

Fig. 7 Fovea localization steps. **a** Original image converted to gray level. **b** Complement of gray image. **c** Binary image from image of the complement image where the brightest region will be the fovea. **d** Localization of fovea in original image

localized based on the intensity of fovea and applying morphological operations. The fovea localization is obtained by using Algorithm 4. Figure 7 shows images starting with original image through obtaining the position of Fovea.

The detection and identification phase, the position of optic disc is used for detecting right and left eye. Algorithm 5 is performed for detection and identification of left and right eye and obtained results are shown in Fig. 8. Another feature for detection

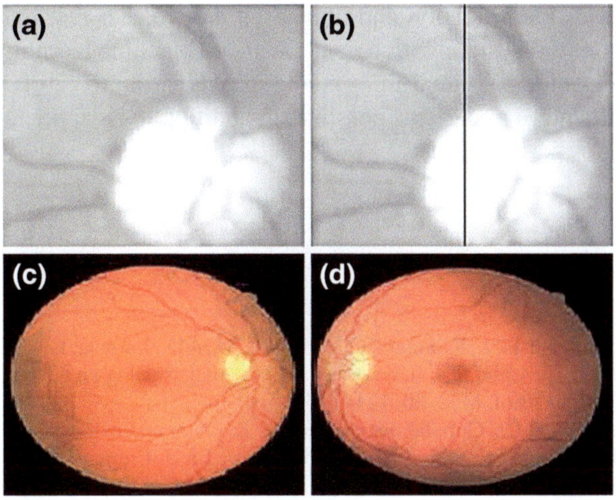

Fig. 8 **a** The region of interest. **b** Logically dividing the ROI into two logical parts *left* and *right*. **c** Optic disc position in *right* eye. **d** Optic disc position in *left* eye

is the position of the fovea, if the fovea is on the right side of the optic disc, then the eye is on the right otherwise the retina is on the left eye, detection will be more accurate when using the two features.

Segmented retina images were evaluated by an expert ophthalmologists regarding missed and false segmented retina blood vessels, optic desk, and fovea as well as the identification of the eye position (left or right). Segmentation results revealed average accuracies of 98 % of correctly identified, and 2 % of falsely indicated. Results indicated that the proposed approach might assist ophthalmologists to detect human retina disease as early as possible to form a successful treatment of disease.

6 Conclusions and Further Research

In this paper an automatic localization and boundary detection of retina image system is presented. It is composed of three main phases: preprocessing, segmentation and detection phase. The pre-processing phase is preformed using three consequently process including correction of non-uniform illumination, color normalization and contrast enhancement. The segmentation phase is applied for main parts of retinal image including blood vessels tree, optic disc and fovea localization. The detection phase is implemented to detect the optic disc based on optic disc localization the infected eye is easily determined weather is right or left eye.

The proposed automatic localization and boundary detection of retina image system is tested and evaluated on retinal images data set [13], the obtained results showed that the proposed system is preformed well and has more accurate results for retinal diseases identification and diagnoses. For further research we intend to integrate the proposed system with recent advanced machine learning techniques to extract features vector of retinal images and classification to retina diseases diagnoses and identification as early as possible; in addition to its application in medical engineering.

References

1. Welfer, D., Scharcanski, J., Kitamura, C.M., Pizzol, M.M.D., Ludwig, L.W.B., Marinho, D.R.: Segmentation of the optic disk in color eye fundus images using an adaptive morphological approach. Comput. Biol. Med. **40**, 124–137 (2010)
2. Siddalingaswamy, P.C.: Gopalakrishna Prabhu, K.: Automatic localization and boundary detection of optic disc using implicit active contours. Int. J. Comput. Appl. **1**(7), 1–5 (2010)
3. Winder, R.J., Morrow, P.J., McRitchie, I.N., Bailie, J.R., Hartd, P.M.: Algorithms for digital image processing in diabetic retinopathy. Comput. Med. Imaging Graph. **33**, 608–622 (2009)
4. Sanchez, C.I., Hornero, R., Lyopez, M.I., Aboy, M., Poza, J., Abasolo, D.: A novel automatic image processing algorithm for detection of hard exudates based on retinal image analysis. Med. Eng. Phys. **30**, 350–357 (2008)
5. Santhi, D., Karkuzhali, S., Manimegalai, D.: Locating optic disc center and segmenting blood vessels in retinal images for screening system of diabetic retinopathy. **1**(1), (2011)
6. Fang, G., Yang, N., Lu H., Li, K.: Automatic segmentation of hard exudates in fundus images based on boosted soft segmentation. In: International Conference on Intelligent Control and Information Processing, Dalian, China, 13–15 Aug 2010
7. Zhang, B., Zhang, L., Zhang, L., Karray, F.: Retinal vessel extraction by matched filter with first-order derivative of Gaussian. Comput. Biol. Med. **40**, 438–445 (2010)
8. Calvo, D., Ortega, M., Penedo, M.G., Rouco, J.: Automatic detection and characterisation of retinal vessel tree bifurcations and crossovers in eye fundus images. J. Comput. Methods Progr. Biomed. **86**, 707–718 (2010)
9. Yun, W.L., Acharya, U.R., Venkatesh, Y.V., Chee, C., Min, L.C., Ngd, E.Y.K.: Identification of different stages of diabetic retinopathy using retinal optical images. Inf. Sci. **178**, 106–121 (2008)
10. Garcia, M., Lopezb, M.I., Alvareza, D., Horneroa, R.: Assessment of four neural network based classifiers to automatically detect red lesions in retinal images. Med. Eng. Phys. **32**, 1085–1093 (2010)
11. Gonzales, R.C., Woods, R.E.: Digital Image Processing, 3rd edn. Prentice Hall, Upper Saddle River (2008)
12. Kwan, H.K.: Fuzzy filters for noisy image filtering. In: Proceedings of the IEEE International Symposium on Circuits and Systems (ISCAS), vol. 4, pp. 161–164 (2003)
13. Niemeijer, M., van Ginneken, B.: http://www.isi.uu.nl/Reseach/Databases/DRIVE/results.php

Integration of a Speech-Analyzer Agent in a Multi Agent System for Remote Healthcare Monitoring

Mohamed Achraf Dhouib, Lamine Bougueroua, Katarzyna Wegrzyn-Wolska and Stefan Todorov

Abstract The age of the population in all societies around the world is increasing. Elderly people prefer to maintain their independence, their autonomy and live at home as long as possible. This research take place in the field of the health care telemonitoring system which proposes software solutions to monitor elderly people in their own home. This study aims to make profit of the technological diversity of the several Decision Support System used to detect distress situation. We propose a multi-Agent approach to federate a collective decision process in which each Agent encapsulates a decision support system. This encapsulation enables the real time combination of the decisions.

1 Introduction

In the coming decades, many European countries will be confronted with aging population issues. It is estimated that by 2020, 28 % of the French population will be over 60 [1]. The traditional solution of keeping the elderly in specialized centers has become too expensive; it also has a negative impact on the independence of the patient. Remote healthcare monitoring systems represent a much more convenient solution which assures the social independence of the elderly as much as their security [2]. Many research projects take place in this context and implement several solutions using various technologies. A corner stone technology in this field is the decision support system (DSS). These systems are able to analyze the data

M. A. Dhouib (✉) · L. Bougueroua · K. Wegrzyn-Wolska · S. Todorov
ESIGETEL, Fontainebleau-Avon, Avon, France
e-mail: mohamed.dhouib@esigetel.fr

L. Bougueroua
e-mail: lamine.bougueroua@esigetel.fr

K. Wegrzyn-Wolska
e-mail: katarzyna.wegrzyn@esigetel.fr

M. Kudělka et al. (eds.), *Proceedings of the Third International Conference on Intelligent Human Computer Interaction (IHCI 2011), Prague, Czech Republic, August, 2011*, Advances in Intelligent Systems and Computing 179, DOI: 10.1007/978-3-642-31603-6_16,

gathered from several ambient captors in order to generate risk detection decisions. These decisions are sent to a remote monitoring center to take action in the case of real threat. Although they use several data modalities (localizational, physiological, actimetric ...), the DSS usually uses a unique type of artificial intelligence (neural network, expert system, fuzzy logic ...). The pertinence of each DSS is determined by the occurrence of false and undetected alarms. A real-time combination of these decisions is able to improve the usage of appropriate resources within an acceptable response time. In order to make best use of this technological diversity, we aim to encapsulate each DSS in an intelligent agent. In fact, the multi-agent system (MAS) architecture enables these DSS to have a uniform view of the decision concept and to exchanges both knowledge and intelligence.

This paper presents a theoretical framework for real-time dispatching of a multi-agent approach for collective decisions. In the first section, we present a survey of three DSS used in health care telemonitoring [2, 3]. In the second section, we present the method for encapsulating a DSS in an intelligent agent. This encapsulation is achieved by transforming the generated decision of the DSS into an abstract form that may thus be used as an environmental fact by other intelligent agents (those agents are probably encapsulating other DSS). For this reason, we propose in this section a semantic alignment approach based on the works of. We then define the 2-phase collective decision algorithm of our MAS and the role of the central agent. In the third section we present the real-time dispatcher plugged into the central agent. The first part of this section is dedicated to explaining the dispatching problems of our MAS. Subsequently we present the different types of message and the classification made by the real-time classifier. In the same section, we present dynamic priority assignment and the scheduling algorithm. In conclusion we will present the advantages of our system and finally outline our future work.

2 A Multi Agent System for Risk Detection in Remote Healthcare Monitoring

2.1 BDI Agent Topology

A remote healthcare monitoring (RHM) network is a complex and dynamic environment. In order to perform an automatic surveillance of a patient, the conventional software techniques are difficult to develop. In addition, they are not enough pertinent to assume a large scale RHM network. In fact, the increasing number of monitored people requires collaboration between the installed system and a highly pertinent decision support system. [4] says: "Agent-oriented systems, based on a radically different view of computational entities, offer prospects for a qualitative change in this position".

There are many classification of agents topologies [5–8] but the one we choose, considering the analyzed problem, is based on the decision process of the agent. We classify these topologies as reactive and cognitive. According to [5], this classification "define a practical evaluation axis of the agent capacity to accomplish individually complex tasks and to schedule its actions". Unlike reactive agent whose life cycle is Perception/Action, cognitive agent's life cycle is perception/ deliberation/ action. In fact, cognitive agents establish action plans according to their global goals. An emblematic approach to implement these agents has been initiated in the eighties [25]. This approach proposes architecture based on three attitudes: Belief, Desire, intention (BDI). It has been the subject of several studies [4, 9–11]. The BDI model is one of the best studied models of practical reasoning agent [12]. It combines a respectable philosophical model of human reasoning. In order to argue this choice of agent topology, we establish the necessity of beliefs, desire and intention for a RHM system to act appropriately in an environment characterized by practical limitation and requirements. Similarly to the works of [4] which first exposed the characteristics of an air traffic management system, we argue this topology choice by exposing the critical characteristics of our RHM system:

- The environment of the system is non-deterministic: it can evolve at any instant of time with many different ways. For example, the patient position, hurt pulses, activity and temperature may change at any time.
- The system is non-deterministic: there are many different actions or procedures that may potentially be executed. For example, the system may generate an alert decision while asking the robot move toward the patient.
- The system may have simultaneously different objectives. For example, it can be asked to maximize the decision pertinence and have a field of vision on the patient within 5 s.
- The choice of actions or procedures that achieve the various objectives of the system depends only on the state of the environment. For example, the choice of actions that make the system elaborate the best decision about the patient situation depends only on the parameters of the patient health.
- A sensing action is not always sufficient for fully determining the state of the environment. For example, the patient heart pulse can be determined by a pulse sensor carried by the patient, unlike the localization which requires the combination of multiple sensor data (RFID, IR ...).
- The rate at which computation and actions can be carried out is within reasonable bounds to the rate at which the environment evolves.

Given characteristic 4, it is essential to the system to have information on the state of the environment. Besides, characteristic 5 implies that this cannot be determined locally. As a result, it is necessary to have a component of the system that represent and update continuously the state of the environment. This component is called beliefs. Otherwise, it is necessary to the system to represent its different objectives as well as the priority given to each one. Given characteristic 3 and 4, these objectives can be generated "instantaneously or functionally, and thus not requiring any state representation". Given this picture, the system needs a deliberative selection function that assigns to generate the course of actions of the system. [4] considered two constraints: the occurrence of significant changes during the execution of the selection function and the failure that may result if we interrupt this computation at each system change. The system intentions are so defined as a component to represent at any instant of time, the chosen course of action. This course may change only if the system detects a considerable change of state. Since we have strong similarities between the system studied by [4], we adopt the same abstract architecture: three dynamic data structures representing belief, desires and intentions with an input queue of events.

2.2 MAS Analysis: Gaia Role Model

Like other software engineering paradigm, agents represent a great advance in developing appropriate high-level abstraction (Procedural abstraction, abstract data types, objects and components) with which to model and develop these complex systems [13]. There are, at present, a great number of methodologies for agent analysis and design: Gaia [13], AUML [14–16], MAS-CommonKADS [17], aSE [18], ZEUS [19], MESSAGE [20], TROPOS [21], etc. Most of these studies are not fully developed or present some deployment limitations [22]. In order to define an appropriate analysis and design of this MAS, the chosen conceptual framework is Gaia. GAIA is an extension of classical software engineering approaches which proposes several conception models for multi-agent systems [13]. This choice is argued by the fact that Gaia is a methodology for very general analysis and design and therefore applicable to a very wide range of multi-agent systems. In addition, the GAIA design is easy to implement. It also allows the user to have a wide knowledge of the multiagent systems both at an organizational (social) level and at a detailed level for each agent. Gaia is particulary convenient to MAS where each agent is susceptible of consuming an important computing resource, which is the case of the DSS that we intend to use in our decisive agents. Otherwise, Gaia represent an abstraction conceptual framework that enables the MAS designer to represent interactions between heteregenous agents (probably developed in different languages). Finally, [13] precise that Gaia should not be used to design MAS where we may have an important number of agents (more than 100) or that potentially can contain intentional conflicts between agents. Our RHM MAS will not contain more than 20 agents. The public ontology introduced in the next paragraph guaranties the inexistence of such conflicts between CODEM agents. Gaia analysis leads to two conceptual models: the role model and the interaction model. In this section, the role model is presented to define the different roles to be played in the MAS. A role is defined by its responsibilities and its associated permission/rights. The roles in this MAS architecture are defined according to the principle functions that are required in this RHM system. RHM consist of communication function, urgency actions and risk detection function. Thus, three main roles categories are to be defined: communication roles, urgency roles and risk detection roles. An agent with a communication role have the responsibilities to perform communication between patient home equipments, tele-operators and distant equipments (probably installed at doctor's offices, hospitals, family ...). It is allowed to access to the required communication equipment as well as the required patient data to perform the appropriate communication (contact data, patient state, patient network ...). An agent with an urgency role is responsible of taking actions in the patient home when a distress situation is detected. When a risk detection agent generates a distress decision, these agents are allowed to access to system actuators, including accompanying robot, doors lock, light switches, cameras position, ...as well as asking for communication agent services. The risk detection roles category is the most concerned by this study. It consist of three roles:

- Informative Agent (IA) role: These agents are responsible for providing the required data for patient diagnosis establishment. They are allowed to access the system data bases as well as the system sensors. For ethic and personal privacy reasons, there are some restrictions on these agents access to the distant data bases of the system (not located at the patient home). An RHM social network involves patient as well as medical stuff, patient's family, friends, neighbors ...The data-mining in these distributed databases implements several artificial intelligence and can be adapted to the patient network evolution. In addition, it needs potentially important computing resources. This justifies the use of BDI agent for this role instead of classical data access services.
- Decision Agent (DA) role: These agents encapsulates Decision Support Systems (DSS). They are responsible for making abstraction of the decisions generated by their corresponding DSS. These decisions are sent to a central agent that orchestrates the collective decision process (detailed in the next section). The decision agents can also be asked for making decisions about the patient state concerning a given time lapse. In order to make real time decisions, response deadlines can be imposed by the central agent. Decision agent are allowed to ask informative agent for patient data. Their DSS are also allowed to directly access system databases.
- Central Agent (CA) role: As its name indicates, this role is central in the collective decision process. In this RHM system, there is only one CA agent. On the one hand, this agent responsibilities are reception, scheduling and fusion of the decision generated by the decision role agents. It also orchesters the decision phases (section II.c) and makes the final decisions. On the other hand, this agent controls the auto-adaptation process of the DA.

Now that the roles required for making the collective decision in the RHM system are completed, the interaction between these agent roles need to be defined. For this purpose, we present formally the collective decision process of CODEM and the public ontology used for the distress detection. These elements enables the definition of the interaction model of CODEM.

2.3 Two-Phases Collective Decision

The collective decision is made in two phases. Each phase consists of a wait window in which the central agent waits for the incoming decision message, and a decision

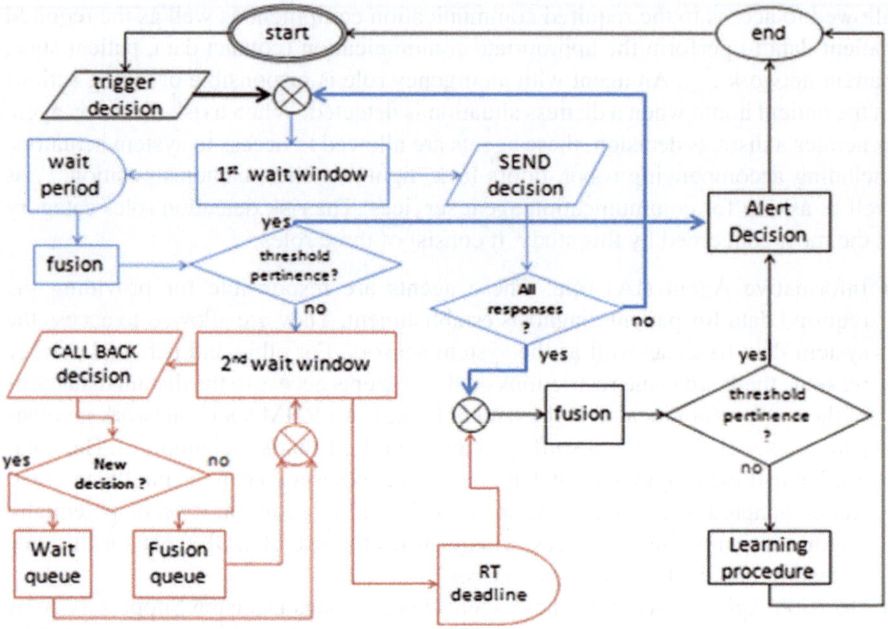

Fig. 1 Two-phases collective decision schema

process in which the central agent fuse the received decision. The procedure starts when a first decision message is received. We call this the trigger decision (Fig. 1).

Phase 1: The CA starts the wait window of phase-1. The duration of the wait window depends on the trigger decision data (explained in section II.d: agent affinity, used modalities ...). In this paper, we do not detail the computing algorithm of the waiting duration. The decision messages received by CA in phase-1 are called SEND decisions. A SEND decision is a spontaneous decision. It is not a response to a previous request. In the case of a trigger decision, a threshold of pertinence is defined by the central agent. During the first wait window, the CA may receive other SENDs from other DA. At the end of the wait window, CA fuse all the received SENDs. If the final decision's pertinence surpasses the threshold, the decision is confirmed as an alert. If the wait window is terminated without reaching the pertinence threshold, CA starts the second phase of decision.

Phase 2: CA starts a new wait window. During this wait window, a real-time consensus is launched among the DAs concerned by the trigger decision modalities. For this purpose, CA assigns to each concerned agent a consensus priority. This is computed as follows:

$$p_i(d) = \sum_{m_j \in d} A_{ij} \cdot c_j \tag{1}$$

where: m_j are the used modalities in the trigger decision d, c_j the corresponding coefficient for each modality, A_{ij} is the affinity coefficient of the agent i for the modality m_j.

This affinity coefficient is determined at the implementation of the agent DSS. It may be changed by the CA according to the DA pertinence. During this second wait window, the received message may be SEND decisions. As they do not concern the launched consensus, they are placed in the wait queue. The response messages are called CALLBCK decisions. At the end of the second wait window, CA computes the global pertinence of the received CALLBCK decisions. If the pertinence threshold is reached, the trigger decision is confirmed. Otherwise it is rejected and a learning procedure is sent to the responsible DA. In this article, we do not detail the inner learning procedure of such an agent.

2.4 Public Ontology of the MAS for Decision Agent

An important aspect of multi-agent systems is communication between different agents. Ontologies represent the technology that supports inter-agent communication by providing a definition of the world to the multi-agent system. Thus, each agent represents his beliefs using the necessary terms of the semantics in order to perform actions and communication with the rest of the system [23]. In our MAS, we consider that communication between two agents takes place via a public ontology. We further assume that all ontologies are encoded in a common language. The problem is therefore limited to the interpretation of this ontology to link it to the private ontology of each agent. Thus, at each message transaction, the agent translates the terminology he intends to use in an area of _shared concepts from the public ontology of the MAS. The terminology in which we are interested in this work concern the risk detection role. In RHM, a DSS uses one or more data modality to detect distress situations. In order to standardize the concept of a decision, we classify the generated decisions according to the modalities used by the DSS. The modalities included in our system are sound, speech, physiological data, actimetric data, video, sensor status, alert call. When a DA receive a decision from its DSS, it must classify it according this list of modalities. For example, if a DSS detect the fall of a patient while located in the corridor, it generates a decision according to two modalities (localisation and fall detection). The corresponding DA considers these two modalities as one because they are both actimetric modalities. Generally, every decision is based on a global pertinence calculated by combining the pertinence affected to each decision modality. For a decision d, the global pertinence is:

$$G_p(d) = \sum_{m_i} p_i(d) \cdot c_i \tag{2}$$

where:

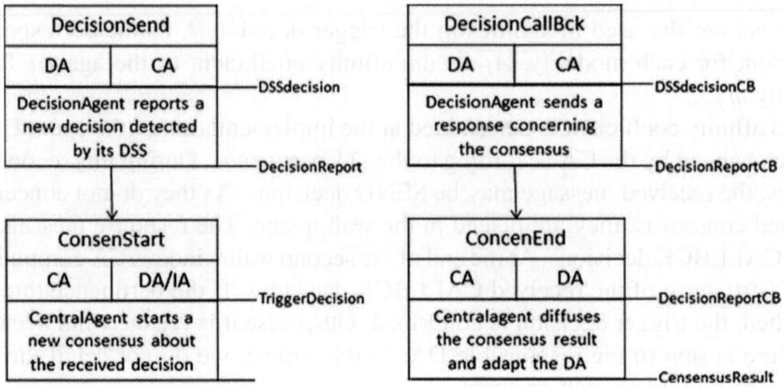

Fig. 2 Gaia interaction model of CODEM

1. m_i: the modalities used for the decision d
2. p_i: the pertinence of the decision d according to the modality m_i
3. c_i: the coefficient of the modality m_i accorded by the DSS

The first public ontology to establish is the decision report. This decision report abstraction is available for SENDs decision as well as CALLBCKs. Once the decision SENDs are received by the CA, they are fused. The resulting decision is not a decision report and cannot be represented in as a decision Report.

The terminology used by CA to manage the collective decision process need to be defined. As explained in Sect. 2.3, CA launches the consensus at the beginning of phase-2 and diffuses the result at its end.

2.5 Interaction Model of the MAS

Gaia interaction model defines for each role the associated protocols. Protocols are the patterns of interaction that occur in the system between the various roles [13]. We present in Fig. 2 the interaction model corresponding to the collective decision protocols:

The Gaia design process aims to allow the system developer to have a sufficiently low level of abstraction that they can easily use the classical design process (especially object oriented techniques) in order to implement the MAS. The Gaia design process consist of three models: Agent model present how the agent roles are instantiated in the MAS at run-time. The services model identifies the main services that are required to realise the agent's role. The acquaintance model documents the lines of communication between the different agents.

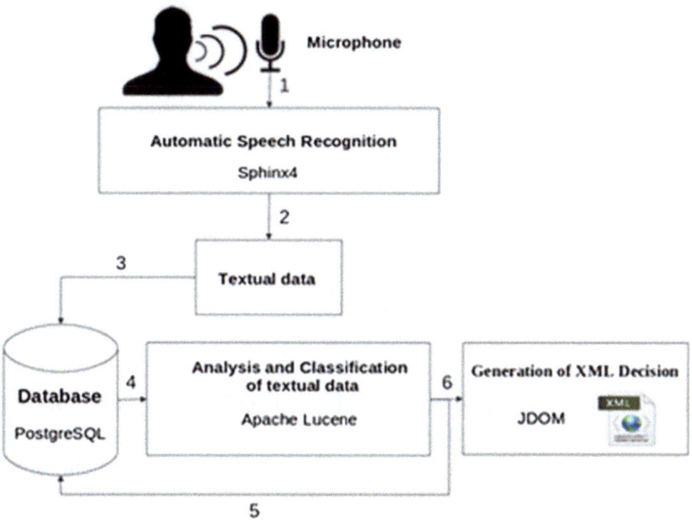

Fig. 3 The workflow of intelligent agent

3 Integration of a Speech-Analyzer Agent in a Jade Multi-Agents Environment

3.1 Speech-Analyzer Agent

The main role of the agent consist on retrieving the textual data from an incoming sound stream then make the necessary classifications, using an existing data base, in order to make the decision about the critical situation of the subject. First, the acoustic signal is recorded and digitized step 1 Fig. 3 which describes the workflow.

In step 2 is used Sphinx-4 which is state-of-the-art speech recognition system written in Java. The beginning and end of the speech signal are then detected, for example by removing the breathing sounds, ambient noise and silence. The result is isolated speech signal. After that is performed acoustic analysis, during which is made the extraction of features (calculation of acoustic parameters or acoustic vectors). The purpose is to extract information which makes it possible to better separate the sounds using the particular fast Fourier transform (FFT), which can process the signal in the frequency domain. The data is stored in a sequence of vectors. These vectors are used with the acoustic model, the language model (also called grammar) and the appropriate dictionary to recognize what was said. First there is a pattern classification using as input the sequence of features that gives as output a sequence of models. Then it is the language processing, or the language models that are used to give words or strings of the recognized words. Next the result of the transcription process is in the form of textual data that is stored in database step 3. For the analysis and classification it is used the Apache Lucene which is a high

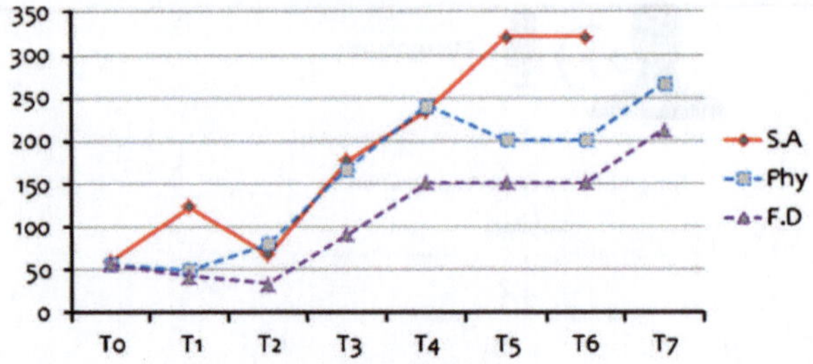

Fig. 4 Evolution of the global pertinence of the agents

performance, scalable and open-source Information Retrieval (IR) library. Lucene is used for the classification process. The list of existing phrases is first extracted from the database and then is used to create Lucene index step 4. The created index is allowing to create quick keyword lookups using various search algorithms to find the most appropriate matching phrase for the textual results coming from Sphinx. One of the most successful algorithms used is the PhraseQuery algorithm which tries to locate phrases where the words are within a certain distance of one another. The maximum allowable positional distance between words to be considered a match is called slop. The distance is the number of positional moves of words to reconstruct the phrase in order. Phrase queries are scored based on the edit distance needed to match the phrase. More exact matches count for more weight than sloppier ones.

3.2 Development Environment

The solution has been implemented on Java Agent Development environment (Jade). Jade is a java software framework for development and execution of multi-agent systems. This platform allows the development of FIPA-compliant agent. It also manages the communication and the deployment of the agents using the standard FIPA-ACL communication language in both SL and XML [24].

3.3 Tests and Discussions

The system used to test the multi-agent collective decision consist of a central agent, a speech analyser agent based on the speech analyser DSS described in III.a and two fictif agent using physiological and fall detection modalities. The test scenario is detailed in Table 1.

The global pertinence of the agent have been evaluated at the end of each decision phase (Fig. 4). This pertinence depends of the agent's affinity to the trigger modalities and on its response to the decision consensus.

Table 1 Test scenario

	Speech analyser agent (S.A)	Physiological agent (Phy)	Fall detector agent (F.D)	Central agent (C.A)
T1, $\Phi1$	TrigDecision(86,8)	–	–	ConsenStart ()
T1, $\Phi2$	Confirm(86,8)	Confirm(90)	Confirm(75)	Confirm()
T2, $\Phi1$	TrigDecision(90)	–	–	ConsenStart ()
T2, $\Phi2$	Confirm(90)	Confirm(98)	Deny(13,3)	Confirm() / Lear()
T3, $\Phi1$	–	TrigDecision(100)	–	ConsenStart ()
T3, $\Phi2$	Deny(0)	Confirm(100)	Confirm(89,1)	Confirm() / Lear()
T4, $\Phi1$	–	TrigDecision(100)	–	ConsenStart ()
T4, $\Phi2$	Confirm(94,5)	Confirm(100)	Confirm(100)	Confirm()
T5, $\Phi1$	TrigDecision(90)	–	–	ConsenStart ()
T5, $\Phi2$	Confirm(90)	Deny(0)	Null	Deny() / Lear()
T6, $\Phi1$	TrigDecision(83)	–	–	ConsenStart ()
T6, $\Phi2$	Confirm(83)	Null	Null	Deny() / Lear()
T7, $\Phi1$	–	TrigDecision(88)	–	ConsenStart ()
T7, $\Phi2$	Null	Confirm(88)	Confirm(91)	Confirm() / Lear()

4 Conclusion

In this paper, we propose a new approach for remote healthcare monitoring risk detection. Based on a Gaia conception of BDI multi-agent system, we presented an architecture that makes profit of the technological diversity of the several DSS used in this field. Our two phases collective decision approach is interesting to make collective decision between heterogeneous DSS which increases the pertinence of the final decision. The first tests made on a real speech analyser agent shows that it's possible to increase the pertinence of each DSS using the responses of the other ones. The next step in this work is to make a better mathematical model to evaluate the global pertinence of an agent.

References

1. McMorrow, K., Roeger, W.: The Economic and Financial Market Consequences of Global Ageing. Springer Series: European and Transatlantic Studies, XIV. ISBN: 978-3-540-40541-2
2. Ali, R., Sylvie, T., Nicolas S., Jean-Marie, P.: An adaptive system for home monitoring using a multiagent classification of patterns. In: International Journal of Telemedicine and Applications, Hindawi Publishing Corporation, vol. 2008. doi:10.1155/2008/136054
3. Souidene, W., Istrate, D., Medjahed, H., Boudy, J., Baldinger, J.L., Belfeki, I., Delavault, F., Steenkeste, F.: Multi-modal platform for in-home healthcare monitoring (EMUTEM). In: Proceedings of the Second International Conference on Health Informatics HEALTHINF, pp. 381–386. Porto, Portugal (2009)
4. Rao, A., Georgeff, M.: BDI Agents: From Theory to Practice Proceedings of the First International Conference on Multi-Agent Systems (ICMAS-95), pp. 312–319. AAAI Press, Menlo Park, California (1995)
5. Ferber, J.: Les systèmes Multi-Agents. Vers une intelligence collective, Paris (1995)

6. Chaib-Draa, B.: Interaction between agents i routine. Familiar and Unfamiliar Situation Int. J. Coop. Inform. Syst. **5**, 1–25 (1996)
7. Huget, M.P., Pinson, S.: Une typologie des Modèles d'Agents Document du Lamsade, n. 110. (1998)
8. Wooldridge, M., Jennings, N.R.: Agents theories, architecture and language: A survey In: Wooldridge, M., Jennings, N.R. (eds.) Intelligent Agents, ECAI'94, LNAI 890, pp. 1–32. Springer, Amsterdam (1994)
9. Kumar, D., Shapiro, S.C.: The OK BDI architecture IJAIT. Int. J. Artif. Intell. Tools **3**, 349–366 (1994)
10. Busetta, P., Ramamohanarao, K.: The BDIM Agent Toolkit Design. Department of Computer Sciences, Victoria, Australia (1998)
11. d'Inverno, M., Kinny, D., Luck, M., Wooldridge, M.: A formal specification of dMARS intelligent agents IV: In: Singh M., Rao A., Wooldridge, M. (eds.) Proceedings of the Fourth International Workshop on Agent Theories, Architectures and Languages, vol. 1365, pp. 155–176. Springer, Heidelberg (1998)
12. Georgeff, M., Pell, B., Pollack, M., Tambe, M., Wooldridge, M.: The Belief-Desire-Intention Model of Agency Intelligent Agents V, In: Müller J. -P., Singh M., Rao, A. (eds.) Lecture notes in AI. vol. 1365, Springler, Berlin (1999)
13. Wooldridge, M., Jennungs, N.R., Kinny, D.: The gaia methodology for agent-oriented analysis and design. J. Auton. Agent. Multi-Agent Syst. **3**(3), 285–312, Kluwer Academic Publishers, Dordrecht (2000)
14. Bauer, B., Huget, M.P.: FIPA Modeling: Agent Class Diagrams (2003)
15. Odell, J., Levy, R., Nodine, M.: FIPA Modeling TC: Agent Class Superstructure Metamodel. FIPA meeting and interim work (2004)
16. Odell, J., Huget, M.P.: FIPA Modeling: Interaction Diagrams (2003)
17. Iglesias, C., Garijo, M., Gonzalez, J.C., Velasco, J.R.: Analysis and Design using MAS-CommonKADS. Inteligent Agents IV LNAI. vol. 1365, Springer, Heidelberg (1998)
18. DeLoach, S.: Anlysis and design using MaSE and AgentTool. In: Proceedings of the 12th Midwest Artificial Intelligence and Cognitive Science Conference (MAICS) (2001)
19. Nwana, H.S., Ndumu, D.T., Lee, L.C., Collins, J.C.: ZEUS: A toolkit for building distributed multi-agent systems. Appl. Artif. Intell. J. **1**(13), 129–185 (1999)
20. EURESCOM MESSAGE: Methodology for engineering systems of software agents. Technical report P907-TI1, EURESCOM (2001)
21. Bresciani, P., Giorgini, P., Giunchiglia, F., Mylopoulos, J., Perini, A.: TROPOS: An Agent-Oriented Software Development Methodology. Technical Report DIT-02-015, Informatica e Telecomunicazioni, University of Trento, Italy (2002)
22. Corchado, J.M., Bajo, J., Tapia, D.I.: ALZ-MAS: Alzheimer's special care multi-agent system. In: Proceedings of the Workshop on Agents Applied in Health Care (ECAI 2006)
23. Huhns MN, Singh MP, Ontologies for Agents, IEEE Internet Computing, **1**(6), pp. 81–83 doi: 10.1109/4236.643942 (1997)
24. Giovanni Caire: WADE User Guide, Telecom Italia (2010)
25. Bratman, M.E., Israel, D.J., Pollack, M.: Plans and resource-bounded practical reasoning. Comput. Intell. **4**, 349–355 (1988)

Part IV
Technological and Theoretical Background

Communication Logic on Multi-Modal Logic S5n

Takashi Matsuhisa

Abstract This article investigates a communication logic based on the multi-modal logic **S5n**, and it is treated from multi-modal logical point of view. We introduce models for the logic as knowledge revision processes on agents' knowledge by communication through messages, and the completeness theorem is shown: the communication logic is determined by the class of all the models.

1 Introduction

This paper aims to give a logical foundation of communication system through messages among agents from multi-modal logical points of view.

Recently, researchers in AI, and computer science, economics become entertained lively concerns about relationships between knowledge and actions. The most interest to us is the situation involving the knowledge of a group of agents rather than of just a single agent. At what point does agents sufficiently know to stop gathering information and make decisions? There are also concerns about the complexity of computing knowledge. In investigating the situation, a communication process among them plays important role to reach consensus among all agents.

Aumann [1] introduced the notion of common-knowledge and applied it to studying the foundation of game theory. By common-knowledge of something we mean the infinite regress of higher levels of mutual knowledge; that is, everybody knows something and everybody knows that everybody knows it and everybody knows that everybody knows that everybody knows it and so on. He showed the 'agreeing to disagree' theorem': If all agents commonly know all the posteriors then these must

T. Matsuhisa (✉)
Department of Natural Sciences, Ibaraki National College of Technology,
Nakane 866, Hitachinaka-shi, Ibaraki312-8508, Japan
e-mail: mathisa@ge.ibaraki-ct.ac.jp

M. Kudělka et al. (eds.), *Proceedings of the Third International Conference on Intelligent Human Computer Interaction (IHCI 2011), Prague, Czech Republic, August, 2011*, Advances in Intelligent Systems and Computing 179, DOI: 10.1007/978-3-642-31603-6_17,

be equal. Since then the notion of common-knowledge can be brought to the foreground on the investigation of decision theoretical problems.

In fact, Aumann and Brandenburger [2] gave an epistemic condition for a Nash equilibrium of a strategic game: They show that if each player's conjecture about the other players' action in a strategic game then the conjectures yields a Nash equilibrium for the game. Matsuhisa [7] applied the theorem of Aumann [1] to a moral hazard problem appeared in a principal-agent model. This is the issue in a principal-agent model whether there exist any sharing rules that resolve both the free ride problem and balance the budget, under which an efficient action is a Nash equilibrium. He gave a possible resolution of it under certain technical assumptions.

However, infinite level of knowledge is very complex, and common-knowledge seems to be problematic regarding to the infinite regression of higher levels of mutual knowledge, and other notions substituting for common-knowledge are sought after as role reversal for common-knowledge.

Parikh and Krasucki [10] focused communication is in pairs. The communication in pairs does not lead to common knowledge for the group. Nevertheless, they showed that conditional probabilities will converge in any fair protocol for communication, and extended a version of agreeing to disagree' theorem in this situation. Matsuhisa [5] and Matsuhisa and Strokan [9] continued the work of Parikh and Krasucki [10] extending the result of Aumann and Brandenburger [2] to the communication system introduced by Parikh and Krasucki [10]. Further a resolution of moral hazard in the principal-agent model uncertainty was given in the communication system by Matsuhisa and Jiang [8].

The communication plays, hence, a reversal role for common-knowledge, and the foundation of communication system is a very important topic brought to the foreground on interactive investigation in the broad areas including Mathematics, Computer Science and AI. Furthermore the investigation of foundation of communication from computer theoretical point of view is expected to lead us a rich store of knowledge on interaction with computer and intelligent human beings.

In this paper a logic of communication among multi-agent is treated with its sound models: We present a communication logic based on the multi-modal logic **S5n**, and we introduce models for the logic as revision processes on agents' knowledge by communication through messages, and show that he completeness theorem; i.e., the communication logic is determined by the models.

This paper is organised as follows. Section 2 recalls the multi-modal logic **S5n** together with the semantics models. In Sect. 3 we present the communication logic **CS5n** based on the logic **S5n**, and introduce the communication model as semantics for the logic **CS5n**. Section 4 gives the proof of the completeness theorem of the logic **CS5n**. In Sect. 5 we conclude with remarks on 'agreeing to disagree theorem' on the finite models for the logic **CS5n**.

2 Multi-Modal Logic

2.1 *Syntax*

Let N be a finite set of *agents* $\{1, 2, \cdots, k, \cdots, n\}$, and T the time horizontal line $\{0, 1, 2, \cdots t, \cdots\}$. The *language* is founded on as follows: The *sentences* of the language form the least set containing each *atomic* sentence $\mathbf{P}_m (m = 0, 1, 2, \dots)$ and closed under the following operations: nullary operators for *falsity* $\bot$ and for *truth* $\top$; unary and binary syntactic operations for *negation* $\neg$, *conditionality* $\rightarrow$ and *conjunction* $\wedge$, *disjunction* $\vee$, respectively; two unary operations for *modalities* $\Box_i$, $\Diamond_i$ together with countably many binary operations $\Box_i^t$, $\odot_i^t$ for $i \in N, t \in T$. Other such operations are defined in terms of those in usual ways.

The intended interpretation of $\Box_i \varphi$ is the sentence that 'agent i knows a sentence φ,' the sentence $\Diamond_i \varphi$ is interpreted as the sentence that 'i thinks a sentence φ is possible,' the sentence $\odot_i^t(\varphi|\psi)$ is meant to express propositions of the *messages* that i sent at time t and $\Box_i^t(\varphi|\psi)$ for $t \in T$ is meant that agent i knows φ when he/she receives ψ on communication through messages at time t.

2.2 *Multi-Modal Logic* S5n

A *multi-modal logic* L is a set of sentences containing all truth-functional tautologies and closed under substitution and modus ponens. A multi-modal logic L' is an *extension* of L if $L \subseteq L'$. A sentence φ in a modal logic L is a *theorem* of L, written by $\vdash_L \varphi$. Other proof-theoretical notions such as *L-deducibility*, *L-consistency*, *L-maximality* are defined in usual ways. (See, Chellas [3].)

A *normal system* of multi-modal logic L is a multi-modal logic containing ($\mathrm{Df}_\Diamond$) and is closed under the rules of inference ($\mathrm{RK}_{\Box_i}$).

($\mathrm{Df}_\Diamond$) $\Diamond_i \varphi \longleftrightarrow \neg \Box_i \neg \varphi$;
($\mathrm{RK}_{\Box_i}$) $\dfrac{(\varphi_1 \wedge \varphi_2 \wedge \cdots \wedge \varphi_k) \longrightarrow \psi}{(\Box_i \varphi_1 \wedge \Box_i \varphi_2 \wedge \cdots \wedge \Box_i \varphi_k) \longrightarrow \Box_i \psi}$ $(k \geqq 0)$

The *logic* **S5n** is the minimal normal system of multi-modal logic containing the following schemas:

(T) $\Box_i \varphi \longrightarrow \varphi$
(5) $\Diamond_i \varphi \longrightarrow \Box_i \Diamond_i \varphi$

The logic **S5n** contains the schema:

(4) $\Box_i \varphi \longrightarrow \Box_i \Box_i \varphi$

2.3 *Knowledge Structure, Model and Truth*

Let Ω be a non-empty set called a *state space* and 2^Ω the field of all subsets of Ω. Each member of 2^Ω is called an *event* and each element of Ω called a *state*.

A *knowledge structure* is a tuple $\langle \Omega, (K_i)_{i\in N}, (\Pi_i)_{i\in N}\rangle$ in which Ω is a state space and $K_i : 2^\Omega \to 2^\Omega$ is agent i's *knowledge operator*. The interpretation of the event $K_i E$ is that 'i knows E.' Π_i is i's *possibility operator* on 2^Ω defined by $\Pi_i E = \Omega \setminus K_i(\Omega \setminus E)$ for every E in 2^Ω. The interpretation of $\Pi_i E$ is that 'E is possible for i.'

A *model on* a knowledge structure is a tuple $\mathcal{M} = \langle \Omega, (K_i)_{i\in N}, (\Pi_i)_{i\in N}, V\rangle$ in which $\langle \Omega, (K_i)_{i\in N}, (\Pi_i)_{i\in N}\rangle$ is a knowledge structure and a mapping V assigns either **true** or **false** to every $\omega \in \Omega$ and to every atomic sentence $\mathbf{P}_m$. The model $\mathcal{M}$ is called *finite* if Ω is a finite set.

Definition 1 By $\models^{\mathcal{M}}_\omega \varphi$, we mean that a sentence φ is *true* at a state ω in a model $\mathcal{M}$. *Truth at a state* ω in $\mathcal{M}$ is defined by the inductive way as usuals:

1. $\models^{\mathcal{M}}_\omega \mathbf{P}_m$ if and only if $V(\omega, \mathbf{P}_m) = \mathbf{true}$, for $m = 0, 1, 2, \ldots$;
2. $\models^{\mathcal{M}}_\omega \top$, and not $\models^{\mathcal{M}}_\omega \bot$;
3. $\models^{\mathcal{M}}_\omega \neg\varphi$ if and only if not $\models^{\mathcal{M}}_\omega \varphi$;
4. $\models^{\mathcal{M}}_\omega \varphi \longrightarrow \psi$ if and only if $\models^{\mathcal{M}}_\omega \varphi$ implies $\models^{\mathcal{M}}_\omega \psi$;
5. $\models^{\mathcal{M}}_\omega \varphi \wedge \psi$ if and only if $\models^{\mathcal{M}}_\omega \varphi$ and $\models^{\mathcal{M}}_\omega \psi$;
6. $\models^{\mathcal{M}}_\omega \varphi \vee \psi$ if and only if $\models^{\mathcal{M}}_\omega \varphi$ or $\models^{\mathcal{M}}_\omega \psi$;
7. $\models^{\mathcal{M}}_\omega \Box_i\varphi$ if and only if $\omega \in K_i(||\varphi||^{\mathcal{M}})$, for $i \in N$;
8. $\models^{\mathcal{M}}_\omega \Diamond_i\varphi$ if and only if $\omega \in \Pi_i(||\varphi||^{\mathcal{M}})$, for $i \in N$:

where $||\varphi||^{\mathcal{M}}$ denotes the set of all the states in $\mathcal{M}$ at which φ is true; this is called the *truth set of* φ.

We say that a sentence φ is *true in the model* $\mathcal{M}$ and write $\models^{\mathcal{M}} \varphi$ if $\models^{\mathcal{M}}_\omega \varphi$ for every state ω in $\mathcal{M}$. A sentence is said to be *valid in* a knowledge structure if it is true in every model on the knowledge structure. Let Γ be a set of sentences. We say that $\mathcal{M}$ is a *model for* Γ if every member of Γ is true in $\mathcal{M}$. A knowledge structure is said to be *for* Γ if every member of Γ is valid in it. Let $\mathbf{C}$ be a class of models on a knowledge structure. A multi-modal logic L is *sound with respect to* $\mathbf{C}$ if every member of $\mathbf{C}$ is a model for L. It is *complete with respect to* $\mathbf{C}$ if every sentence valid in all members of $\mathbf{C}$ is a theorem of L. A multi-modal logic L is said to be *determined by* $\mathbf{C}$ if L is sound and complete with respect to C. It is said to have the *finite model property* if it is sound and complete with respect to the class of all finite models in $\mathbf{C}$. We denote by $\mathbf{C}_F$ the class of all finite models in $\mathbf{C}$.

Theorem 1 *A normal system of modal logic L has the finite model property; i.e.,* $\vdash_L \varphi$ *if and only if* $\models^{\mathcal{M}} \varphi$ *for all* $\mathcal{M} \in \mathbf{C}_F$*. In particular,* **S5n** *has the finite model property.*

Proof Chellas [3].

3 Communication

We assume that the players communicate by sending *messages*. A *protocol* is a mapping $\mathrm{Pr} : T \to N \times N, t \mapsto (s(t), r(t))$ such that $s(t) \neq r(t)$. Here t stands for *time* and $s(t)$ and $r(t)$ are, respectively, the *sender* and the *recipient* of the communication which takes place at time t. We consider the protocol as the directed graph whose vertices are the set of all players N and such that there is an edge (or an arc) from i to j if and only if there are infinitely many t such that $s(t) = i$ and $r(t) = j$.

A protocol is said to be *fair* if the graph is strongly connected; in words, every player in this protocol communicates directly or indirectly with every other player infinitely many often. It is said to contain a *cycle* if there are players $i_1, i_2, \dots, i_k$ with $k \geq 3$ such that for all $m < k$, i_m communicates directly with i_{m+1}, and such that i_k communicates directly with i_1. The communication is assumed to proceed in *rounds*[1]. The protocol is said to be *acyclic* if the graph has no cycle. (C.f.: [4, 10])

3.1 Communication Logic CS5n

Into the language of the logic **S5n** we introduce sentences of the forms $\odot_i^t(\varphi|\psi)$ and $\Box_i^t(\varphi|\psi)$ for $i \in N, t \in T$. The sentence $\odot_i^t(\varphi|\psi)$ is meant to express propositions of the *messages* that i sent at time t and $\Box_i^t(\varphi|\psi)$ expresses propositions of the form that agent i knows φ when he/she receives ψ on communication through messages at time t.

A *normal system of communication logic* based on the logic **S5n** according to a fair protocol $Pr : T \to N \times N$ is an extension of **S5n** is closed under the rules of inference (RCE_μ) and (RCK_μ): For $\mu = \odot, \Box$ and each $t \in T$,

$$(\mathrm{RCE}_\mu)\ \frac{\psi_1 \longleftrightarrow \psi_2}{\mu_i^t(\varphi|\psi_1) \longleftrightarrow \mu_i^t(\varphi|\psi_2)};$$

$$(\mathrm{RCK}_\mu)\ \frac{\varphi_1 \wedge \varphi_2 \wedge \cdots \wedge \varphi_k \longrightarrow \varphi}{\mu_i^t(\varphi_1|\psi) \wedge \mu_i^t(\varphi_1|\psi) \wedge \cdots \wedge \mu_i(\varphi_k|\psi) \longrightarrow \mu_i^t(\varphi|\psi)} \quad (k \geqq 0)$$

Definition 2 The *communication logic* **CS5n** is the smallest normal system of communication logic based on the logic **S5n** which contains the schemas: For every $t \in T$ with $t \geqq 1$,

(CDef) $\Box_i \varphi \longleftrightarrow \Box_i^0(\varphi|\psi)$
(UN) $\Box_i^t(\varphi|\bot) \longrightarrow \bot$;
(CS5) $\mu(*|\psi)$ gives a unary modal operator for **S5n** for $\mu = \odot_i^t, \Box_i^t$;
(MESDef)

[1] There exists a time m such that for all t, $\mathrm{Pr}(t) = \mathrm{Pr}(t+m)$. The *period* of the protocol is the minimal number of all m such that for every t, $\mathrm{Pr}(t+m) = \mathrm{Pr}(t)$.

$$\Box_j^{t+1}(\varphi|\psi) \longleftrightarrow (\Box_j^t \wedge \odot_i^t)(\varphi|\psi) \quad \text{for } (i, j) = \Pr(t);$$

$$\Box_j^{t+1}(\varphi|\psi) \longleftrightarrow \Box_j^t(\varphi|\psi) \quad \text{for } (i, j) \neq \Pr(t);$$

(COH) $\Box_i^t(\varphi|\psi) \longrightarrow \Box_i^{t+1}(\varphi|\psi)$ for $\psi \neq \bot$;
(COHMES) $\odot_i^t(\varphi|\psi) \longrightarrow \Box_i^t(\varphi|\psi)$

3.2 Communication Model

Let Z be a set of decisions, which set is common for all agents. By a *system design function* (decision function) we mean a mapping f of $2^{\Omega} \times 2^{\Omega}$ into the set of decisions Z. We refer the following property of the function f: Let X be an event.

DUC (Disjoint Union Consistency): For every pair of disjoint events S and T, if $f(X; S) = f(X; T) = d$ then $f(X; S \cup T) = d$;

Remark 1 If f is intended to be a posterior probability, we assume given a probability measure μ on a state-space Ω which is common for all agents; precisely, for some event X of Ω, $f(X; \cdot)$ is given by $f(X; \cdot) = \mu(X|\cdot)$. This function satisfies (DUC).

Definition 3 Let $\mathcal{M} = \langle \Omega, (K_i)_{i\in N}, (\Pi_i)_{i\in N}, V\rangle$ be a model for the logic **S5n**. A *communication model* $\mathcal{C}$ is a triple

$$\langle \Omega, (K_i)_{i\in N}, (\Pi_i)_{i\in N}, V, \Pr, f, (Q_i^t), (W_i^t) \mid (i, t) \in N \times T\rangle$$

in which $\Pr(t) = (s(t), r(t))$ is a fair protocol such that for every t, $r(t) = s(t + 1)$, f is a system design function, and $Q_i^t(*|*)$ is the mapping of $2^{\Omega} \times \Omega$ into 2^{Ω} for i at time t that is defined inductively in the following way: We assume given a mapping $Q_i^0(X; \omega) := \Pi_i(\{\omega\})$. Suppose Q_i^t is defined. Let

$$W_i^t(X; \omega) = \{\xi \in \Omega \mid f(X; Q_i^t(X; \{\xi\})) = f(X; Q_i^t(X; \{\omega\}))\}.$$

This is interpreted as the message that i sends to j at time t. Then $Q_j^{t+1}(X|\omega)$ is defined as follows:

- If j is not a recipient of a message at time $t + 1$, then $Q_j^{t+1}(X; \omega) = Q_j^t(X; \omega)$.
- If j is a recipient of a message at time $t + 1$, then $Q_j^{t+1}(X; \omega) = Q_j^t(X; \omega) \cap W_{s(t)}^t(X; \omega)$.

Truth at a state ω in $\mathcal{C}$ is defined by the usual way. In particular,

$$\models_{\omega}^{\mathcal{C}} \Box_i^t(\varphi|\psi) \quad \text{if and only if} \quad Q_i^t(||\psi||^{\mathcal{C}}; \omega) \subseteq ||\varphi||^{\mathcal{C}}$$

$$\models_{\omega}^{\mathcal{C}} \odot_i^t(\varphi|\psi) \quad \text{if and only if} \quad W_i^t(||\psi||^{\mathcal{C}}; \omega) \subseteq ||\varphi||^{\mathcal{C}}$$

We can show that

Theorem 2 *The logic* **CS5n** *is determined by all communication models.*

4 Proof of Theorem 2

The proof will be given in the same line as described in Chellas [3].

4.1 Proof Sets

Let L be a system of communication logic based on the logic **S5n** according to a fair protocol Pr. We recall the *Lindenbaum's lemma*:

Lemma 1 *Every L-consistency set of sentences has an L-maximal extension.*

This is because L includes the ordinary propositional logic.

We call the extension an *L-maximally consistent set*. As a consequence, *a sentence in L is deducible from a set of sentences Γ if and only if it belongs to every L-maximally consistent set of Γ*, and thus we can observe

Theorem 3 *Let L be a system of communication logic on the logic* **S5n** *according Pr. A sentence is a theorem of L if and only if it is a member of every L-maximally consistent set of sentences.*

We denote by $|\varphi|_L$ the class of L-maximally consistent sets of sentences containing the sentence φ; this is called the *proof set* of φ. We note that

Corollary 1 *Let L be a system of communication logic on the logic* **S5n** *according Pr.*

(i) *A sentence φ is a theorem of L if and only if $|\varphi|_L = \Omega_L$;*
(ii) *A sentence $\varphi \to \psi$ is a theorem of L if and only if $|\varphi|_L \subseteq |\psi|_L$.*

4.2 Canonical Model

Let L be a system of communication logic on the logic **S5n** according Pr.

Definition 4 The *canonical model* for L is the model

$$\mathcal{C}_L = \langle N, \Omega_L, (K_{Li}), (\Pi_{Li}), V_L, (Q_{L_i}^t), (W_{L_i}^t) \mid (i,t) \in N \times T \rangle$$

for L consisting of:

1. Ω_L is the set of all the L-maximally consistent sets of sentences;
2. $K_{Li} : 2^{\Omega_L} \to 2^{\Omega_L}$ is given by

$$\omega \in K_{Li}(E) \text{ if and only if there exists } \varphi \in L \text{ such that } |\varphi|_L = E \text{ and } \Box_i \varphi \in \omega$$

3. $\Pi_{Li} : 2^{\Omega_L} \to 2^{\Omega_L}$ is given by

$$\omega \in \Pi_{Li}(E) \text{ if and only if there exists } \varphi \in L \text{ such that } |\varphi|_L = E \text{ and } \diamond_i \varphi \in \omega$$

4. V_L is the mapping defined by $V_L(\omega, \mathbf{P}_m) = \{\omega \in \Omega^L \mid \mathbf{P}_m \in \omega\}$ for $m = 0, 1, 2, \dots$.
5. $Q_{L_i}^{\,t} : 2^{\Omega_L} \times \Omega_L \to 2^{\Omega_L}$ is given as follows: For every $\omega \in \Omega_L$,

$$\Box_i^t(\phi|\psi) \in \omega \quad \text{if and only if} \quad Q_{L_i}^{\,t}(|\psi|_L; \omega) \subseteq |\varphi|_L$$

6. $W_{L_i}^{\,t} : 2^{\Omega_L} \times \Omega_L \to 2^{\Omega_L}$ is given as follows:

$$W_i^t(|\psi|_L; \omega) = \{\xi \in \Omega_L \mid f(|\psi|_L;\, Q_i^t(|\psi|_L; \{\xi\})) = f(|\psi|_L;\, Q_i^t(|\psi|_L; \{\omega\}))\}.$$

We can easily observe that

Proposition 1 *The canonical model $\mathcal{C}_L$ is a model for a system L of communication on the multi-modal logic* **S5n**.

The important result about a canonical model is the following:

Basic theorem 1 *Let $\mathcal{C}_L$ be the canonical model for a system of communication logic L. Then for every sentence φ, $\models^{\mathcal{C}_L} \varphi$ if and only if $\vdash_L \varphi$. In other words, $||\varphi||^{\mathcal{C}_L} = |\varphi|_{\mathcal{C}_L}$.*

Proof By induction on the complexity of φ. We treat only that φ is $\Box_i\psi$ and φ is $\Box_i^t(\psi|\chi)$. As an inductive hypothesis we assume that $||\psi||^{\mathcal{C}_L} = |\psi|_{\mathcal{C}_L}$ and $||\chi||^{\mathcal{C}_L} = |\chi|_{\mathcal{C}_L}$. Then

$$\begin{aligned}
\models^{\mathcal{C}_L} \Box_i\varphi \quad & \text{if and only if} \quad \omega \in K_{Li}(||\psi||^{\mathcal{C}_L}) \text{ for every } \omega \in \Omega_L, \\
& \qquad \text{by the definition of validity;} \\
& \text{if and only if} \quad \omega \in K_{Li}(|\psi|^{\mathcal{C}_L}) \text{ for every } \omega \in \Omega_L, \\
& \qquad \text{by the inductive hypothesis as above;} \\
& \text{if and only if} \quad \vdash_L \Box_i\psi \\
& \qquad \text{by the definition of canonical model.} \\
\models^{\mathcal{C}_L} \Box_i^t(\psi|\chi) \quad & \text{if and only if} \quad Q_{L_i}^{\,t}(||\chi||^{\mathcal{C}_L}; \omega) \subseteq ||\psi||^{\mathcal{C}_L} \text{ for every } \omega \in \Omega_L, \\
& \qquad \text{by the definition of validity;} \\
& \text{if and only if} \quad Q_{L_i}^{\,t}(|\chi|_L; \omega) \subseteq |\psi|_L \text{ for every } \omega \in \Omega_L, \\
& \qquad \text{by the inductive hypothesis as above;}
\end{aligned}$$

$$\text{if and only if} \quad \vdash_L \Box_i^t(\psi|\chi)$$
$$\text{by the definition of canonical model.} \qquad \Box$$

Proof of Theorem 2 Soundness has been already observed in Proposition 4.1. The completeness will be shown as follows: Suppose that some sentence φ is not a theorem in L. In view of Basic theorem, it follows that φ is not valid for a canonical model $\mathcal{C}_L$. □

5 Concluding Remarks

This article present a multi-modal logic of communication **CS5n** based on the logic **S5n** as a deontic logic. We give sound models for the logic, and show the complete theorem for the logic **CS5n**.

However it is not yet settled whether the logic **CS5n** has a finite model property or not. For a finite model for the logic **CS5n** we can show the agreeing to disagree theorem ([1, 4]) that every agents can be reached consensus after long run communication: If Pr is acyclic then for sufficient large $t \in T$ and if f satisfies (**PUD**), $f(||\varphi||^{\mathcal{C}}; Q_i^t(||\varphi||^{\mathcal{C}}; \{\omega\})) = f(||\varphi||^{\mathcal{C}}; Q_j^t(X; \{\omega\}))$ for any $\omega \in \Omega$.

Worthy noting that the communication model can be treated from Fuzzy theoretical point of view, and in the fuzzy communication model the Aumann' theorem is still valid (Matsuhisa [6]). The logic **CS5n** is expected to give a base for investigations of interactive decision theory with incomplete information in broad sense.

Acknowledgments The research was supported by Grant-in-Aid for Scientific Research (C) (No. 23540178) in the Japan Society for the Promotion of Sciences and it was also supported by the BUSAIKU-BUHI Foundation for Scientific Researches.

References

1. Aumann, R.J.: Agreeing to disagree. Ann. Stat. **4**, 1236–1239 (1976)
2. Aumann, R.J., Brandenburger, A.: Epistemic conditions for mixed strategy Nash equilibrium. Econometrica **63**, 1161–1180 (1995)
3. Chellas, B.F.: Modal logic: An introduction. Cambridge University Press, Cambridge (1980)
4. Krasucki, P.: Protocol forcing consensus. J. Econ. Theor. **70**, 266–272 (1996)
5. Matsuhisa, T.: Communication leading to Nash equilibrium through robust messages: S5-knowledge model case. Combinatorial optimization and applications, Lecture Notes in Computer Science **4616**. pp. 136–145. Springer, Berlin (2007)
6. Matsuhisa, T.: Fuzzy communication reaching consensus under acyclic condition. PRICAI 2008: Trends in Artificial Intelligence, Proceedings, Lecture Notes in Computer Science **5351**. pp. 760–767. Springer, Berlin (2007)
7. Matsuhisa, T.: Moral hazard resolved by common-knowledge in principal-agent model. Int. J. Intell. Inform. Database Syst. **6**(3), 220–230 (2012)
8. Matsuhisa, T., Jiang, D.Y.: Moral hazard resolved in communication network. World J. Soc Sci. **1**(3), 100–115 (2011)

9. Matsuhisa, T., Strokan, P.: Bayesian communication leading to a Nash equilibrium. Int. J. Math. Game Theor. Algebra **18**(4–5), 427–434 (2009)
10. Parikh, R., Krasucki, P.: Communication, consensus, and knowledge. J. Econ. Theor. **52**, 178–189 (1990)

Handling Possibly Conflicting Preferences

Radim Nedbal

Abstract There is a need for handling preferences in relational query languages that arises naturally in real-world applications dealing with possible choices generated by the current state of the world captured in the relational data model. To address this problem, we propose a fully declarative language for encoding preferences conditional on the current state of the world represented as a relation database instance. The language has constructs for various kinds of preferences, and we show how to interpret (sets of) its formulae; even sets of formulae that encode conflicting preferences. This leads to a flexible approach for specifying the most desirable choices of autonomous systems that act on behalf of their designers. Throughout the paper, we use an example of a control support system for a bank surveillance to motivate the need for our framework and to illustrate it.

1 Introduction

The Problem and the Main Goal

Complex autonomous systems make many decisions during their run time in environment where they act. The decisions are driven by designer's desires, and the environment is usually dynamic. Representing and processing desires in terms of preferences, which order possible choices (decision outcomes) so that a more desirable choice precedes a less desirable one, appears to be especially appealing as it allows one to specify desires in a declarative way and to deal with inconsistencies and exceptions in a quite flexible manner. Indeed, a preference-based approach is amenable to customization both by allowing specification of additional desires

R. Nedbal (✉)
Institute of Computer Science of Academy of Sciences of the Czech Republic,
Pod Vodárenskou věží 2, 182 07Prague 8, Czech Republic
e-mail: radned@seznam.cz

M. Kudělka et al. (eds.), *Proceedings of the Third International Conference on Intelligent Human Computer Interaction (IHCI 2011), Prague, Czech Republic, August, 2011*, Advances in Intelligent Systems and Computing 179, DOI: 10.1007/978-3-642-31603-6_18,

and providing additional information about the environment. As the relational data model (RDM) allows to describe a generic model of a dynamic environment, where the number and properties of concrete object instances can change, our main goal is to develop a general framework for preference handling in relational query languages (RQL) to support the user-friendly design of autonomous systems that can act in a dynamic environment.

To achieve our goal, we propose a fully declarative language for encoding preferences conditional on the current state of the world represented as a relation database (DB) instance. The main objective of this paper is to define its semantics so that we can interpret any set of (possibly conflicting) formulae of the language.

- After introducing the illustrating example to motivate the need for our language, we give the syntax (Sect. 2) of the language whose signature contains symbols for eight various kinds of preferences to support compact specifying of the model of designer's preferences.
- Then we recall the main concepts of the so-called logics of preferences (Sect. 3.1), introduced in [9, 10], which support describing the model of designer's preferences in up to sixteen different, qualitative (both precise and imprecise) ways.
- Elaborating on the idea of the so-called minimal logic of preference (Sect. 3.2), introduced in [2], we define the semantics of our language in terms of the satisfaction relation (Sect. 3.2) between models of preferences and sets of language's formulae so that every set of formulae is satisfied by at least one preference model (Sect. 3.2).
- Last (Sect. 3.3), we discuss the need for a non-monotonic reasoning mechanism (NMRM) that will support efficient algorithms for selection most desirable choices.

Motivation: System Configuration and Design

To see the need for our framework, consider a command and control room for real-time operations. According to [6], this can be a control room for handling large-scale emergencies such as earthquakes, flooding, terrorist attacks; a control center for complex missions such as NASA shuttle missions; an army command center; or a monitoring center for some company that receives a lot of real time data that requires the attention of a decision maker. To be concrete, let us focus on the setting of a bank surveillance system with infrared (IR) and non-infrared cameras located in various places. For the sake of illustration, see Fig. 1, depicting a room with a single gate, two light sources, four non-IR cameras A1, …, A4, and a single IR camera A5. In a realistic scenario, we are talking about tens of rooms and hundreds of information streams to which the decision maker in the control room might have access. Clearly, no decision maker can handle this volume of information. Therefore we aim at a feasible solution that can have enormous impact on the quality of decisions made. As information can be displayed in various modes; e.g., a video can be shown in different

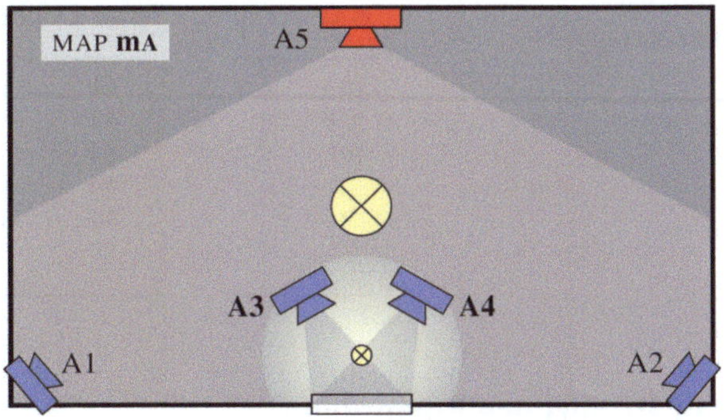

Fig. 1 Room equipped with one IR camera *A5* and four non-IR cameras *A1*, *A2*, *A3*, *A4*, where the gate area is lit and shot by both *A3* and *A4*

sizes and different places on a limited number of the decision maker's screens, we seek a tool that will automatically control the information sources displayed to a decision maker based on the current state of the world and principles of camera surveillance for banks. It is usual that such principles are expressed in a natural language by the decision maker. For example:

(p-1) Streams from non-IR cameras shooting a lit area are more desirable than streams from IR cameras.
(p-2) For unlit areas, streams from IR cameras are more desirable than streams from non-IR cameras.
(p-3) A camera stream of a gate area is always desirable.
(p-4) Displaying interactive digital maps with non-IR cameras shooting a lit area is more desirable than displaying each single stream from those non-IR cameras.

Consider the characteristics of such a system. On the one hand, the set of cameras is known ahead of time, and some of their properties are also known, e.g., camera types and proportions (from the interval [0, 1]) of the gate area in their viewing fields. On the other hand, lighting of the room changes dynamically. Besides, we consider interactive digital maps of rooms that can be displayed on a special, decision maker's touch screen to enable displaying a desired camera stream on a control screen upon touching a respective camera icon. We can imagine such a map as the one (MAP mA) depicted in Fig. 1. Altogether, note that all the above information on possible choices (camera streams and interactive maps) can be represented in the RDM (Fig. 2). Accordingly, the aforementioned changes in the room lightening are captured by updating values (from the interval [0, 1]) of the attribute LIT, triggering new display choices for a limited number of screens available in the control room; these most desirable choices must be made quickly, i.e., in a matter of seconds.

Fig. 2 Relation DB instance representing the bank surveillance system

MAP	ROOM
mA	A
mB	B
mC	C
⋮	⋮

CAMERA	ROOM	IR	LIT	GATE
A1	A	N	0.2	0
A2	A	N	0.2	0
A3	A	N	1	1
A4	A	N	1	1
A5	A	Y	0.3	0.04
⋮	⋮	⋮	⋮	⋮

2 Language of Preference Formulae

Preferences ought to be specified in a formal language in which a designer (in the decision analyst role) can encode the decision maker's statements representing main principles of his or her domain of knowledge. As a system designer's willingness to adapt a new tool to effectively encode knowledge is likely to depend greatly on the convenience and intuitive appeal of this tool, there is a need for a language that mediates comfortable encoding a specific type of knowledge while remaining sufficiently expressive. Such a language must, among others, be based on information that designers find cognitively easy to reflect upon and to express, and the information communicated in such a language should be reasonably easy to interpret. Considering this expectations, it appears unavoidable [5] that such a language be based on pieces of qualitative information about the (preference) model. Indeed, while quantitative information appears much harder to introspect upon (e.g., "the utility of Linda is 12, and the utility of Nora is 5"), qualitative language statements (e.g., "I like Linda more that Nora") appear quite often in our daily conversation. In particular, note that also the statements (p-1) through (p-4), listed on the preceding page, are of qualitative nature, each of them expressing a more intens desire for (some or all) choices from some set than from some other set, and observe that

- information on all the possible choices (in domains of our interest) is represented as a DB instance, $\boldsymbol{J}$, and recall that DB queries are system interpretable specifications of user's intents in engineering applications, where they denote sets by properties (rather than by names) of their elements;
- the respective sets of choices are denoted intensionally, and their *intension* (denoted by properties), in contrast to their *extension* (instantiation), does not depend on the actual state of the DB instance;
- (p-1) and (p-2) provide precise while (p-3) and (p-4) imprecise pieces of information on (the model of user) preferences in a specific, compact way (for more details refer to Example 2).

Example 1 (Comparative preference statements) Consider the statement (p-1) introduced on the previous page. Basically, it denotes two sets of streams (intensionally) and describes the "desirability relation" between them. The two sets can be denoted in the relational algebra as follows:

$$\varphi_1(\boldsymbol{J}) = \delta_{\text{CAMERA}\to\text{INPUT}}\Big(\pi_{\text{CAMERA}}\big(\sigma_{\substack{\text{IR} = \text{“N”} \\ \text{LIT} \geq 0.5}}(I_2)\big)\Big),$$

$$\psi_1(\boldsymbol{J}) = \delta_{\text{CAMERA}\to\text{INPUT}}\Big(\pi_{\text{CAMERA}}\big(\sigma_{\text{IR}=\text{“Y”}}(I_2)\big)\Big),$$

where $\boldsymbol{J}$ includes the relation instances depicted in Fig. 2 to which we refer as I_1 and I_2, and δ, π, and σ are (unary) relational algebra operators *renaming*, *projection*, and *selection*, respectively (for more details about RQLs, refer to, e.g., [1]). As a result, the designer can encode (p-1) as $\varphi_1 \rhd \psi_1$ where $\rhd$ denotes a specific kind of preference of φ over ψ, i.e., a specific kind of information on the "desirability relation" between the two sets.

Definition 1 (*Preference formula and specification*) A *preference formula* is an expression of the form $\varphi \rhd \psi$ where φ, ψ are *schema equivalent* (from the same DB schema to relation schemata of the same type, i.e, with an identical attribution) queries in a RQL, and $\rhd$ represents a distinct *preference symbol* for a respective kind of a preference. A set of preference formulae where all the queries in a RQL are schema equivalent is termed *preference specification*, denoted $\mathcal{P}$.

We borrow the preference symbols (to be substituted for $\rhd$) from [9]. However, relaxing their semantics (outlined in Sect. 3.2), the respective kinds of preferences they denote are termed as follows:

- (strict) *negatively opportunistic*, $({}^{M}{>}^{m})$ ${}^{M}{\geq}^{m}$,
- (strict) *negatively careful*, $({}^{m}{>}^{M})$ ${}^{m}{\geq}^{M}$,
- (strict) *negatively locally optimistic*, $({}^{M}{>}^{M})$ ${}^{M}{\geq}^{M}$,
- (strict) *negatively locally pessimistic*, $({}^{m}{>}^{m})$ ${}^{m}{\geq}^{m}$,

where the left hand side and the right hand side of the preference symbol is indexed by either the small *m* for the "pessimistic" minimal (worst) choices or the capital *M* for the "optimistic" maximal (best) choices, as introduced in [9] and explained in Sect. 3.1.

Remark 1 (*Queries in RQLs*) In this paper, φ, ψ, queries in a RQL, are (query) mappings from the family of all instances of a(n) (input) DB schema to the family of instances of a(n) (output) relation schema. For example, $\boldsymbol{J}$ being an instance of an input DB schema, the image $\varphi(\boldsymbol{J})$ of $\boldsymbol{J}$ under φ is an instance of an output relation schema. Moreover, we use the syntactic abbreviations $\varphi \wedge \psi$ and $\varphi \wedge \neg\psi$ that we view as follows:

$$[\varphi \wedge \psi](\boldsymbol{J}) = \varphi(\boldsymbol{J}) \cap \psi(\boldsymbol{J}), \qquad [\varphi \wedge \neg\psi](\boldsymbol{J}) = \varphi(\boldsymbol{J}) \setminus \psi(\boldsymbol{J}).$$

3 Interpretation

Interpretation gives exact meaning to language's formulae. As a good preference system should allow users to express preferences under different interpretations at the same time, each preference formula (Def. 1) encodes a comparative preference statement whose interpretation is given by the symbol substituted for $\rhd$. Such an approach is consistent with psychological results as argued in [12].

We base ourselves on the main concepts of the languages presented in [9, 10] (Sect. 3.1). Elaborating on the idea of minimal logic of preference (Sect. 3.2), we weaken the definitions of respective kinds of preferences so that our interpretation maps any preference specification (possibly including a mixture of preference formulae of all the above eight kinds) to a nonempty set of models. To narrow down this set to a subset of preference models that give a most accurate picture of a preference specification, we discuss desirability of a NMRM (Sect. 3.3).

3.1 Logics of Preferences

A drawback of the logic of (a single) preference is formalization of preference of one kind only. Consequently, when expressing preferences, one has to choose which kind of preference formulae are used to encode all preferences under consideration.

To study the interaction among diverse kinds of preferences, a non-monotonic preference logic for various kinds of preferences, *logic of preferences*—in contrast to the usual reference to the *logic of preference*, has been introduced in [9] and extended in [10]. The authors show that we can differentiate between four distinct kinds of preferences from the viewpoint of optimism and pessimism. Accordingly, the preferences are termed: *locally optimistic*, *locally pessimistic*, *opportunistic*, and *careful*. In brief, a preference of "φ over ψ" is defined as a preference of "$\varphi \wedge \neg\psi$ over $\neg\varphi \wedge \psi$," and its semantics is based on *total pre-order* (i.e., a transitive and total relation) on the set of possible choices. Specifically, by maximal and minimal referring respectively to maximal and minimal choices in terms of the total pre-order, a preference of "φ over ψ" is interpreted in accordance with its kind as follows:

(Strict) opportunistic (we are ptimistic about φ and pessimistic about ψ):
maximal choices from $\varphi(\boldsymbol{J}) \setminus \psi(\boldsymbol{J})$ are (strictly) preferred to minimal choices from $\psi(\boldsymbol{J}) \setminus \varphi(\boldsymbol{J})$.

(Strict) careful (we are pessimistic about φ and optimistic about ψ):
minimal choices from $\varphi(\boldsymbol{J}) \setminus \psi(\boldsymbol{J})$ are (strictly) preferred to maximal choices from $\psi(\boldsymbol{J}) \setminus \varphi(\boldsymbol{J})$.

(Strict) locally optimistic (we are optimistic about both φ and ψ):
maximal choices from $\varphi(\boldsymbol{J}) \setminus \psi(\boldsymbol{J})$ are (strictly) preferred to maximal choices from $\psi(\boldsymbol{J}) \setminus \varphi(\boldsymbol{J})$.

(Strict) locally pessimistic (we are pessimistic about both φ and ψ): minimal choices from $\varphi(\boldsymbol{J}) \setminus \psi(\boldsymbol{J})$ are (strictly) preferred to minimal choices from $\psi(\boldsymbol{J}) \setminus \varphi(\boldsymbol{J})$.

Furthermore, in [10], the authors distinguish between preferences *with* and *without ceteris paribus proviso*, which we don't consider here, leaving them for the future research.

3.2 The Idea of Minimal Logic of Preference

The idea of minimal logic of preference [2] may be used to represent any set of preferences in a consistent way, which is achieved in two steps. First, a *preference model* is defined as a partial pre-order (i.e., a reflexive and transitive relation) over a (finite) set of choices, where incomparable choices indicate some kind of conflict between these choices. Second, a preference of choices denoted (intensionally) by φ over choices denoted (intensionally) by ψ is defined as the absence of a preference of choices denoted by ψ over choices denoted by φ.

Definition 2 (*Preference model*) Let Ω denote the set of all and only the so-called acceptable choices, which fulfill given mandatory requirements encoded in a RQL as a DB query, say q, that takes a DB instance $\boldsymbol{J}$ as input. That is, Ω is abstracted to $q(\boldsymbol{J})$. Then we say that a partial pre-order over Ω, denoted $\langle \Omega, \succeq \rangle$, is a preference model for mandatory requirements denoted as q.

Remark 2 (*Preference notation*) For precise identification, we use the notation $\langle \Omega, \succeq \rangle$, where Ω is both the *domain* and *codomain*, and $\succeq$ is the *graph* of the binary relation on Ω, i.e., a subset of $\Omega \times \Omega$. Abusing the notation, we write $w \succeq w'$ instead of $\langle w, w' \rangle \in \succeq$. For brevity, we use the usual notational shorthand and, write $w \succ w'$ for $w \succeq w'$ without $w' \succeq w$ and $w \not\succeq w'$ for negation of $w \succeq w'$. Finally, $w \succeq w'$ ($w \succ w'$) reads: "w is (strictly) preferred to w'."

In general, each preference formula provides a piece of information on the preference model. Formally, given a preference specification $\mathcal{P} = \{\varphi_1 \rhd \psi_1, \ldots, \varphi_n \rhd \psi_n\}$, preference formulae from $\mathcal{P}$ define respective preference models $\langle \Omega, \succeq_1 \rangle$, ..., $\langle \Omega, \succeq_n \rangle$, from which one should derive the global preference model, say $\langle \Omega, \succeq_{\mathcal{P}} \rangle$. Unfortunately, some pieces of information given by two distinct preference models, in general, may be conflicting, leading to cyclic preferences, e.g., $w \succ w'$ and $w' \succ w$. To overcome this problem, we propose relaxing the semantics of preference formulae introduced in [9] to pieces of information that are never inconsistent. To this end, not only do we define a preference model as a partial pre-order, but, moreover, we elaborate on the idea of minimal logic of preference and introduce semantic rules, defining the meaning of respective preference formulae, that, as will be seen in Sect. 3.2 are satisfiable in any combination.

Satisfaction Relation and Semantic Rules

Formally, the semantics of the language is given via the *satisfaction relation* between preference models and sets of preference formulae. The following *semantic rules* define the *satisfaction relation w.r.t.* $\boldsymbol{J}$ between a preference model and a respective kind of a preference formula $\varphi \rhd \psi$ in the so-called *normal form* w.r.t. $\boldsymbol{J}$.

Remark 3 (*Normal form*) The notion of normal form is based on the so-called *conjunctive expansion principle*, introduced in [13], a book that is usually taken to be the seminal work in preference logic. In brief, $\varphi \wedge \neg\psi \rhd \neg\varphi \wedge \psi$ is the normal form of a preference formula $\varphi \rhd \psi$ w.r.t. to any DB instance, and $\varphi \rhd \psi$ is in normal form w.r.t. $\boldsymbol{J}$ iff $\varphi(\boldsymbol{J}) \cap \psi(\boldsymbol{J})$ is empty.

(Strict) negatively opportunistic
(we are optimistic about φ and pessimistic about ψ):
✔ some acceptable choice in $\psi(\boldsymbol{J})$ is not (non-)strictly preferred to some acceptable choice from $\varphi(\boldsymbol{J})$, or
✗ there is no acceptable choice in $\psi(\boldsymbol{J})$, or
✗ there is no acceptable choice in $\varphi(\boldsymbol{J})$.
(Strict) negatively careful (we are pessimistic about φ and optimistic about ψ):
✔ no acceptable choice from $\psi(\boldsymbol{J})$ is (non-)strictly preferred to an acceptable choice from $\varphi(\boldsymbol{J})$.
(Strict) negatively locally optimistic (we are optimistic about both φ and ψ):
✔ there is an acceptable choice in $\varphi(\boldsymbol{J})$ such that no acceptable choice from $\psi(\boldsymbol{J})$ is (non-)strictly preferred to it, or
✗ there is no acceptable choice in $\varphi(\boldsymbol{J})$.
(Strict) negatively locally pessimistic (we are pessimistic about both φ and ψ):
✔ some acceptable choice in $\psi(\boldsymbol{J})$ is (non-)strictly preferred to no acceptable choice from $\varphi(\boldsymbol{J})$, or
✗ there is no acceptable choice in $\psi(\boldsymbol{J})$.

Example 2 (*Preference formulae*) Observe that the statements (p-1) and (p-4), listed on page 3, can be encoded as preference formulae

$$\varphi_1 \,{}^{m}\!\!>^{M} \psi_1, \quad \varphi_4 \,{}^{m}\!\!>^{m} \varphi_1$$

where

$$\varphi_4(\boldsymbol{J}) = \delta_{\text{MAP}\to\text{INPUT}}\Big(\pi_{\text{MAP}}\big(I_1 \bowtie \sigma_{\substack{\text{IR} = \text{"N"} \\ \text{LIT} \geq 0.5}}(I_2)\big)\Big)\,.$$

To see this, recall Example 1 and note that

- both the above formulae are in normal form w.r.t. any DB instance,
- we interpret "lit area" as that with lighting evaluated with the value 0.5 or more,

(p-1) "streams from $\varphi_1(\boldsymbol{J})$ are more desirable than streams from $\psi_1(\boldsymbol{J})$" is understood as "no acceptable stream from $\psi_1(\boldsymbol{J})$ is non-strictly more desirable than an acceptable stream from $\varphi_1(\boldsymbol{J})$," (p-4) "maps from $\varphi_4(\boldsymbol{J})$ are more desirable than the set $\varphi_1(\boldsymbol{J})$ (of streams) as whole" is understood as "some acceptable stream from $\varphi_1(\boldsymbol{J})$ is non-strictly more desirable than no acceptable map from $\varphi_4(\boldsymbol{J})$, or there is no acceptable stream in $\varphi_1(\boldsymbol{J})$,"

where "acceptable streams and maps" are denoted as

$$\delta_{\text{MAP}\to\text{INPUT}}\Big(\pi_{\text{MAP}}\big(\sigma_{\text{ROOM}=\text{“A”}}(I_1)\big)\Big) \bigcup \delta_{\text{CAMERA}\to\text{INPUT}}\Big(\pi_{\text{CAMERA}}\big(\sigma_{\text{ROOM}=\text{“A”}}(I_2)\big)\Big). \quad (1)$$

Accordingly, (p-1) and (p-4) are encoded respectively as a strict negatively careful and a strict negatively locally pessimistic preference formulae if we read the word "desirable" as "preferred." Also, note that (p-1) communicates a precise piece of information on the preference model while (p-4) an imprecise one because "some acceptable stream from $\varphi_1(\boldsymbol{J})$" is an "imprecise" reference to (a stream from) $\varphi_1(\boldsymbol{J})$.

Following the same train of thoughts, observe that

$$\varphi_2 \,{}^{m}\!\!>^{M} \psi_2 \,, \qquad\qquad \varphi_3 \,{}^{M}\!\!>^{M} \psi_3$$

encode the other respective statements (p-2) and (p-3), and

$$\varphi_2(\boldsymbol{J}) = \delta_{\text{CAMERA}\to\text{INPUT}}\left(\pi_{\text{CAMERA}}\left(\sigma_{\substack{\text{IR} = \text{“Y”} \\ \text{LIT} \le 0.5}}(I_2)\right)\right),$$

$$\psi_2(\boldsymbol{J}) = \delta_{\text{CAMERA}\to\text{INPUT}}\left(\pi_{\text{CAMERA}}\left(\sigma_{\substack{\text{IR} = \text{“N”} \\ \text{LIT} \le 0.5}}(I_2)\right)\right),$$

$$\varphi_3(\boldsymbol{J}) = \delta_{\text{CAMERA}\to\text{INPUT}}\Big(\pi_{\text{CAMERA}}\big(\sigma_{\text{GATE}=1}(I_2)\big)\Big),$$

$$\psi_3(\boldsymbol{J}) = \delta_{\text{CAMERA}\to\text{INPUT}}\Big(\pi_{\text{CAMERA}}\big(\sigma_{\text{GATE}<1}(I_2)\big)\Big).$$

Definition 3 (*Satisfaction relation*) We say that $\langle \Omega, \succeq \rangle$ satisfies a preference formula $\varphi \rhd \psi$ w.r.t. $\boldsymbol{J}$, denoted $\langle \Omega, \succeq \rangle \models_{\mathbf{J}} \varphi \rhd \psi$, iff the semantic rule defining the satisfaction relation w.r.t. $\boldsymbol{J}$ between $\langle \Omega, \succeq \rangle$, on the one hand, and the normal form of $\varphi \rhd \psi$ w.r.t. $\boldsymbol{J}$, on the other hand, holds true.[1] Furthermore, we say that $\langle \Omega, \succeq \rangle$ *satisfies a preference specification* $\mathcal{P}$ w.r.t. $\boldsymbol{J}$, $\langle \Omega, \succeq \rangle \models_{\mathbf{J}} \mathcal{P}$, iff $\langle \Omega, \succeq \rangle$ satisfies every preference formula from $\mathcal{P}$ w.r.t. $\boldsymbol{J}$:

$$\langle \Omega, \succeq \rangle \models_{\mathbf{J}} \mathcal{P} \iff \forall [\varphi \rhd \psi] \in \mathcal{P} \colon \langle \Omega, \succeq \rangle \models_{\mathbf{J}} \varphi \rhd \psi \,. \quad (2)$$

Example 3 (*Preference models satisfying* $\mathcal{P}_{\text{bank}}$) Observe that Fig. 3 depicts preference models for the mandatory requirements denoted as (1), and all the preference

[1] To be strictly rigorous, we should write $\langle \Omega, \succeq \rangle \models_{\boldsymbol{J}} \{\varphi \rhd \psi\}$ as the satisfaction relation holds between preference models and sets of preference formulae.

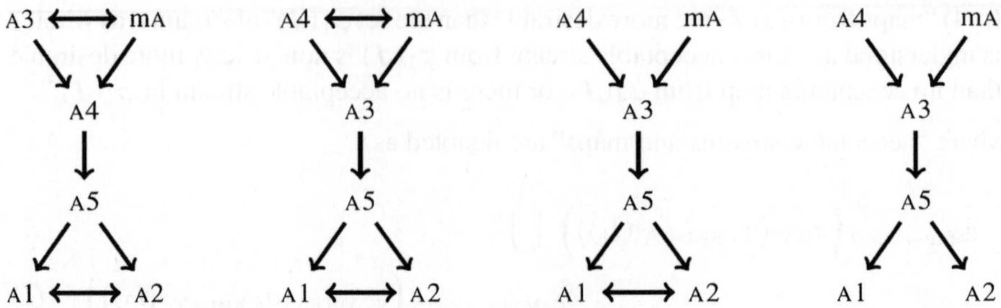

Fig. 3 Preference models for the bank surveillance (a directed edge from a to b means $a \succeq b$)

models w.r.t. $\boldsymbol{J}$ satisfy the preference specification $\mathcal{P}_{\text{bank}}$,

$$\mathcal{P}_{\text{bank}} = \{\varphi_1 \;{}^{m}{>}^{M}\; \psi_1,\ \varphi_2 \;{}^{m}{>}^{M}\; \psi_2,\ \varphi_3 \;{}^{M}{>}^{M}\; \psi_3,\ \varphi_4 \;{}^{m}{>}^{m}\; \varphi_1\}\ , \tag{3}$$

which includes the preference formulae from Example 2.

Totality of Satisfaction Relation

The subtlety of the semantics is in applying the idea of minimal logic of preference, yielding a very special class of semantic rules ensuring that any set of preference formulae, defined by them, is satisfied by at least one preference model, namely an identity relation. To see this, note that an identity relation $\langle \Omega, \succeq_\Delta \rangle$, whose graph $\succeq_\Delta = \{\langle w, w \rangle \mid w \in \Omega\}$, is a trivial instance of the partial pre-order over Ω (the least preference model in terms of set inclusion). Then the claim follows from the case-by-case analysis of the semantic rules. As a result, we get the so-called *right-totality of satisfaction relation*, which states that every preference specification is satisfied by at least one preference model (w.r.t. a given DB instance).

3.3 Discussion on Non-Monotonic Reasoning

As a preference specification, in general, may be satisfied in terms of Def. 3 by multiple preference models, we have to either

1. propose a NMRM to identify the so-called *distinguished preference models* (DPMs) with desired properties,
2. or employ the concept of preference compilation [5] to transform DPMs into a utility function (suitable in the more general, decision theoretic setting)

in order to capture the intended meaning of (mandatory, hard requirements encoded as) q and (soft requirements encoded as) $\mathcal{P}$ w.r.t. $\boldsymbol{J}$. Opting for the first path, we have to define a suitable order on the set of preference models that satisfy $\mathcal{P}$. Such an order may be defined in various ways [2, 7, 9]. For example, it may model a preference[2] on the set of the preference models so that we can identify DPM(s) with its (preferably the single) maximal elements (preference models). Due to the right-totality of satisfaction relation and because of the finiteness of the set of preference models that satisfy a preference specification under consideration, such a mechanism will always return at least one DPM if the set Ω of acceptable choices is nonempty. This property is referred to as *totality of interpretation*.

Another desirable feature of the DPM semantics concerns the concept of preference equivalence w.r.t. $\boldsymbol{J}$: we say that choices w, w' are *$\mathcal{P}$-equivalent w.r.t.* $\boldsymbol{J}$ if and only if $[w \in \varphi(\boldsymbol{J})$ iff $w' \in \varphi(\boldsymbol{J})]$ and $[w \in \psi(\boldsymbol{J})$ iff $w' \in \psi(\boldsymbol{J})]$ for each normal form $\varphi \rhd \psi$ of a preference formula from $\mathcal{P}$ and desire that choices from Ω that are $\mathcal{P}$-equivalent w.r.t. $\boldsymbol{J}$ should also be *equivalent w.r.t.* DPMs (the set of DPMs should be invariant w.r.t. the interchange of those choices). That is to say, we desire that, given a DB instance $\boldsymbol{J}$ containing all the information on possible choices, interpretation of user preferences (over the possible choices) encoded as a preference specification $\mathcal{P}$ depends on information provided by $\boldsymbol{J}$ and $\mathcal{P}$ only. The idea is that, from the perspective of preferences, possible choices have no properties other than the relationships determined by $\boldsymbol{J}$ and $\mathcal{P}$. Thus the information on the choices is essentially unchanged if all the choices are consistently renamed. This desired *invariance of DPM semantics*, which is in line with the intuition, is crucial for development efficient algorithms for selection most desirable choices.

4 Related Work

It is worth noticing that representing and reasoning about comparative preference statements has been widely investigated in AI [3, 10]. Unfortunately, these works have not been used in DB preference queries except [4, 12]. As a result, two main preference semantics of comparative preference statements have been defined in DB framework: *careful* (also called *totalitarian* in [4] or *strong* in other literature) and *ceteris paribus*. Authors in DB community seem to implicitly favour the careful semantics. However, as shown on different examples in [4], this semantics may be inappropriate, yielding unintuitive or empty results even for very simple queries.

To trigger an action, **rule-based systems** have been pursued, e.g., in AI. However, in settings where the total volume of choices to select is constrained by resource limitations, e.g., a number of screens available in the control room or decision maker's attention, using a traditional rule-based system is unlikely to be feasible as the resource constraints introduce strong interdependence into the system.

[2] Here, in the context of non-monotonic reasoning, the term preferred has a different meaning than that ascribed to it in decision theory: it refers to models preferred in reasoning and not to models we would like to be true.

With a motivation similar to ours, a rule-based preference specification language, extending regular rule-based languages, has been introduced in [6]. The language's *preference rules* allow more flexible specification of designer's preferences for autonomous systems by specifying the relative value of some property as a function of a (hopefully) small number of other properties that influence this property.

Similar approaches to incorporating a single kind of preferences (with careful semantics) in RQLs are pursued independently in [8, 11], where a preference model capturing intensity of desire for various tuples is specified directly, using operators for combining binary relations: Pareto or lexicographic composition, etc. By contrast to our approach, specifying the preference model in an imprecise way is not supported and, moreover, the exact sequence of preference combinators to specify the preference model is always required, which might be cognitively quite hard. In fact, this is the price we pay for the preference specification consistency.

5 Conclusions

The importance of preference handling techniques for many areas of AI and decision support systems is apparent. As, there is a great (and illustrated) need for such techniques, we have outlined a novel, flexible approach based on the language that is capable of encoding preferences conditional on the current state of the world in a purely qualitative way. As the language facilitates specification of the preference model intensionally and in an imprecise way: its formulae do not require complete information about the system, it is suitable for control of dynamic systems, where both the state and number of objects changes. The main contribution of the paper lies in the semantics that gives meaning to any set of preference formulae; even those encoding conflicting preferences. Because of the lack of space, we have only very briefly discussed the need and desirable properties of a NMRM that support efficient algorithms for selection most desirable choices.

There is much that needs to be done to demonstrate the effectiveness of the outlined framework. The first and best demonstration would be a working prototype: nothing succeeds like a successful implementation. However, such an implementation, among others, requires not only a NMRM that should be tailored down to the specifics of a concrete application, but also addressing the issue of compact *representation* to capture preference formulae in a framework suitable for efficient query-answering algorithms. Here, we note that our latest research findings show that such a representation is feasible provided that the NMRM has the desirable properties discussed in Sect. 3.3. Besides, theoretical questions concerning expressivity (e.g., other preference constructs and more general, the so-called *ceteris paribus* semantics [9, 13]) ask for further investigation.

Acknowledgments The work supported by the project GAP202/10/0761\Web Semantization.

References

1. Abiteboul, S., Hull, R., Vianu, V. (eds.): Foundations of Databases: The Logical Level. Addison-Wesley Longman Publishing Co., Inc., Boston (1995)
2. Boella, G., van der Torre, L.W.N.: A non-monotonic logic for specifying and querying preferences. In: Kaelbling, L.P., Saffiotti, A. (eds.) IJCAI, pp. 1549–1550. Professional Book Center (2005)
3. Boutilier, C., Brafman, R.I., Domshlak, C., Hoos, H.H., Poole, D.: Cp-nets: a tool for representing and reasoning with conditional ceteris paribus preference statements. J. Artif. Intell. Res. **21**, 135–191 (2004). http://dblp.uni-trier.de/db/journals/jair/jair21.html#BoutilierBDHP04
4. Brafman, R.I., Domshlak, C.: Database preference querie revisited. Technical Report, TR2004-1934, Cornell University, Computing and Information Science (2004)
5. Brafman, R.I., Domshlak, C.: Preference handling—an introductory tutorial. Technical Report, 08–04, Computer Science Department, Ben-Gurion University, Negev Beer-Sheva, Israel 84105 (2007)
6. Brafman, R.I.: Relational preference rules for control. In: Brewka, G., Lang J. (eds.) KR, pp. 552–559. AAAI Press, Menlo Park (2008)
7. Brewka, G., Niemelä, I., Truszczynski, M.: Preferences and nonmonotonic reasoning. AI Magazine **29**(4), 69–78 (2008)
8. Chomicki, J.: Preference formulas in relational queries. ACM Trans. Database Syst. **28**(4), 427–466 (2003). http://doi.acm.org/10.1145/958942.958946
9. Kaci, S., van der Torre, L.W.N.: Algorithms for a non-monotonic logic of preferences. In: Godo, L. (ed.) ECSQARU, Lecture Notes in Computer Science, vol. 3571, pp. 281–292. Springer, Berlin (2005)
10. Kaci, S., van der Torre, L.W.N.: Non-monotonic reasoning with various kinds of preferences. In: Brafman, R.I., Junker, U. (eds.) IJCAI-05 Multidisciplinary Workshop on Advances in Preference Handling, pp. 112–117. Edinburgh, Scotland (2005)
11. Kießling, W.: Foundations of preferences in database systems. In: Proceedings of the 28th VLDB Conference, pp. 311–322. Hong Kong, China (2002)
12. Neves, R.D.S., Kaci, S.: Combining totalitarian and ceteris paribus semantics in database preference queries. Logic J. IGPL **18**(3), 464–483 (2010)
13. von Wright, G.H.: The Logic of Preference. Edinburgh University Press, Edinburgh (1963)

Evaluation of Models for Semantic Information Filtering

Ivo Lašek and Peter Vojtáš

Abstract In this paper we evaluate various approaches to a user profile modelling for news recommendation. We represent a user profile as a bag of real world entities, the user is interested in. News articles are thus recommended based on its contained concepts and not based on a text similarity. We propose several ways of such a user profile construction based on a user feedback. Different ways of a user feedback collection are compared. This paper addresses the problem of precise user modelling for information filtering.

1 Introduction

Nowadays we have access to a wide range of information sources from all around the world. These include online news articles, blog entries, short news published on social networks. With any such a new information source, it becomes more difficult to manage the amount of data, we are facing every day. Today, anyone can become a publisher.

With the easiness of text publishing, it is much more difficult to maintain an elementary order of news. We can not read all the news published every minute. So, we need to consume them selectively. But there is a little help in selecting unstructured texts of news articles. Online news servers maintain some basic categorization, but this is only a rough filtering. Even with the categories we need, at least quickly, walk through many irrelevant texts to find an interesting information. Thus we are limited

I. Lašek (✉)
Faculty of Information Technology, Czech Technical University in Prague, Prague, Czech Republic
e-mail: lasekivo@fit.cvut.cz

P. Vojtáš · I. Lašek
Faculty of Mathematics and Physics, Charles University in Prague, Prague, Czech Republic
e-mail: vojtas@ksi.mff.cuni.cz

M. Kudělka et al. (eds.), *Proceedings of the Third International Conference on Intelligent Human Computer Interaction (IHCI 2011), Prague, Czech Republic, August, 2011*, Advances in Intelligent Systems and Computing 179, DOI: 10.1007/978-3-642-31603-6_19, © Springer-Verlag Berlin Heidelberg 2013

to only a small amount of information sources (our favourite news server, couple of blogs of our friends, a Twitter and a Facebook account), leaving many potentially much more valuable information sources unnoticed.

As a consequent, we often consume a large amount of general information at the expense of not seeing information particularly interesting or important for us personally. For example lets assume, we are interested in news about a particular company. We would have to follow all the news from a business section of our favourite news sources. Even so, it is not guaranteed that we do not miss something important. With this approach, we are limited to only a small number of information sources.

We address the problem of locating relevant news articles based on their semantic content. We build on our previous work, where we introduced a concept of a semantic information filtering [1]. We proposed a framework enabling news articles filtering on a much more detailed level than a basic categorization. We use named entities extracted from articles to characterize its semantic meaning. This approach gives us the possibility to offer a real personalised news stream.

In this paper, we further elaborate on models exploiting named entities to characterize a semantic content of news articles. We compare several ways of the entities scoring in order to automatically match articles describing a same topic. This should serve us to more precisely recommend articles fitting even specific user needs.

The rest of the paper is organized as follows. First, we recall related work in Sect. 2. Section 3 introduces the proposed semantic information filtering approach. Section 4 describes various ways of user feedback collection. In Sect. 5 we briefly describe the implementation details of the proposed solution. Section 6 presents our experimental results. Finally, we conclude in Sect. 7.

2 Related Work

We focus on news filtering. There are two basic approaches to news filtering: Collaborative filtering and Content based filtering.

Collaborative filtering is a general concept not necessarily specific for the news domain. It draws on the idea, that people who agreed in their subjective evaluation of past items are likely to agree again in the future. Famous implementation include product domain [2] or movies and tv shows domain [3]. The earlier representative of the collaborative filtering in news domain is GroupLens [4] using this techniques to reduce the information overload by filtering Netnews.

More recent work on this topic is presented by Google [5]. By Google News[1] MinHash and Probabilistic Latent Semantic Indexing algorithms are used as an adaptation of the collaborative filtering technique. Interesting news are recommended to registered users based on their clicks history.

---

[1] http://news.google.com/

Collaborative filtering has its advantages as a universal method—to a certain extent domain independent. However, there are its drawbacks as well. We need to keep track of many users with overlapping interests. The news corpus changes very often. But for the collaborative filtering it may take hours to learn the recommendations from other users [6]. Also the recommendation process remains hidden from the user. The user can not directly influence her or his own profile.

The other approach is content based filtering. We filter the information based on its properties. In the case of articles, this means in most cases its text (either a full text, only a title or a short description).

Very often, there are used information retrieval techniques [7]. In such cases the similarity of the texts of articles is compared in order to find similar articles. Used models vary from simple keyword matching or the vector space model with basic tf-idf weighting to the Okapi BM25 retrieval function [8] or language modeling framework [9]. A nice comparison of the latter two is provided in [10].

Recently, Google incorporated content based recommendations to its algorithm as well [6]. Text classifiers are used to determine a category, which an article belongs to. Users get recommendations based on their preference of particular categories. To model user's preferences a Bayesian framework is proposed using her or his past click history.

In this paper, we focus on the evaluation of various models for identification of related articles. The idea is that the user is interested in several topics that he wants to read about. It is thus useful, to offer him articles related to the topics contained in articles, he liked.

The relatedness [11] of articles should not be considered only as similarity of their texts. Such a system can tend to recommend very similar articles every time. However, users are interested rather in new information about the same fact, than in same information in small variations. In [11] the relatedness of articles is characterized from multiple aspects: relevance and novelty, connection clarity and transition smoothness.

We believe that recommendation based on more abstract content of the article, represented by named entities can overcome the mentioned problems.

An automatic annotation of unstructured text coming from news articles is necessary for our system to work. Our inspiration comes mainly from two projects. SemTag [12] developed in IBM Almaden Research Center introduces the idea of automated semantic annotations. They show it is applicable in the large scale as well. SemTag was applied to a collection of 264 million web pages producing 434 million annotations.

The other project is KIM platform [13]. KIM platform focuses at the named entity recognition and the recognition of a limited amount of phrases.

Further inspiration can be found in the survey papers about web information extraction [14] and semantic annotation platforms [15]. In the more recent survey [16] a wide range of Semantic APIs are introduced. Many of these APIs can be used for named entity recognition including Dapper [17] concentrating rather at the product domain, Zemanta [18] or OpenCalais [19].

Our aim is not to create another semantic tagging or a named entity recognition system. We rather concentrate on using the listed existing solutions and adding our information filtering logic.

Both the mentioned projects SemTag and KIM platform use their own well defined knowledge bases. In the case of SemTag it is TAP [20] in case of KIM the data are collected from several structured databases. These data are clean, but tend to become obsolete in time and not ideally extensible. From this perspective, more flexible and extensible seem to be semantic APIs like Zemanta or OpenCalais. These provide unique identifiers of extracted entities in the form of Uniform Resource Identifiers (URIs). Additionally both mentioned services provide links to additional information sources for the extracted entities. This is a great advantage, when we consider collecting additional information from semantic web resources.

OpeanCalais also provides an information about the score of each entity in the context of an annotated article. The score represents the importance of the entity in the text.

3 Semantic Information Filtering

In the framework, we introduced in [1], a user interest is characterized by her or his profile. The profile composes of named entities extracted from news articles rated by the user. Each entity is characterized by its significance. The significance is a numeric measure of the user interest in a particular topic (represented by the entity).

In order to compute the significance of entities, we applied a variant of the ordinary tf-idf metric. But instead of term frequencies, we use entity frequencies. The entity frequency indicates, how often is an entity mentioned in the text of an article. The second part of the metric–Inverse Document Frequency is replaced with the Inverse Document Entity Frequency and again instead of terms we calculate inverse frequencies of occurrences of entities in documents in the whole news articles corpus. For the clarity, we denote our variant ef-idef as Entity Frequency–Inverse Document Entity Frequency.

The significance is then computed as the sum of ef-idef weights multiplied by user rating of articles containing this entity.

$$s_i = \frac{\sum_{j=0}^{n}(r_{i,j} \times ef\text{-}idef_{i,j})}{n \times \max(ef\text{-}idef)} \tag{1}$$

With the Eq. 1, we can obtain a significance of an entity i. In the equation, $r_{i,j}$ is a user rating of an article a_j that contains an entity i. It turned out to be advantageous to normalize the value of significance. Thus, we are able to determine a uniform value of the threshold independently of the count of news articles in the corpus.

When a new article comes to the framework, we determine its score ar in the context of a particular user profile as follows:

$$ar = \frac{\sum_{i=0}^{m} s_i}{m} \tag{2}$$

The significances of entities that are present in the article and also in the user profile are summed. Again the score is normalized.

If the score exceeds a certain threshold, the article is considered as interesting for the user.

4 User Rating

In Eq. 1 the rating $r_{i,j}$ can get following values:

- -1 in case of a negative rating.
- 1 in case of a positive rating.
- 0 if the user does not vote for an article j that contains an entity i.

We propose three possible ways to incorporate a user rating:

- Positive rating only
- Positive and negative rating
- Positive, negative and neutral rating

By positive rating a user is given the opportunity to rate articles only positive. This can be done either explicitly (for example by pressing a voting button) or implicitly—simply by monitoring a user activity (for example by monitoring user clicks [5]).

By positive and negative rating we can again use an explicit or implicit feedback. The user has the same opportunity as in the previous case, but the interpretation differs. We consider not voting or not clicking as a negative feedback.

Finally, for the positive, negative and neutral rating, we give a user the opportunity to vote explicitly either negative or positive for an article. If the user does not vote for some article, we consider it as a neutral rating (i.e. $r_{i,j} = 0$).

5 Implementation

We implemented a framework for evaluation of the proposed approaches to the user feedback collection. The whole framework is implemented in Java and collected articles are stored in MySQL database.

For named entity extraction, we use a web service provided by OpenCalais [19]. For parsing the output of the web service, we use Jena Framework [21]. OpenCalais disposes of a comprehensive database of companies, people, products and places. It performs the named entity extraction. The entities are additionally disambiguated, if needed. For example, when resolving a company name, clues in the text as what the company does or where it is located may help to resolve among several possible identities of the company.

OpenCalais uses natural language processing and machine learning methods to analyse the posted documents. It is able to locate mentions of entities also from the context (e.g. pronoun as a reference to an entity). Each entity is assigned a score based on the frequency, the entity is mentioned in the text. We use this score as an entity frequency to compute ef-idef. In our case, the OpenCalais score is normalised in order to take values from 0 to 1.

6 Evaluation

6.1 Used Data

For the evaluation, we used the dataset provided by the authors of [10]. They built a dataset using the Reuters RCV1 collection [22]. The dataset consists of 4,800 articles. 50 of them were chosen as base articles and for each of them 100 arbitrarily selected articles were manually assigned a score (0–4) according to their relatedness. Authors claim the articles with the score 3 and 4 to be rather related, while articles with the score 1 and 0 to be rather unrelated.

6.2 Evaluation Results

Our aim during the evaluation was to compare proposed approaches to the user rating collection. We built several training and testing sets of equal size to simulate various user demands. Each training set contained a group of related articles. User profiles were created based on the related articles from the training set. The framework should then find remaining related articles from the testing set.

In the first phase, we evaluated the collection of a positive rating only. We built three training and testing sets. In the first case, we considered only the articles related with the score of 4 as correctly matched. In the second phase, the articles with the scores 3 and 4 were considered as correctly matched. Finally, articles with scores 2, 3 and 4 were considered as related.

The results of the evaluation are shown in Fig. 1. The figure shows precision and recall for three training and testing groups. As expected, the precision and recall improves with the growth of the training and testing group size. The training group composed of only articles related with the score 4 was too specific and the framework was not able to satisfactorily distinguish between the degree of 3 and 4.

Much better was the situation, when we incorporated the negative rating of articles, too (Fig. 2). We initialized the user profile with entities from articles related with the score of 4 rated as positive. The entities from the remaining articles from the training

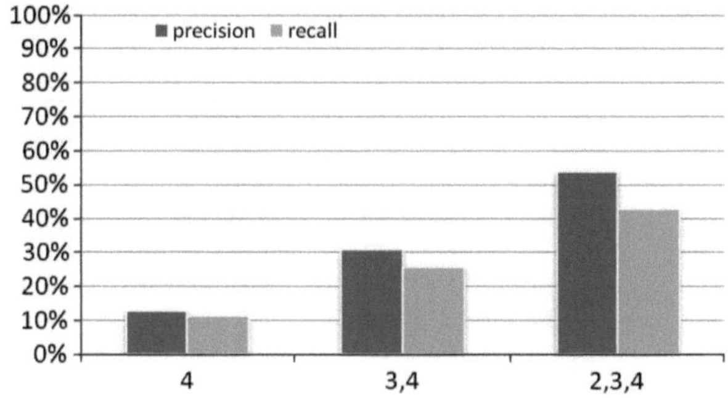

Fig. 1 Articles recommended based on a positive rating only. Training and testing groups composed of articles related with a score 4, a score 3, 4 and a score 2, 3, 4

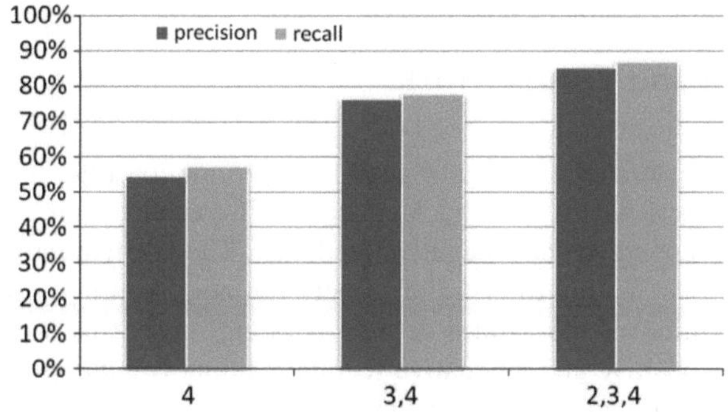

Fig. 2 Articles recommended based on a positive and a negative rating. Articles related with a score 4, a score 3, 4 and a score 2, 3, 4 rated as positive, the rest of the training set rated as negative

set were assigned a negative rating. Then we extended the related articles with the score 3 and 4 to be rated positively and finally also the articles with the scores of 2, 3 and 4. We can see a massive improvement of precision and recall, especially in the first group, representing users focused only on a specific topic. The results improve again, as the user needs are looser.

Finally, we performed a set of tests with the neutral rating as well. We formed two training and testing groups. In the first one, we considered articles related with the score of 4 as positive ratings, articles with the score of 3 as neutral and the rest as negative. Then we extended the group of positive to the score of 3 and 4, the score of 2 was considered as neutral and the rest as negative. The results are shown in Fig. 3.

In the last set of experiments, the result for the group '0, 1, n, 3, 4' was a little surprising. The increase of correctly recommended articles was not so high, which caused the recall to fall down. In all the experiments, we tried to determine an optimal threshold. In all the cases, we selected the threshold that would filter approximately same amount of articles from training groups as was the count of positively rated articles. In this particular case, the threshold could be probably better balanced.

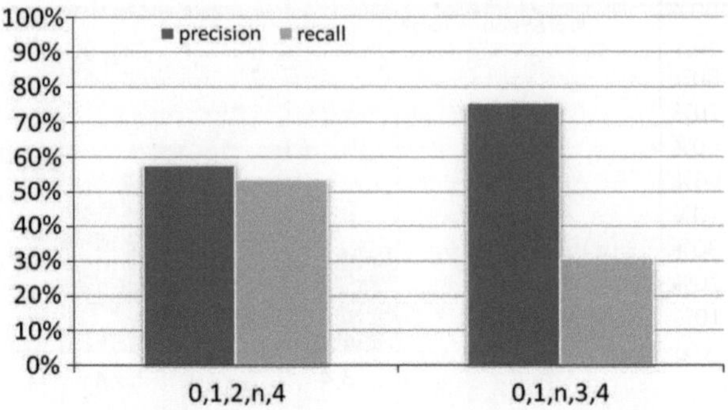

Fig. 3 Articles recommended based on a positive, a negative and a neutral rating. In the graph 'n' denotes neutrally rated articles. The higher scores were rated as positive, lower scores as negative

7 Conclusion and Future Work

In this paper, we extended our idea of semantic information filtering by evaluating various approaches to the user feedback collection. We achieved the best results with the collection of a positive and a negative feedback. The positive rating only did not provide enough information to distinguish between neighbouring groups of related articles. The addition of the neutral degree of rating did not provide a significant improvement. A positive and a negative feedback remains quite flexible, as it does not necessarily require an explicit user interaction.

However, the method of a user feedback collection should be further examined in a real world scenario with a real users. For example not rating or not clicking at an article does not have to mean that the user is not interested in it. It is possible that she or he simply does not have the time to read it right now. Thus the incorporation of the neutral degree of rating may still be beneficial in such cases.

In our future work, we would like to extend the user model and incorporate also the properties of extracted entities. We would thus be able to identify more abstract concepts than mere presence of an entity in the text.

Acknowledgments This work has been partially supported by the grant of The Czech Science Foundation (GAČR) P202/10/0761 and by the grant of Czech Technical University in Prague registration number SGS11/085/OHK3/1T/18.

References

1. Lašek, I., Vojtáš, P.: Semantic information filtering—beyond collaborative filtering. In: 4th International Semantic Search Workshop. http://km.aifb.kit.edu/ws/semsearch11/11.pdf (2011). Accessed 13 June 2011
2. Linden, G., Smith, B., York, J.: Amazon.com recommendations: item-to-item collaborative filtering. IEEE Internet Comput. **7**(1), 76–80 (2003)

3. Breese, J.S., Heckerman, D., Kadie, C.: Empirical analysis of predictive algorithms for collaborative filtering. In: Cooper, G.F., Moral, S. (eds.) Proceedings of the 14th Conference on Uncertainty in Artificial Intelligence (UAI-98), pp. 43–52. Morgan-Kaufmann, San Francisco, Calif (1998)
4. Resnick, P., Iacovou, N., Suchak, M., Bergstrom, P., Riedl J.: GroupLens: an open architecture for collaborative filtering of netnews. In Proceedings of the: ACM Conference on Computer supported Cooperative Work (CSCW '94), pp. 175–186. ACM, New York, NY, USA (1994)
5. Das, A.S., Datar, M., Garg, A., Rajaram, S.: Google news personalization: scalable online collaborative filtering. In: Proceedings of the 16th International Conference on World Wide Web (WWW '07), pp. 271–280. ACM, New York, NY, USA (2007)
6. Liu, J., Dolan, P., Pedersen, E.R.: Personalized news recommendation based on click behavior. In: IUI'10: Proceedings of the: International Conference on Intelligent User. Interfaces, pp. 31–40 (2010)
7. Belkin, N.J., Croft, W.B.: Information filtering and information retrieval: two sides of the same coin? Commun. ACM **35**(12), 29–38 (1992)
8. Robertson, S., Walker, S.: Some simple effective approximations to the 2-poisson model for probabilistic weighted retrieval. In: Proceedings of SIGIR '94, pp. 232–241. ACM Press, New York, NY (1994)
9. Ponte, J.M., Croft, W.B.: A language modeling approach to information retrieval. In: Proceedings of SIGIR '99, pp. 275–281. ACM Press, New York, NY (1998)
10. Bogers, T., Bosch, A.: Comparing and evaluating information retrieval algorithms for news recommendation. In: Proceedings of the ACM Conference on Recommender Systems. pp. 141–144. (2007)
11. Lv, Y., Moon, T., Kolari, P., Zheng, Z., Wang, X., Chang, Y.: Learning to model relatedness for news recommendation. In: Proceedings of the 20th International Conference on World Wide Web, WWW 2011, pp. 57–66. ACM Press (2011)
12. Dill, S., Eiron, N., Gibson, D., Gruhl, D., Guha, R., Jhingran, A., Kanungo, T., Rajagopalan, S., Tomkins, A., Tomlin, J. A., Zien, J.Y.: Semtag and seeker: bootstrapping the semantic web via automated semantic annotation. In: Proceedings of the 12th International Conference on World Wide Web, WWW'03, pp. 178–186. ACM, New York, NY, USA, 2003
13. Popov, B., Kiryakov, A., Ognyanoff, D., Manov, D., Kirilov, A.: Kim—a semantic platform for information extraction and retrieval. Nat. Lang. Eng. **10**, 375–392 (2004)
14. Chang, C.-H., Kayed, M., Girgis, M.R., Shaalan, K.: A survey of web information extraction systems. IEEE Trans. Knowl. Data Eng. **18**, 1411–1428 (2006)
15. Reeve, L., Han, H.: Survey of semantic annotation platforms. In: Proceedings of the 2005 ACM Symposium on Applied Computing, SAC '05, pp. 1634–1638. ACM, New York, NY, USA (2005)
16. Dotsika, F.: Semantic apis: scaling up towards the semantic web. Int. J. Inf. Manag. **30**(4), 335–342 (2010)
17. Dapper. http://www.dapper.net (2005). Accessed 13 June 2011
18. Zemanta. http://www.zemanta.com/api/ (2009). Accessed 13 June 2011
19. OpenCalais. http://www.opencalais.com/ (2008). Accessed 13 June 2011
20. TAP. http://ksl.stanford.edu/projects/TAP/ (2005). Accessed 13 June 2011
21. Jena. http://jena.sourceforge.net/. Accessed 13 June 2011
22. Lewis, D., Yang, Y., Rose, T., Li, F.: Rcv1: a new benchmark collection for text categorization research. J. Mach. Learn. Res. (JMLR) **5**, 361–397 (2004)

GPU Based Parallelism for Self-Organizing Map

Petr Gajdoš and Jan Platoš

Abstract Modern graphics cards take role of powerful computation hardware. This hardware becomes more popular due to purchasing costs and its availability. The advantages of Graphics Processor Unit (GPU) in parallel computation of Self-Organizing Network are described in this paper including a comparison with multi-threaded CPU. The parallelism on GPU is explained in a separated section. Mentioned section is divided into parts with respect to different forms of parallelism. The results of experiments at the end confirmed, that the utilization of GPU brings significant improvements in time of computation in case of large data sets.

1 Power of GPU

Modern graphics hardware play an important role in the area of parallel computing. Graphics cards have been used to accelerate gaming and 3D graphics applications, and now, they are also used to accelerate computations with relatively distant topics, e.g. remote sensing, environmental monitoring, business forecasting, medical applications or physical simulations etc. Architecture of Graphics Processing Units (GPU) is suitable for vector and matrix algebra operations. That leads to the wide usage of GPUs in the area of information retrieval, data mining, image processing, data compression, etc. [10]. Nowadays, one does not need to be an expert on graphics hardware because of existence of various Application Programming Interfaces (API), which help programmers to implement their software faster. Nevertheless, it

P. Gajdoš (✉) · J. Platoš
Department of Computer Science, FEECS VSB-Technical University of Ostrava,
17. listopadu 15, 708 00Ostrava Poruba, Czech Republic
e-mail: petr.gajdos@vsb.cz

J. Platoš
e-mail: jan.platos@vsb.cz

M. Kudělka et al. (eds.), *Proceedings of the Third International Conference on Intelligent Human Computer Interaction (IHCI 2011), Prague, Czech Republic, August, 2011*, Advances in Intelligent Systems and Computing 179, DOI: 10.1007/978-3-642-31603-6_20, © Springer-Verlag Berlin Heidelberg 2013

will be always necessary to keep basic rules of GPU programming to write required code more effective.

There are two graphics hardware leaders, vendors (ATI and nVIDIA) that prefer they own solution before any other. ATI developed technology called ATI Stream and nVIDIA presented nVIDIA CUDA. Comparison of these two APIs is not a goal of this article. We refer to [10] for more information. CUDA is an acronym for Compute Unified Device Architecture and it was used in our experiments because of hardware availability. It is a general purpose parallel computing architecture that leverages the parallel compute engine in nVIDIA graphics processing units.

1.1 nVIDIA CUDA

A schematic view of the nVIDIA G80 GPU can be seen in the Fig. 1. The simplified general description in this paragraph was taken from [10]. GeForce GTX 280 was used in our experiments. The GPU processor is composed of 30 multiprocessors. Each Single Instruction Multiple Data (SIMD) multiprocessor drives eight arithmetic logic units (ALU) which process the data, thus each ALU of a multiprocessor executes the same operations on different data, lying in the registers. In contrast to standard CPUs which can reschedule operations (out-of-order execution), the selected GPU is an in-order architecture. This drawback is overcome by using multiple threads as described by Wellein et al. in [3] . Current general purpose CPUs with clock rates of 3 GHz outrun a single ALU of the multiprocessors with its rather slow 1.3 GHz. The huge number of 240 parallel processors on a single chip compensates this drawback. The processing is optimized for floating point calculations and a Fused Multiply Add (FMA) is four step pipelined, thus its latency is four clock cycles. Additional operations have different specifications and therefore require different numbers of clock cycles to complete. GPU producers prefer double precision in later GPUs because of their wide utilization in scientific computations. CUDA Capability 1.3 and above is necessary to have double precision support. The Table 1 contains a detailed description of selected GPU.

Andrecut [1] in 2008 described computing based on CUDA on two variants of Principal Component Analysis (PCA). The usage of parallel computing on GPU improved efficiency of the algorithm more than 12 times in comparison with CPU. Preis et al. [15] applied GPU on methods of fluctuation analysis, which includes determination of scaling behavior of a particular stochastic process and equilibrium autocorrelation function in financial markets. The speed up was more than 80 times than the previous version running on CPU. Patnaik et al. [14] used GPU in the area of temporal data mining in neuroscience. They analyzed spike train data with the aid of a novel frequent episode discovery algorithm. Achievement of more than 430 speed up is described in the mentioned paper. We refer to [11] for more nVidia CUDA examples.

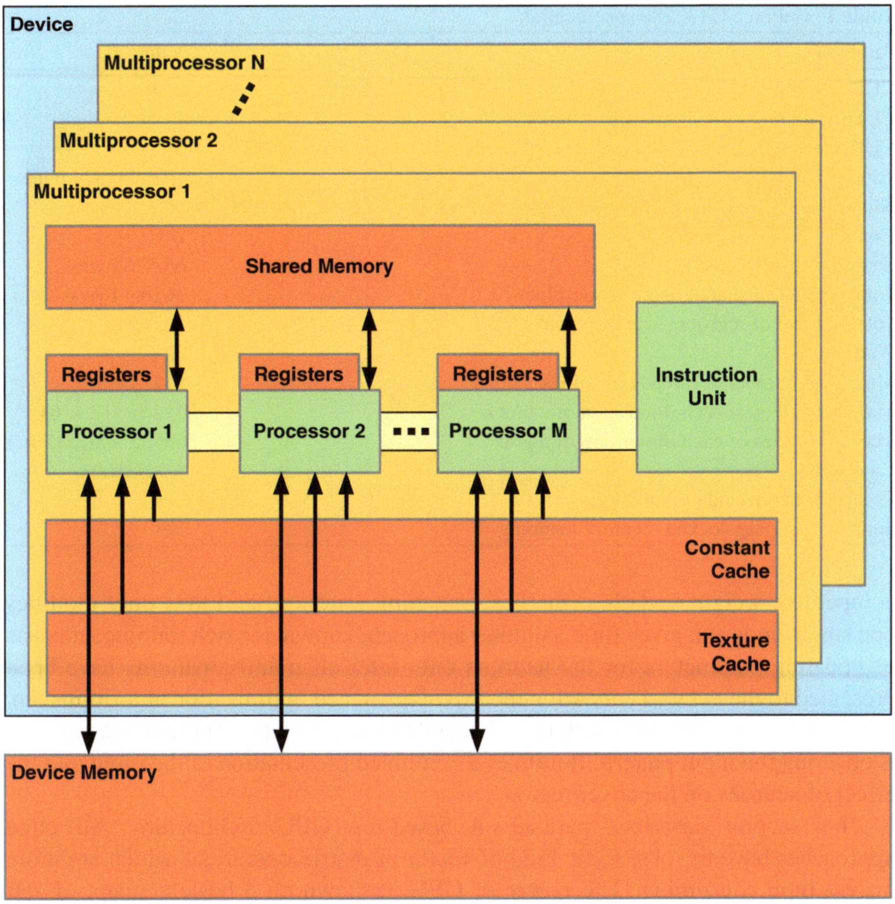

Fig. 1 nVIDIA G80 GPU architecture (taken from [10])

2 GPU Parallelism

Several different implementations of parallel Kohonen maps have already been presented [7, 8, 13, 19] and there are also studies on how the computer architecture could be modified in order to support highly parallel calculations especially for SOM training [9]. These approaches focus mainly on time efficiency issues and how the mathematical operations can be distributed on different machines to speed up the self-organization process. Different approach was presented by I. Valova [16, 17], where the amount of parallelism is not determined by the number of available hardware resources but rather by the size of the input pattern. More convenient approach for a completely parallel SOM has been presented by Weigang [18]. Here in fact, the network processes the whole input space at once, but the algorithm still uses

Table 1 GeForce GTX 280 specification

Parameter	Value
CUDA runtime version	2.30
CUDA capability major revision number	1
CUDA capability minor revision number	3
Total amount of global memory	1073741824 bytes
Number of multiprocessors	30
Number of cores	240
Total amount of constant memory	65,536 bytes
Total amount of shared memory per block	16,384 bytes
Total number of registers available per block	16384
Warp size	32
Maximum number of threads per block	512
Maximum sizes of each dimension of a block	$512 \times 512 \times 64$
Maximum sizes of each dimension of a grid	$65535 \times 65535 \times 1$
Clock rate	1.30 GHz
Concurrent copy and execution	Yes
Support host page-locked memory mapping	Yes

competitive weight updating (the best-matching criterion) and thus only modifies one single node at a given time. Another approach, known as batch training, consists of updating parameters for the neurons only after all training patterns have been presented to the network, allowing the use of advanced optimization algorithms, i.e. gradient descent. Our approach uses a cover function to represent how the network is covering the input pattern. It utilizes a combined presentation of the input and the effect of neurons on the cover region.

This section describes parallelism based on GPU architecture. All cited approaches have to solve some kind of vector or matrix operations which are more or less time consuming. The power of GPU is shown on a basic concept of self organizing maps, although proposed improvements can be applied on any kind of SOM.

2.1 SOM Definition

Presented in the early 1980s by Kohonen [5], the self-organizing map is one of the common approaches on how to represent and visualize data and how to map the original dimensionality and architecture of the input space onto another, usually lower dimensional, architecture in the output space. Further, some classification or clustering tools use SOM to find data which are more closer to each other [2]. All experiments and examples in this paper follow the specification of SOM presented below (see also the Fig. 2):

- The SOM is initialized as a network of fixed topology. The variables $dimX$ and $dimY$ are dimensions of such 2-dimensional topology.

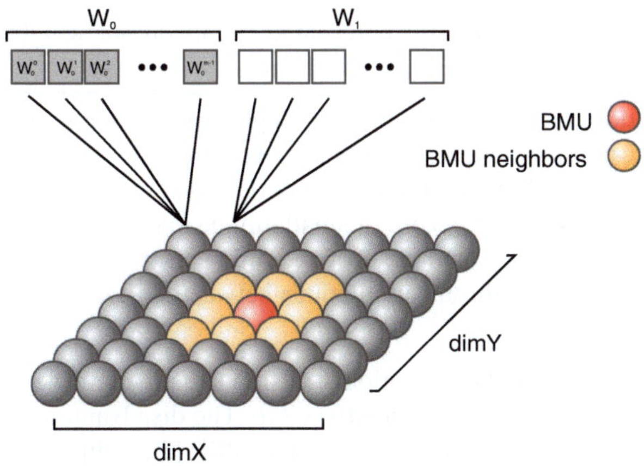

Fig. 2 Illustrative example of the SOM

- V^m represents an m-dimensional input vector.
- W^m represents an m-dimensional weight vector.
- The number of neurons is defined as $N = dimX * dimY$ and every neuron $n \in< 0, N-1 >$ has its weight vector W_n^m.
- The neighborhood radius r is initialized to the value $min(dimX, dimY)/2$ and will be systematically reduced to a unit distance.
- All weights vectors are updated after particular input vector is processed.
- The number of epochs e is know at the beginning.

2.2 CUDA Streams and Input Data Segmentation

The SOM can be updated in two ways; sequential (or incremental) and batch training. In case of sequential training, the SOM is updated just after every input vector has been processed. Such training is more time-consuming in comparison with batch training where the SOM is updated after a set of input vectors (batch) has been processed and the final weight vector has been computed. The sequential training has been chosen to better demonstrate advantages of parallel computation on GPU.

A stream is a sequence of commands that execute in order. Different streams, on the other hand, may execute their commands out of order with respect to one another or concurrently. Following points can be included into advantages of CUDA streams:

- Every stream initializes the same SOM with respect to the first setting of weight vectors, epochs, etc.
- Required memory is allocated for every stream which reduces memory conflicts.
- Streams are independent on each other.

Nevertheless, the utilization of CUDA streams leads to additional computation after all input vectors have been processed. All partial results of each streams (output weight vectors) should be merged together which usually invoke another computation of SOM, where input vectors are represented by all output weight vectors from previous streams. Moreover, there are other limits:

- If kernel functions are designed to process single input vector, loops over all inputs and epochs must be implemented and executed on the CPU. Another loop runs over all streams. Finally, additional loop and superimposed calling of kernel functions prolong computations time in case of small dimension of input vectors.
- The number of streams expand memory requirements.
- There must be some synchronization points between CUDA streams and CPU in the code.

The experiments show, that the utilization of CUDA streams can save the computation time in case of large dimension of input vectors. The disadvantage associated with inner CPU loops can be partially solved by OpenMP [12] which enables to divide CPU code into different threads to ensure better performance with CUDA streams [6].

2.3 Parallel Reduction

A correct design of CUDA kernel function is necessary to get good performance during runtime. Kernels functions, when called, are executed N times in parallel by N different CUDA threads, as opposed to only once like regular functions. Thus it is necessary to design a kernel function so that every thread access different memory block. We refer to [4] for more information on CUDA memory management.

A kernel is defined using the *__global__* declaration specifier and the number of CUDA threads for each call is specified using a new <<< ... >>> syntax. The brackets <<< and >>> hold grid size, block size, amount of shared memory and stream index. The setting of these parameters affect the distribution of threads over GPU. We refer to [10] for more information.

The principle of parallel reduction (PR) brings a significant time savings. A modified version of PR was used in a phase of computation of vector differences and Euclidean distances. The Fig. 3 shows an illustrative example of settings of grid, blocks and threads which were applied on SOM, where the number of neurons N = 9 and vector dimension m = 8. The GPU grid contains 3 blocks. Every block has 32 threads, therefore it can cover at most four weights vectors. The setting of kernel function itself play an important role in CUDA. It depends on GPU hardware parameters as well. The parallel reduction is shown in the middle part of the Fig. 3 and has following steps:

(A) The differences between all elements of a given input vector and an appropriate weight vector are computed and stored in the shared memory on GPU. The elements can be powered by 2 in case of Euclidean distance.
(B) A half of previously used threads computes partial sums of elements of difference vector.
(C) Last thread computes square root of the final sum and stores the result output into distance vector, which is stored in global GPU memory.

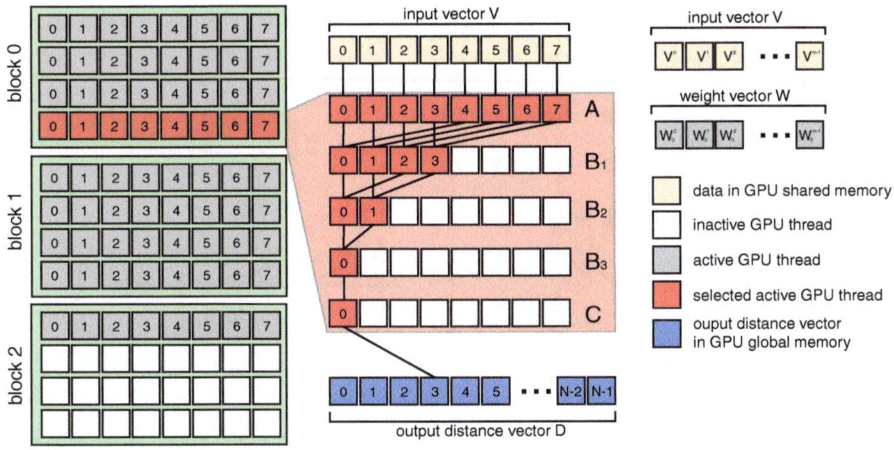

Fig. 3 Processing of a single input vector by parallel GPU threads

When all blocks finish, the process of parallel reduction is used to find a minimum value in the distance vector. An index of the minimal element represents Best Matching Unit (BMU).

2.4 Parallel SOM Update

Some fix data is precomputed in our application. Such data is stored in the global memory and then partially copied into shared memory by all thread blocks to reduce computational time [4] during SOM update. First, it is a distance matrix which holds the distances among neurons in a SOM. In the Fig. 4 on the left, there is an example of the SOM with rectangular topology of dimension 7×7. The maximum distance in this topology is $maxDist = (dimX - 1) + (dimY - 1) = (7-1) + (7-1) = 12$. Just for illustration, the black node in the middle represents BMU and all nodes with the same gray scale have the same distance to the BMU with respect to the rectangular topology.

A vector of learning factors F makes the second fix data. In the illustrative Fig. 4 on the right, there are curves of learning functions for all epochs ($e = 4$). The vector F has a dimension equal to $(maxDist + 1) * e = (12 + 1) * 4 = 48$ in our example and represents a set of values $f_i(d)$, where $f_i(d)$ is the evaluated learning function (learning factors) for a given epoch i, and a distance d, where $d \in < 0, maxDist >$ and $i \in < 0, e >$.

The process of updating of SOM runs in a few steps:

1. The GPU kernel function, that is responsible for updating, is set so that one neuron is processed by m threads, where m is a dimension of weight vector.
2. Fixed distance vector and learning factors for a given epoch are copied into GPU shared memory to ensure better performance.

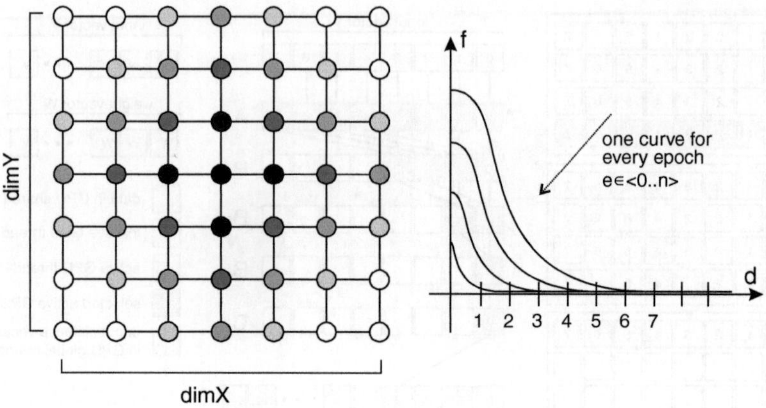

Fig. 4 Neuron distances in the SOM for a given BMU in the center of the SOM (on the *left*) and curves of learning functions for all epochs

3. Every thread updates a part of its weight vector with respect to the learning factor and the distance to BMU.

2.5 Kernel Example

Although there is no place to show the whole code of GPU implementation of SOM because it contains hundreds of lines, following part of the code illustrates the philosophy of parallel programming with CUDA. The kernel function *updateCompleteSOM* is called from the main program when BMU is already known. A basic description of all used variables and keywords is in the Table 2.

```
__global__ void updateCompleteSOM(
    float* weightVectors,
    const unsigned int wSize,
    const unsigned int* distanceMatrix,
    const unsigned int* BMUindex,
    const float *distances,
    const float *learingFactors)
{
    unsigned int tid = threadIdx.x;
    __shared__ unsigned int sBMUindex;

    if (tid == 0) sBMUindex = *bmuIndex;
    __syncthreads();

    unsigned int wid = blockDim.x * blockIdx.x + tid;
    if (wid >= wSize) return;
    unsigned int nid = (wid / vectorDim)
          * dimX * dimY + sBMUindex;
    weightVectors[wid] +=
                          learingFactors[distanceMatrix[nid]]
                          * distances[wid];
}
```

Table 2 The description of the method called updateCompleteSOM

Row	Keyword/Variable	Description
1	__global__	CUDA keyword to define kernel function
2	weightVectors	Pointer to the array of all weight vectors
3	wSize	The size (number of elements) of weightVectors weightVectors = dimX ∗ dimY ∗ vectorDimension
4	distanceMatrix	The distance matrix with respect to the SOM topology (see the Sect. 2.4)
5	BMUindex	Index of the Best Matching Unit
6	distances	Vector of distances of all neurons to the input vector (e.g. Euclidean distances)
7	learningFactors	Vector of learning factors for a given epoch (see the Sect. 2.4)
9	tid	Index of the CUDA thread (threadIdx is a variable of CUDA framework [10])
10	__shared__	CUDA keyword which says that a variable will be stored in the shared memory for all threads of a given thread block
12	sBMUindex	Just the first thread of block copies the BMUindex into the shared memory
13	__sybcthreads ()	CUDA function which synchronizes all threads in the given block
15	wid	Index of the weight vector related to the CUDA thread
17	nid	Index of the neuron related to the CUDA thread
18-19	weightVectors [wid]	Every thread updates one element of the weightVectors (parallelism)

3 Experiments

All experiments include comparison between GPU and CPU implementation of SOM. The Table 3 includes detailed specifications named GPU, CPU-A and CPU-B witch respect different hardware configuration used for experiments.

A SOM with a fix network size is computed in the first experiment. Dimensions of input vectors are fix as well and they are set to four in this case. This experiment reveals hidden disadvantages of GPU utilization, as can be seen in the Fig. 5 or in the corresponding table. The computation time increases near linear, however, GPU time is approximately 2 times greater in comparison with CPU-A multi-threaded implementation. This is due to additional time costs associated with transactions between RAM and GPU memory. Next, calling *kernels* (GPU functions) and inner thread indexing take some time. These costs make GPU implementation not suitable for SOM in case of small data.

In the Fig. 6, there are results of the second experiment which deals with increasing dimension of input and weight vectors. Again, GPU implementation seems to be not

Table 3 Hardware specification

	GPU	CPU-A / CPU-B
CPU	Intel Core 2 Duo 2,2 GHz	4 × AMD Opteron 1,8 GHz
RAM	4 GB	32 GB
GPU	GeForce 280 GTX, 1 GB	-
Threads	Depends of GPU	2 (CPU-A), 8 (CPU-B)

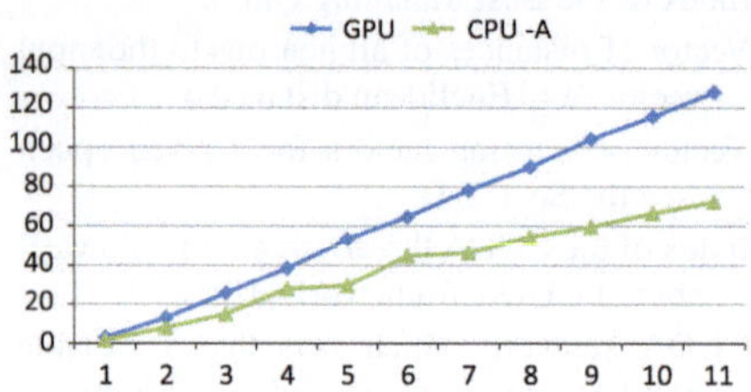

ID	number of inputs	vector dim.	SOM network		time in seconds	
			dimX	dimY	GPU	CPU-A
1	1000	4	10	10	3,11	1,64
2	5000	4	10	10	13,11	8,05
3	10000	4	10	10	25,68	14,98
4	15000	4	10	10	38,49	27,80
5	20000	4	10	10	53,23	29,43
6	25000	4	10	10	64,56	44,91
7	30000	4	10	10	78,02	46,41
8	35000	4	10	10	89,95	54,44
9	40000	4	10	10	104,11	59,46
10	45000	4	10	10	115,47	66,38
11	50000	4	10	10	127,90	72,36

Fig. 5 Variable number of inputs

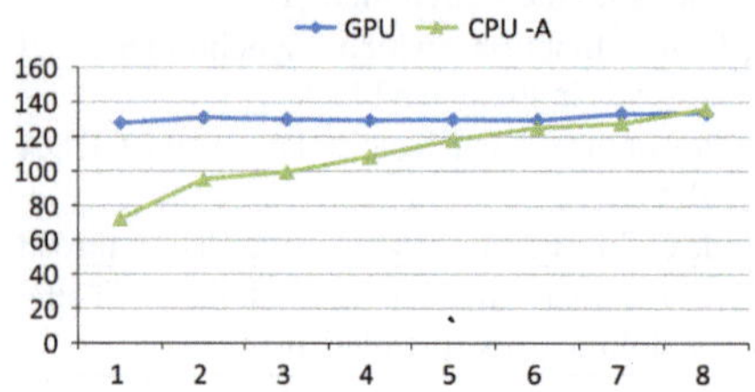

ID	number of inputs	vector dim.	SOM network		time in seconds	
			dimX	dimY	GPU	CPU-A
1	50000	4	10	10	127,90	72,36
2	50000	8	10	10	131,06	95,37
3	50000	12	10	10	130,03	99,51
4	50000	16	10	10	129,44	108,37
5	50000	20	10	10	129,79	118,06
6	50000	24	10	10	129,62	125,15
7	50000	28	10	10	133,25	127,58
8	50000	32	10	10	133,53	136,18

Fig. 6 Variable dimension of input vectors

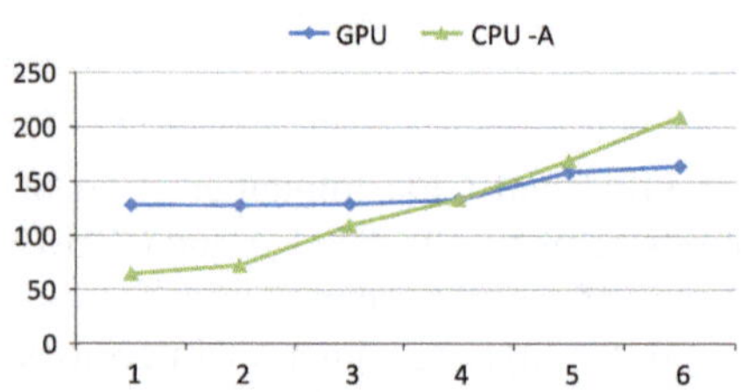

ID	number of inputs	vector dim.	SOM network		time in seconds	
			dimX	dimY	GPU	CPU-A
1	50000	4	5	5	128,21	64,66
2	50000	4	10	10	127,90	72,36
3	50000	4	15	15	129,11	109,10
4	50000	4	20	20	133,20	133,11
5	50000	4	25	25	158,70	169,38
6	50000	4	30	30	164,21	209,22

Fig. 7 Variable dimension of SOM network

effective until dimension 32 because of the same reasons as in the first experiment. However, the trends of graphs predict future computation times and it is clear, that GPU time increases very slowly in comparison with CPU-A.

The third experiment and its results in the Fig. 7 show the power of parallelism on neurons (we refer to the Sect. 2.4), whereas the previous two experiments illustrate parallelism on particular vectors (we refer to the Sect. 2.3). The contribution of GPU implementation is perceptible in higher dimension of SOM network.

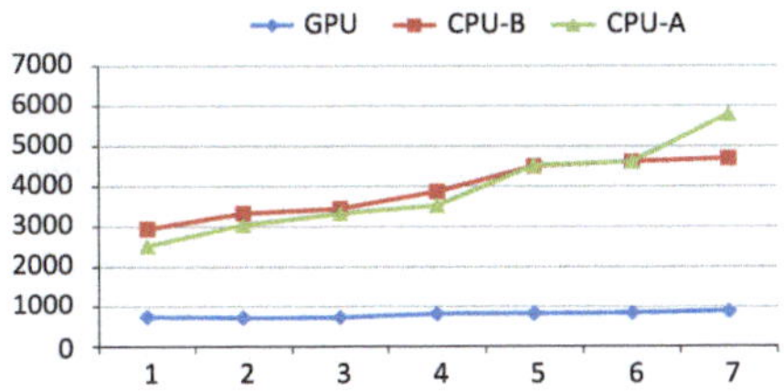

ID	number of inputs	vector dim.	SOM network		time in seconds		
			dimX	dimY	GPU	CPU-A	CPU-B
1	250000	40	10	10	741,41	2518,67	2943,67
2	250000	50	10	10	716,44	3038,69	3329,97
3	250000	60	10	10	722,97	3325,90	3447,45
4	250000	70	10	10	815,01	3525,53	3871,43
5	250000	80	10	10	819,56	4520,88	4503,06
6	250000	90	10	10	841,52	4617,80	4614,17
7	250000	100	10	10	885,38	5793,50	4689,08

Fig. 8 Variable dimension of input vectors

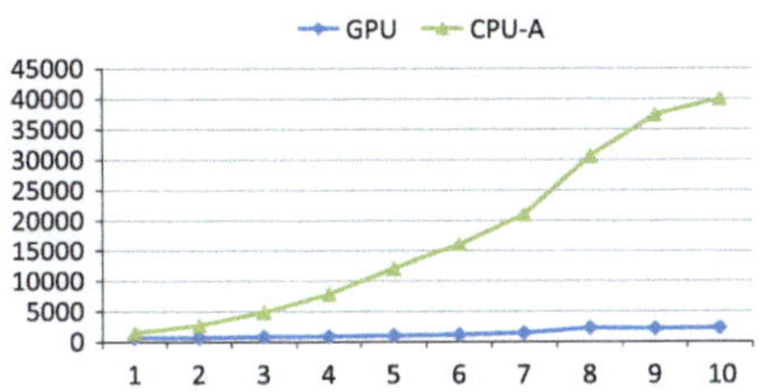

ID	number of inputs	vector dim.	SOM network		time in seconds	
			dimX	dimY	GPU	CPU-A
1	250000	4	10	10	692,37	1513,03
2	250000	4	20	20	698,44	2724,38
3	250000	4	30	30	818,93	4846,69
4	250000	4	40	40	905,82	7858,48
5	250000	4	50	50	1042,55	12048,28
6	250000	4	60	60	1176,00	16104,82
7	250000	4	70	70	1459,97	20990,82
8	250000	4	80	80	2268,49	30564,54
9	250000	4	90	90	2199,91	37387,30
10	250000	4	100	100	2307,67	39984,57

Fig. 9 Variable dimension of SOM network

Previous demonstrations confirm, that the power of GPU utilization increases with the size of neural network and with the dimensions of input and weights vectors, respectively. The last two experiments (see the Figs. 8 and 9) show the computation times in such cases.

4 Conclusion

Evergrowing bulk data and demand for their processing force software developers to look for new solutions. The trends in the areas of Information Retrieval, Data mining, Stream processing, etc. show, that utilization of modern hardware will play more remarkable role in the future. Hardware producers reflect such demand for their products. This article illustrated power of GPU in connection with self-organized network. Experiments confirmed, that the contribution of GPU is significant and can reduce computation time in case of large data sets.

Acknowledgments This work was supported by the Ministry of Industry and Trade of the Czech Republic, under the grant no. FR-TI1/420.

References

1. Andrecut, M.: Parallel GPU implementation of iterative PCA algorithms 2008. J. Comput. Biol. **16**(11), 1593–1599 (2009)
2. Flexer A.: On the use of self-organizing maps for clustering and visualization. In: Principles of Data Mining and Knowledge Discovery 1999. pp. 80–88
3. Hager, G., Zeiser, T., Wellein, G.: Data access optimizations for highly threaded multi-core CPUs with multiple memory controllers. In: Workshop on Large-Scale Parallel Processing 2008 (IPDPS2008), Miami, 18 April 2008
4. Kirk, D.B. Hwu, W.W.: Programming Massively Parallel Processors A Hands-on Approach 2010. ISBN: 978-0-12-381472-2
5. Kohonen, T.: Self-Organizing Maps. Springer, Berlin (1995)
6. Lee, S., Min, S-j., Eigenmann, R.: OpenMP to GPGPU: a compiler framework for automatic translation and optimization. In: Proceedings of the 14th ACM SIGPLAN Symposium on Principles and Practice of Parallel Programming, Raleigh, 2009. pp. 101–110
7. Mann, R., Haykin, S.: A parallel implementation of Kohonen's feature maps on the warp systolic computer. In: Proceedings of the International Joint Conference on Neural Networks (IJCNN-90), Washington, DC 1990. Vol. II, pp. 84–87
8. Myklebust, G., Solheim, J.G., Steen, E.: Wavefront implementation of self-organizing maps on RENNS. In: International Conference on Digital Signal Processing 1995. pp. 268–273
9. Nordström, T.: Designing parallel computers for self organizing maps. Forth Swedish Workshop on Computer System Architecture, Linköping (1992)
10. nVIDIA®: Cuda Programming Guide. http://developer.download.nvidia.com/compute/cuda/2_3/toolkit/docs/NVIDIA_CUDA_BestPracticesGuide_2.3.pdf (2010). Jan 2010
11. nVIDIA®: Cuda Zone. http://www.nvidia.com/object/cuda_home_new.html (2010). Jan 2010
12. OpenMP: API specification for parallel programming, http://openmp.org/wp/, (2010). Jan 2010
13. Openshaw, S.: Turton I.:A parallel Kohonen algorithm for the classification of large spatial datasets. Comput. Geosci. **22**, 1019–1026 (1996)
14. Patnaik, D., Ponce, S.P., Cao, Y., Ramakrishnan, N.: Accelerator-oriented algorithm transformation for temporal data mining. CoRR, abs/0905.2203, (2009)
15. Preis, T., Virnau, P., Paul, W., Schneider, J.J.: Accelerated fluctuation analysis by graphic cards and complex pattern formation in financial markets. New J. Phys. **11**(9), 093024 (21pp) (2009)
16. Valova, I., MacLean, D., Beaton, D.: Identification of patterns via region-growing parallel SOM neural network. In: Seventh International Conference on Machine Learning and Applications, ICMLA 2008. pp. 853–858
17. Valova, I., Szer, D., Gueorguieva, N., Buer, A.: A parallel growing architecture for self-organizing maps with unsupervised learning. Neurocomputing **68**, 177–195 (2005)
18. Weigang, L.: A study of parallel self-organizing map. In: Proceedings of the International Joint Conference on Neural Networks, Washington, DC 1999
19. Wu, C.H., Hodges, R.E., Wang, C.J.: Parallelizing the self-organization feature map on multi-processor systems. Parallel Comput. **17**(6–7), 821–832 (1991)

Boolean Factor Analysis by Expectation-Maximization Method

Alexander A. Frolov, Dušan Húsek and Pavel Yu. Polyakov

Abstract Boolean factor analysis is one of the most efficient methods to reveal and to overcome informational redundancy of high-dimensional binary signals. In the present study, we introduce new Expectation-Maximization method which maximizes the likelihood of Boolean factor analysis solution. Using the so-called bars problem benchmark, we compare efficiencies of the proposed method with Dendritic Inhibition neural network.

1 Introduction

Factor analysis is one of the most powerful statistical methods to reveal and reduce information redundancy in high dimensional signals. Boolean Factor Analysis (BFA) as a special case of factor analysis implies that components of original signals, factor loadings and factor scores are binary values. Each binary signal can be interpreted as a representation of presence or absence of attributes (binary signal components) in the pattern. The number of considered attributes is the dimensionality of the signal space, the appearance of an attribute is encoded as One, and absence as Zero. The patterns are assumed to be composed of many "objects" in different combinations.

A. A. Frolov (✉)
Institute of Higher Nervous Activity and Neurophysiology,
Russian Academy of Sciences,, Moscow, , Russia
e-mail: aafrolov@mail.ru

D. Húsek
Institute of Computer Science, Academy of Sciences of the Czech Republic,
Prague, Czech Republic
e-mail: pavel.mipt@mail.ru

P. Yu. Polyakov
VSB Technical University of Ostrava, Ostrava, Czech Republic
e-mail: dusan@cs.cas.cz

M. Kudělka et al. (eds.), *Proceedings of the Third International Conference on Intelligent Human Computer Interaction (IHCI 2011), Prague, Czech Republic, August, 2011*, Advances in Intelligent Systems and Computing 179, DOI: 10.1007/978-3-642-31603-6_21,

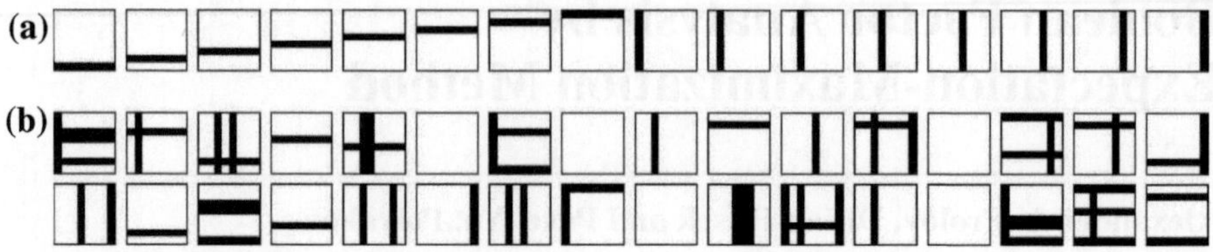

Fig. 1 **a** Sixteen vertical and horizontal bars in 8-by-8 pixel images. **b** Examples of images in the standard bars problem. Each image contains two bars on average

We define an object as a collection of highly correlated attributes and suppose that objects are relatively independent of one another. Hence the attributes of different objects are only slightly correlated. In terms of BFA, objects are factors, attributes constituting the object are factor loadings, and the presence or absence of an object in the pattern is identified by the value of the factor score (One or Zero). Correlation between the attributes constituting each factor can be revealed by statistics over large data set constituted by patterns that contain each factor many times in different combinations with other factors. The aim of BFA is to detect this hidden structure of the signal space and to form a representation in which these independent objects are presented explicitly.

The well-known benchmark for learning of objects from complex patterns is the Bars Problem (BP) [2]. In this problem, each pattern of the data set is an n-by-n binary pixel image containing several of $L = 2n$ possible (one-pixel wide) horizontal and vertical bars (Fig. 1). Pixels belonging and not belonging to the bar take values 1 and 0, respectively. For each image each bar could be chosen with a probability C/L, where C is the mean number of bars mixed in an image. In the point of intersection of vertical and horizontal bars, pixel takes the value 1. The Boolean summation of pixels belonging to different bars simulates the occlusion of objects. The task is to recognize all bars as individual objects on the basis of a data set containing M images consisting of bar mixtures. In most papers where the BP was used as benchmark, C was set to 2 and $n \leq 8$.

In spite of the fact that binary data representation is typical for many fields, including social science, marketing, zoology, genetics and medicine, BFA methods are rather moderately developed. To overcome this lack we propose an Expectation-Maximization (EM) approach to BFA [5]. This method maximizes likelihood of BFA solution appropriate to general BFA generative model.

Using BP as the benchmark, we compare the performance of the proposed method with Dendritic Inhibition (DI) neural network [7] which supposed to be most efficient in solving the bars problem [7–9]. We used information gain provided by BFA as a measure of performance of BFA methods. It was shown previously [3] that this measure is sensitive to both noise in signals and errors in BFA results, and thus is a reliable basis for comparison of different BFA methods.

1.1 Boolean Factor Analysis Generative Model

In terms of BFA, each pattern of a signal space is defined by a binary row vector $\mathbf{x}$ with dimensionality N equal to the total number of attributes. Every component of $\mathbf{x}$ takes value One or Zero, depending on the presence or absence of the related attribute. Each factor $\mathbf{f}_i$ is a binary row vector of dimensionality N whose One valued entries correspond to highly correlated attributes of the ith object. Although the probability of the object's attribute to appear in a pattern simultaneously with its other attributes is high, it is not obligatory equal to 1. We denote this probability as p_{ij}, where j is the index of an attribute and i is the index of a factor. For attributes constituting the factor, that is for attributes with $f_{ij} = 1$, probability p_{ij} is high, and for other attributes (with $f_{ij} = 0$) it is zero.

As in linear factor analysis, we suppose that additionally to common factors $\mathbf{f}_i$ that influence more than one attribute, each signal also contains some specific or unique factors that influence only particular attributes. The contribution of specific factors is defined by a binary row vector $\boldsymbol{\eta}$, which we call "specific noise". Each specific factor η_j is characterized by a probability q_j that jth component of vector η takes One.

As a result, any vector $\mathbf{x}$ can be presented in the form

$$\mathbf{x} = \left[\bigvee_{i=1}^{L} s_i \wedge \mathbf{f}'_i \right] \vee \boldsymbol{\eta}, \tag{1}$$

where $\mathbf{s}$ is a binary row vector of factor scores of dimensionality L, L is the total number of factors, $\mathbf{f}'_i$ is a distorted version of factor $\mathbf{f}_i$ and η is a specific noise defining the influence of specific factors. Factor distortion implies that some Ones of the ith factor become Zeros with probability $1 - p_{ij}$, that is components constituting the factor can take Zeros but neither component not constituting the factor (for components not constituting the factor $p_{ij} = 0$) can take One in distorted version of factor. Nevertheless, it can be One in the input signal due to the presence of other factors or specific noise. We assume that factors appear in patterns (that is related scores s_i take Ones) independently with probabilities π_i $(i = 1, \ldots, L)$.

The aim of Boolean Factor Analysis is to find the parameters of generative model $\boldsymbol{\Theta} = (p_{ij}, q_j, \pi_i, i = 1, \ldots, L, j = 1, \ldots, N)$ and the factor scores $s_{mi}, m = 1, \ldots, M$ for each of patterns of the observed data set $\mathcal{X}$ containing M signals. However, it is supposed that the found factors could also be detected in any arbitrary pattern $\mathbf{x} \notin \mathcal{X}$, if generated by the same model.

We suppose that each component of the common factor is distorted independently of the presence of other factors in the input signal and independently of specific noise.

Thus, given generative parameters $\boldsymbol{\Theta}$ and factor scores $\mathbf{s}$, the probability of the jth component of $\mathbf{x}$ to take the value x_j is

$$P(x_j|\boldsymbol{\Theta},\mathbf{s}) = x_j - (2x_j - 1)(1 - q_j)\prod_{i=1}^{L}(1 - p_{ij})^{s_i}, \tag{2}$$

We suppose also that, given generative parameters and factor scores, different components of $\mathbf{x}$ are also statistically independent. Thus,

$$P(\mathbf{x}|\boldsymbol{\Theta},\mathbf{s}) = \prod_{j=1}^{N} P(x_j|\boldsymbol{\Theta},\mathbf{s}). \tag{3}$$

Since we assume that factors appear in patterns independently then

$$P(\mathbf{s}|\boldsymbol{\Theta}) = \prod_{i=1,L} \pi_i^{s_i}(1 - \pi_i)^{1-s_i}. \tag{4}$$

2 Expectation-Maximization Method: EMBFA

The EM method [1] allows for finding parameters $\boldsymbol{\Theta}$ of a given probabilistic generative model to maximize the likelihood of the observed dataset $\mathcal{X}$. Since it is applied here to the BFA generative model we call it EMBFA. The EM method maximizes the likelihood of the observed data by maximizing the free energy

$$\mathcal{F}(\boldsymbol{\Theta},\mathbf{g}) = \sum_{m=1}^{M}\sum_{\mathbf{s}} g_m(\mathbf{s})[\log P(\mathbf{x}_m|\mathbf{s},\boldsymbol{\Theta}) + \log P(\mathbf{s}|\boldsymbol{\Theta})] + \sum_{m=1}^{M} H(g_m(\mathbf{s})), \tag{5}$$

where $g_m(\mathbf{s})$ is the expected distribution of factor scores for the mth pattern, $H()$ denotes Shannon entropy. The iterations of EM alternatively increase $\mathcal{F}$ with respect to the distributions g_m, while holding $\boldsymbol{\Theta}$ fixed (the E-step), and with respect to parameters of the model $\boldsymbol{\Theta}$, while holding g_m fixed (the M-step).

At the E-step, when $\boldsymbol{\Theta}$ is fixed, the distributions g_m which maximize $\mathcal{F}(\boldsymbol{\Theta},\mathbf{g})$ are calculated according to the following equation

$$g_m(\mathbf{s}|\boldsymbol{\Theta}) = \frac{P(\mathbf{s}|\boldsymbol{\Theta})P(\mathbf{x}_m|\mathbf{s},\boldsymbol{\Theta})}{\sum_{\mathbf{s}} P(\mathbf{s}|\boldsymbol{\Theta})P(\mathbf{x}_m|\mathbf{s},\boldsymbol{\Theta})}, \tag{6}$$

where $P(\mathbf{x}_m|\mathbf{s}, \mathbf{\Theta})$ is given by (3) and $P(\mathbf{s}|\mathbf{\Theta})$ is given by (4). The obtained distributions g_m provide the expected likelihood of the observed data over factor scores given parameters of the generative model [6].

At the M-step, when distributions g_m are fixed, π_i can be estimated as $(1/M) \sum_{m=1}^{M} s_{mi}$, where $s_{mi} = \sum_{\mathbf{s}} g_m(\mathbf{s}|\Theta)s_i$. Respectively, p_{ij} and q_j can be found by steepest ascent maximization of $\mathcal{F}(\mathbf{\Theta}, \mathbf{g})$:

$$\Delta p_{ij} = \gamma \frac{\partial \mathcal{F}}{\partial p_{ij}}, \quad \Delta q_j = \gamma \frac{\partial \mathcal{F}}{\partial q_j}, \tag{7}$$

where γ is a learning rate.

Since we assume that probabilities p_{ij} are sufficiently high for components constituting the ith factor ($f_{ij} = 1$) and equal to zero for other components ($f_{ij} = 0$), at each iteration step we put $p_{ij} = 0$ if the jth component satisfies inequality

$$p_{ij} < 1 - \prod_{l \neq i}(1 - \pi_l p_{lj}), \tag{8}$$

where the right side of the inequality is the probability that the jth attribute appears in the pattern due to other factors except $\mathbf{f}_i$. It is interesting to note that without this threshold truncation of p_{ij}, EM does not converge because of uncertainties arising from the competition between factors defined by p_{ij} and noise defined by q_j.

The iterative procedure (7) at each step M continues until $\sum_{ij} |\Delta p_{ij}|$ become smaller than $\epsilon_2 = 1$. The obtained values of p_{ij}, q_j and π_i are used as the input for the next E-step. EM iterative procedure terminates once all values $\sqrt{\sum_j (p_{ij}^{old} - p_{ij}^{new})^2 / \sum_j p_{ij}^{old}}$ $(i = 1 \dots L)$ remained smaller than $0.25 \cdot 10^{-3}$.

When procedure converges, the final values s_{mi} are estimates of the factor scores. To satisfy the generative model, we binarized those values. The binarization threshold was chosen to maximize BFA information gain [3].

We restricted the EMBFA algorithm to the case of sparse scores, when only a small number of factors (no more than 3) are supposed to be mixed in the observed patterns. In this case, summation over $\mathbf{s}$ in the above formulas is reduced to

$$\sum_{\mathbf{s}}(\dots) = (\dots)_{\mathbf{s}=0} + \sum_{i}(\dots)_{\mathbf{s}=\mathbf{s}_i} + \sum_{i<j}(\dots)_{\mathbf{s}=\mathbf{s}_{ij}} + \sum_{i<j<k}(\dots)_{\mathbf{s}=\mathbf{s}_{ijk}}, \tag{9}$$

where $\mathbf{s}_i$ is the vector of factor scores with all zeros except s_i, $\mathbf{s}_{ij}$ is the vector of factor scores with all zeros except s_i and s_j, and $\mathbf{s}_{ijk}$ is the vector of factor scores with all zeros except s_i, s_j and s_k. An increase of the number of terms in (9) leads to a considerable rise in computational complexity.

The algorithm is given in Appendix. Note the parameter γ in the algorithm is not fixed. It is equal to $p_{ij}(1 - p_{ij})$ while computing Δp_{ij} and to $q_j(1 - q_j)$ while computing Δq_j.

3 Performance of BFA Methods in Solving Bars Problem

In this section, we compare efficiencies of EMBFA and DI. DI was developed in [8] for finding parts-based decompositions of images. The method is based on a feed-forward neural network with lateral inhibition. The main idea of the method is to use lateral inhibition of individual synapses instead of total inhibition of a neuron. As a result of network learning, neurons of the output layer acquire specific sensitivity to factors constituting patterns of the data set: the appearance of a factor in the pattern presented to the input layer leads to the strong activation of the related neuron at the output layer. Output neurons can be activated when factors are partially distorted and in the presence of noise. Thus, activity of each neuron at the output layer provides a gradual estimation of the confidence that this pattern contains the related factor. To transform gradual activity of output layer to binary vector of factor scores we used the same binarization procedure as for EMBFA. Namely, we used the binarization threshold maximizing BFA information gain.

The performance of the methods are compared using information gain G [4]. The calculation of G for given dataset $\mathcal{X}$ and set of factor scores S can be performed using special procedure [3] which also derives the parameters of generative model $\boldsymbol{\Theta}$.

For EMBFA, as well as for DI, the number of desired hidden factors has to be set in advance. In the majority of computer experiments performed by Spratling [7], this number was taken twice higher than the actual number of factors. In our experiments with EMBFA and DI, the predefined number of hidden factors was also taken twice higher than the actual number of factors.

To start the EM procedure, we set $\pi_i = 1/32$, $q_j = 0$ and initialized p_{ij} with random values uniformly distributed in the range from 0.3 to 0.8.

Figure 2 demonstrates sensitivity of BFA methods to noise. The noise was assumed to be distributed uniformly over signal components and factors so that $q_j = q$ for any j, and $p_{ij} = p$ for any i and j. We tested BFA methods using data sets of the size $M = 800$.

In Fig. 2 "ideal" solution corresponds to the case when all scores are exactly the same as those used in the generated data set. As expected, information gain decreases when noise increases even for "ideal" solution. And for very noisy data G tends to zero that means BFA has no sense.

As shown in Fig. 2a the performance of EMBFA is better for large q (regarding "ideal" solution) and slightly worse for small q. In contrast, DI demonstrates strong sensitivity to noise: its performance is almost ideal for $q = 0$ but it is getting worse for noisy data. EMBFA and DI exhibits similar sensitivity to factor distortion as to specific noise (Fig. 2b).

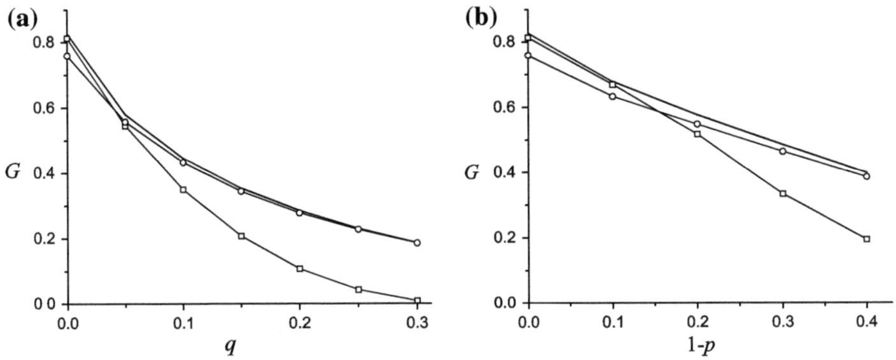

Fig. 2 Information gain G in dependence on q for $p = 1$ (**a**) and in dependence on p for $q = 0$ (**b**). ○—EMBFA, □—DI. *Thick line*—"ideal" solution

To investigate sensitivity of BFA methods to the mean number of bars mixed in an image, which we call as complexity C, we increased the size of images to the grid of 16-by-16 pixels. Each of 32 bars is chosen for each image independently with probability $C/32$. The increased image size allowed us to study the effects of increasing C up to $C = 10$. Here, we investigated only noiseless case, that is we put $p = 1, q = 0$.

Since EMBFA is restricted to the case of sparse scores ($C \leq 3$ according to (9)), one would expect that this method is most sensitive to the increase of C. However, it amazingly gives reasonable results until $C = 8$ (Fig. 3), although with less efficiency. The reason is the peculiarity of EMBFA to treat some redundant bars as noise when the number of bars mixed in the image exceeds three.

When the number of active neurons at the input of the DI network becomes relatively large due to large C, DI fails. This disadvantage can be overcome by repeating factors search with other initial synaptic weights. We randomly changed synaptic weights until more than a half of bars were found, but made maximally 10 attempts. As shown in Fig. 3, this modification essentially improved DI performance.

It seems that the presence of the images with few mixed factors in the data set facilitated factors search by DI. To check this hypothesis, we studied DI performance for the case when the number of mixed bars in each image was exactly C (i.e there were no images containing less than exactly C bars). As shown in Fig. 3, this actually worsened DI performance for $C > 4$.

4 Conclusion

The proposed EMBFA method is proved to be suitable for Boolean Factor Analysis. In comparison with DI neural network [7], it demonstrates slightly worse performance when noise is absent. But in contrast to DI, it exhibits a low sensitivity to noise (both factor distortion and specific noise).

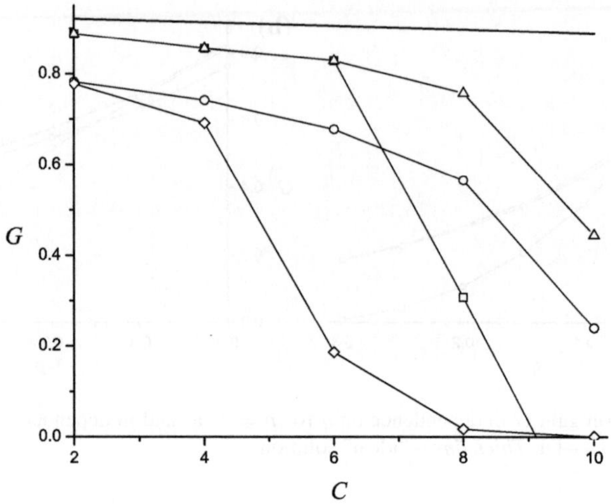

Fig. 3 Information gain G in dependence on mean number of bars mixed in an image C for 16-by-16 pixel images at $M = 800$. ○—EMBFA, □—DI, ◇—DI for fixed number of mixed bars in patterns, △—DI with repeated initializations of factor search

Both methods are shown to be sensitive to number of factors mixed in a pattern (task complexity C). It is expectable for EMBFA since according to (9) this method is restricted to the case of sparse scores ($C \leq 3$). However, it gave reasonable results even when $C = 6$ (Fig. 3) and demonstrates less sensitivity to the rise of C in comparison with DI. When C increases, DI becomes unstable in the sense that its operation strongly depends on the realization of initial synaptic weights of the basic network. For a set of initial weights DI may converge to true factors, for other initial weight realizations it converges to a random solution. It is especially evident for large C (see Fig. 3). Another peculiarity of DI is its sensitivity to hints in data set, that is to patterns with only a few mixed factors, which seems to facilitate significantly the search for true factors. So, DI fails when the number of mixed factors in each pattern is large and fixed (Fig. 3). Another peculiarity of DI is the sensitivity of factors search performed by DI to the presence of the images with few mixed factors in a pattern.

The peculiarity of EMBFA is the competition between two kinds of noise. When specific noise is large, EMBFA prescribes additional pixels to bars. When factor distortion is large, EMBFA deletes some pixels from bars, adjusting itself to the specific noise realization in the data set. Low sensitivity of EMBFA to both kinds of noise and its overall high efficiency is due to using powerful EM approach together with the proposed BFA generative model.

5 Appendix

Algorithm 1: Expectation-Maximization algorithm for Boolean Factor Analysis

input : $\mathcal{X}$ is a set of M binary signals $\mathbf{x}_m$ of dimension N with components x_{mj}, $m = 1 \ldots M$, $j = 1 \ldots N$
output: $\mathbf{\Theta} = (p_{ij}, q_j, \pi_i, i = 1, \ldots, L, j = 1, \ldots, N)$ are parameters of generative model, $\mathcal{S}$ is a set of factor scores with components s_{mi}, $m = 1 \ldots M$, $i = 1 \ldots L$

begin
    foreach $\pi_i \in \mathbf{\Theta}$ **do** $\pi_i \leftarrow 1/L$
    foreach $p_{ij} \in \mathbf{\Theta}$ **do** $p_{ij} \leftarrow$ random number form [0.3, 0.8]
    foreach $q_j \in \mathbf{\Theta}$ **do** $q_j \leftarrow 0$
    repeat
        for $i \leftarrow 1$ **to** L **do for** $j \leftarrow 1$ **to** N **do** $p_{ij}^{prev} \leftarrow p_{ij}$
        $[\mathbf{Q}, \mathbf{A}, \mathbf{B}, \mathbf{C}] \leftarrow$ *ComputerScoresExpectedDistribution*$(\mathcal{X}, \mathbf{\Theta})$
        for $m \leftarrow 1$ **to** M **do**
            for $i \leftarrow 1$ **to** L **do**
                $s_{mi} \leftarrow A_i^m + \sum_{j=1,\neq i}^{L} (B_{ij}^m + \sum_{k=1,\neq i,\neq j}^{L} C_{ijk}^m)$
            end
        end
        for $i \leftarrow 1$ **to** L **do** $\pi_i \leftarrow \sum_{m=1}^{M} s_{mi}/M$
        $\sigma \leftarrow \sum_{m=1}^{M} \left(Q^m + \sum_{i=1}^{N} [A_i^m + \sum_{j<i}^{L} (B_{ij}^m + \sum_{k<j}^{L} C_{ijk}^m)]\right)$
        repeat
            $\mathbf{\Theta}' \leftarrow \mathbf{\Theta}$
            $\mathbf{\Theta} \leftarrow$ *Estep*$(\mathcal{X}, \mathbf{\Theta}', \mathbf{Q}, \mathbf{A}, \mathbf{B}, \mathbf{C}, \sigma)$
            foreach $p_{ij} \in \mathbf{\Theta}$ **do**
                if $p_{ij} < 1 - \exp\left(-\frac{1}{M} \sum_{l \neq i} (p_{lj} \sum_{m=1}^{M} s_{ml})\right)$ **then** $p_{ij} \leftarrow 0$
            end
        until $\sum_{i=1}^{L} \sum_{j=1}^{N} |p'_{ij} - p_{ij}| < 1$
        foreach $\pi_i \in \mathbf{\Theta} : \pi_i = 0$ **do**
            for $j \leftarrow 1$ **to** N **do** $p_{ij} \leftarrow p_{Lj}$
            for $m \leftarrow 1$ **to** M **do** $s_{mi} \leftarrow s_{mL}$
            $\pi_i \leftarrow \pi_L$
            $L \leftarrow L - 1$
            if $L = 0$ **then** break algorithm
        end
        for $i \leftarrow 1$ **to** L **do** $d_i \leftarrow \sqrt{\sum_{j=1}^{N} (p_{ij}^{prev} - p_{ij})^2 / \sum_{j=1}^{N} p_{ij}}$
    until $\max_{i=1 \ldots L} d_i < 0.25 \cdot 10^{-3}$
end

Procedure ComputeScoresExpectedDistributions($\mathcal{X}$, Θ)

input : $\mathcal{X}$ is a set of M binary signals with components $x_{mj}, m = 1 \dots M, j = 1 \dots N$,
$\mathbf{\Theta} = (p_{ij}, q_j, \pi_i, i = 1, \dots, L, j = 1, \dots, N)$ are parameters of generative model
output: **Q**, **A**, **B**, **C** are vectors of dimension M with components of dimension $\mathbb{R}^0, \mathbb{R}^1, \mathbb{R}^2, \mathbb{R}^3$, correspondingly: $Q^m = g_m(\mathbf{s}|\mathbf{\Theta})_{\mathbf{s}=0}$, $A_i^m = g_m(\mathbf{s}|\mathbf{\Theta})_{\mathbf{s}=\mathbf{s}_i}$, $B_{ij}^m = g_m(\mathbf{s}|\mathbf{\Theta})_{\mathbf{s}=\mathbf{s}_{ij}}$, $C_{ijk}^m = g_m(\mathbf{s}|\mathbf{\Theta})_{\mathbf{s}=\mathbf{s}_{ijk}}$

```
begin
  for m ← 1 to M do
    Q^m ← ∏_{n=1}^{N} (x_mn − (2x_mn − 1)(1 − q_n))
    for i ← 1 to L do
      A_i^m ← π_i/(1−π_i) ∏_{n=1}^{N} (x_mn − (2x_mn − 1)(1 − q_n)(1 − p_in))
      foreach j < i do
        B_ij^m ← π_i π_j/((1−π_i)(1−π_j)) ∏_{n=1}^{N} (x_mn−
                 (2x_mn−1)(1−q_n)(1−p_in)(1−p_jn))
        foreach k < j do
          C_ijk^m ← π_i π_j π_k/((1−π_i)(1−π_j)(1−π_k)) ∏_{n=1}^{N} (x_mn−
                    (2x_mn−1)(1−q_n)(1−p_in)(1−p_jn)(1−p_kn))
        end
      end
    end
    v^norm ← Q^m + Σ_{i=1}^{L} [A_i^m + Σ_{j<i} (B_ij^m + Σ_{k<j} C_ijk^m)]
    if v^norm ≠ 0 then
      Q^m ← Q^m / v^norm
      for i ← 1 to L do
        A_i^m ← A_i^m / v^norm
        foreach j < i do
          B_ij^m ← B_ij^m / v^norm
          foreach k < j do
            C_ijk^m ← C_ijk^m / v^norm
          end
        end
      end
    end
  end
end
```

Procedure Estep($\mathcal{X}$, $\mathbf{\Theta}'$, **Q**, **A**, **B**, **C**, σ)

input : $\mathcal{X}$ is a set of M binary signals, $\mathbf{\Theta}' = (p'_{ij}, q'_j, \pi'_i)$ are parameters of the generative model for previous iteration step, **Q**, **A**, **B**, **C** are vectors of expected distributions of factor scores, and σ is a normalization factor

output: optimized parameters of the generative model $\mathbf{\Theta} = (p_{ij}, q_j, \pi_i)$

```
begin
    foreach p_ij ∈ Θ do p_ij ← 0
    foreach q_j ∈ Θ do q_j ← 0
    for m ← 1 to M do
        foreach n = 1, ..., N : x_mn = 1 do
            q_n ← q_n + Q^m
            v_0 ← 1 − q'_n
            for i ← 1 to L do
                v_1 ← v_0(1 − p'_in)
                if 1 − v_1 = 0 then
                    q_n ← q_n + A_i^m,   p_in ← p_in + A_i^m
                end
                else
                    q_n ← q_n + q'_n A_i^m/(1−v_1)
                    p_in ← p_in + p'_in A_i^m/(1−v_1)
                end
                foreach j < i do
                    v_2 ← v_1(1 − p'_jn)
                    if 1 − v_2 = 0 then
                        q_n ← q_n + B_ij^m,   p_in ← p_in + B_ij^m
                        p_jn ← p_jn + B_ij^m
                    end
                    else
                        q_n ← q_n + q'_n B_ij^m/(1−v_2)
                        p_in ← p_in + p'_in B_ij^m/(1−v_2)
                        p_jn ← p_jn + p'_jn B_ij^m/(1−v_2)
                    end
                    foreach k < j do
                        v_3 ← v_2(1 − p'_kn)
                        if 1 − v_3 = 0 then
                            q_n ← q_n + C_ijk^m,   p_in ← p_in + C_ijk^m
                            p_jn ← p_jn + C_ijk^m,   p_kn ← p_kn + C_ijk^m
                        end
                        else
                            q_n ← q_n + q'_n C_ijk^m/(1−v_3)
                            p_in ← p_in + p'_in C_ijk^m/(1−v_3)
                            p_jn ← p_jn + p'_jn C_ijk^m/(1−v_3)
                            p_kn ← p_kn + p'_kn C_ijk^m/(1−v_3)
                        end
                    end
                end
            end
        end
    end
    foreach p_ij ∈ Θ do p_ij ← p_ij/(M π'_i)
    foreach q_j ∈ Θ do q_j ← q_j/σ
    for i ← 1 to L do π_i ← π'_i
end
```

Acknowledgments This paper has been partly elaborated in the framework of the IT4Innovations Centre of Excellence project, reg. no. CZ.1.05/1.1.00/02.0070, supported by Operational Programme 'Research and Development for Innovations' funded by Structural Funds of the European Union and state budget of the Czech Republic and partly supported by the projects AV0Z10300504, GACR P202/10/0262, 205/09/1079.

References

1. Dempster, A.P., Laird, N.M., Rubin, D.B.: Maximum likelihood from incomplete data via the EM algorithm. J. Royal Stat. Soc. Ser. B (Methodol.) **39**(1), 1–38 (1977)
2. Foldiak, P.: Forming sparse representations by local anti-hebbian learning. Biol. Cybern. **64**, 165–170 (1990)
3. Frolov, A.A., Husek, D., Polyakov, P.Y.: New measure of boolean factor analysis quality. Adapt. Nat. Comput. Algorithms **6593**, 100–109 (2011a)
4. Frolov, A.A., Husek, D., Polyakov, P.Y.: Bulevskij faktornyj analiz na osnove attraktornoj nejronnoj seti i nekotoryje ego prilozenija (Boolean factor analysis by means of attractor neural network and some its applications). Neirokomputery: Razrabotka, Primenenie (in Russian ISSN 1999–8554) **1**, 25–46 (2011b)
5. Frolov, A.A., Husek, D., Polyakov, P.Y.: Expectation-maximization approach to boolean factor analysis. In: Proceedings of International Joint Conference on Neural Network. San Jose, California (July 2011c)
6. Neal, R.M., Hinton, G.E.: A view of the EM algorithm that justifies incremental, sparse, and other variants. Learn. graph. model. **89**, 355–368 (1998)
7. Spratling, M.W.: Learning image components for object recognition. J. Mach. Learn. Res. **7**, 793–815 (2006)
8. Spratling, M.W., Johnson, M.H.: Preintegration lateral inhibition enhances unsupervised learning. Neural Comput. **14**(9), 2157–2179 (2002)
9. Spratling, M.W., Johnson, M.H.: Exploring the functional significance of dendritic inhibition in cortical pyramidal cells. Neurocomputing **52**(54), 389–395 (2003)

Part V
Knowledge Discovery II

A Cognitive Interactive Framework for Multi-Document Summarizer

Anupam Srivastava, Divij Vaidya, Malay Singh, Pranjal Singh and U. S. Tiwary

Abstract In this paper, we present a generic interactive framework based on human cognition, where the system can learn continuously from the Internet and from its interaction with the users. To show the utilization of this framework, Iintelli, an agent based application for multiple text document summarization is developed and compared with the MEAD on the Cran Data Set. Mead is a natural language processing-based summarizer, which provides summary by extracting sentences from a cluster of related documents and Cran is a data set maintained by Information Retrieval Group at University of Glasgow. The human knowledge and experience are represented through fuzzy logic-based word-mesh and sentence-mesh, which can learn. Learning is performed using the competitive models, namely, Maxnet and Mexican Hat Models. As the result shows, the framework performs well as a multi-document summarizer. Though we have tested the framework for multi-document summarization, we believe that it can be extended to develop interactive applications for other domains and tasks.

A. Srivastava (✉) · D. Vaidya · M. Singh · P. Singh · U. S. Tiwary
Indian Institute of Information Technology, Allahabad, UP, India
e-mail: anupam328@gmail.com

D. Vaidya
e-mail: iit2008077@iiita.ac.in

M. Singh
e-mail: iit2008001@iiita.ac.in

P. Singh
e-mail: iit2008008@iiita.ac.in

U. S. Tiwary
e-mail: ust@iiita.ac.in

M. Kudělka et al. (eds.), *Proceedings of the Third International Conference on Intelligent Human Computer Interaction (IHCI 2011), Prague, Czech Republic, August, 2011*, Advances in Intelligent Systems and Computing 179, DOI: 10.1007/978-3-642-31603-6_22, © Springer-Verlag Berlin Heidelberg 2013

1 Introduction

We live in a world where the computational capability of machines is increasing at a tremendous rate. However, these machines are still constrained due to lack of human-like intelligence and cognition. Active research in the area of machine learning and Human Computer Interaction has led to the next generation of machines imbibing cognitive capabilities in them. These capabilities represent a significant step towards improving the efficiency and relevance of the results of these systems.

In this paper, we present a generic interactive framework modeled on the human behavior and thinking. The system aims towards blurring the lines of distinction between the data filtered by a human and that done by a machine. The knowledge base used by the system is an integration of data retrieved from the Internet and user knowledge and experience gained through interaction. The prolonged interaction is responsible for dynamic changes in the knowledge base that enables the system to adopt characteristics of its user resulting in understanding the human perception. Human like reasoning and learning abilities have been imbibed in the system using various concepts like competitive learning, fuzzy knowledge representation etc., each represented through a separate agent. All these agents collectively form the framework, which can subsequently be used to develop interactive applications for various domains.

We have found through experiments that use of fuzzy logic is more appropriate to represent human knowledge and experience in our cognitive system. The primary source of information acquisition is the Internet. Moreover, we have used a new form of information gathering method in the task of multi-document summarization i.e. through interaction with the users. This ability to store the episodic memory is realized by the use of Short Term Memory (STM) and Long Term Memory (LTM). This helps in relating to human experiences and memory.

Using cognitive learning techniques the system is able to filter out relevant information from the LTM and Internet, providing its users a personalized information portal, which adapts to their needs. The learning approaches used to replicate human thinking are mainly based on reinforced learning coupled with a lateral inhibitive approach as seen in competitive learning. The storage framework of the system based on LTM and STM, assists in providing the reward/penalty based system, which dynamically modifies the knowledge base and incorporates human-like thinking and reasoning.

The system has been developed using the concepts of self-sustaining software agents. The agents are independent pieces of code that can execute independently. Each agent can be used again in developing an application where there is a need to represent human knowledge and experience. Further, the agents given in this work could easily be clubbed with other new agents to improve upon the information summarization capability of the system.

Various attempts at developing cognitive architecture [1, 2] capable of inference and reasoning on the basis of their knowledge base have been made in the past.

RASCALLI [3–5] is a system developed under a project sponsored by the European Commission. It is capable of performing the above mentioned features. The system tries to model the human metaphor i.e. the system tries to define the mind, various sensors and effectors and the environment from which it continuously learns. The system continuously searches the internet to expand its knowledge base, and works with two forms of memory—LTM and STM.

TextRank [6], developed by Rada Mihalcea and Paul Tarau, is a graph based ranking model for text processing and can be applied for both word abstraction and sentence extraction . In their work, they have used graphs with each node representing a unique lexical unit. The strength of relationship between two nodes is determined by the co-occurrence relation, which is defined in terms of the distance between the two lexical units in the original text. Extending the same principle to sentence extraction, the nodes are used to represent sentences. Here, the similarity between the two sentences was calculated as a function of their content overlap. Content overlap can be found out by use of syntactic filters, or by number of overlapping tokens etc. The sentence similarity was calculated as:

$$Similarity\ (S_i, S_j) = \frac{|\{w_k \in S_i \& w_k \in S_j\}|}{log(|S_i|) + log(|S_j|)} \tag{1}$$

where:

$$S_i = w_i^1, w_i^2, \ldots\ldots, w_i^N. \tag{2}$$

The LexRank [7], developed by Erkan and Radev also performs sentence extraction to create a summary for multiple documents. It is again a graph based method where the nodes represent unique sentences. The level of similarity of two sentences is calculated using cosine similarity of the feature vector formed for the sentences using the inverse frequency—term frequency product.

$$idf\text{-}modified\text{-}cosine(x, y) = \frac{\sum_{w \in x,y} tf_{w,x} tf_{w,y} (idf_w)^2}{\sqrt{\sum_{x_i \in x} (tf_{x_i,x} idf_{x_i})^2} \sqrt{\sum_{y_i \in y} (tf_{y_i,y} idf_{y_i})^2}} \tag{3}$$

Once they have the level of similarity, they calculate the relative importance of a sentence based on the concept of eigenvector centrality in the graph.

$$p(u) = \sum_{v \in adj[u]} \frac{p(v)}{deg(v)} \tag{4}$$

where $p(u)$ denotes the centrality of the node, $adj[u]$ represents the nodes adjacent to node u and $deg(v)$ represent the degree of node v.

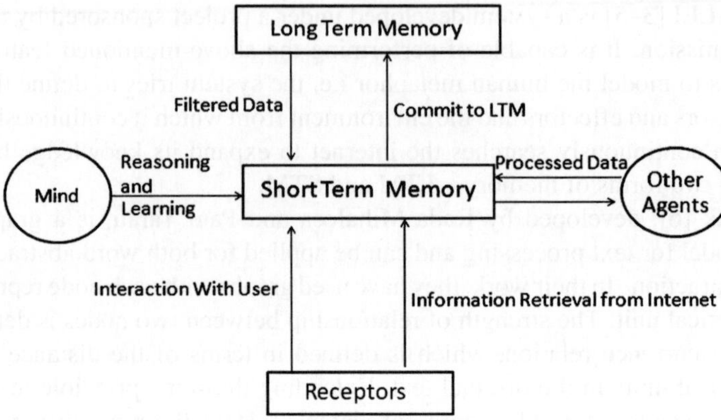

Fig. 1 Cognitive architecture

2 Cognitive Architecture

The proposed architecture models human thinking process. Broadly, the architecture is divided into receptors, agent modules, memory and the mind, which performs reasoning and learning. Each module works in collaboration to produce a result which is similar to the way a human mind would have reasoned and learned to produce the result. Figure 1 shows the interaction between these components.

2.1 Receptors (Sensors)

Receptors are the agents which act as the interface between the external environment and the system. These are the primary source of raw data and essentially contribute towards building and enriching the knowledge base. The agent we have used has its primary source of data as the Internet. Another agent works as an interface between the user and the cognition system. Its primary aim is to interact with the user so that learning can be performed based upon the user inputs and feedbacks. The data gathered by both the receptors is stored in the STM for further processing.

2.2 Memory

The architecture requires both volatile and non-volatile memory to serve different purposes.

The non-volatile form of the memory for permanent storage is called the LTM and the volatile memory is called the STM. LTM serves as the durable centralised database where the information from all the episodes (instances) is stored and updated. It is stored on physical media and is updated through every instance of the application. The data structure used for internal storage is in the form of an undirected graph with nodes and edges. However, the usage of these components varies from one agent to another. The information stored in the LTM models the preferences of its respective user and is used as the knowledge base for perception, reasoning and learning for further episodes of interaction with user.

STM is the temporary work space for the application, where the filtering and processing of data take place. It is directly connected to the various agents which take the raw data, process it and give it to the STM. During each instance, STM is allocated for that particular session, which at the end of the session filters its data and appropriate commits are made to the LTM which includes forming a perception like the user.

2.3 Application Specific Agents

Once we have the raw data acquired by the receptors and stored in the STM, specific agents are invoked to process and filter out the data in user understandable form. Due to the agent based programming methodology of the framework, the number of such agents can be varied as per the needs of the application. In this work, we use a number of agents for tasks like, word abstraction, sentence extraction, etc. for text summarizing, which collaborate with each other and mind agent in carrying out the cognitive processes. At the moment, the agents are designed to perform the task of text summarization. However, the same architecture can be used for other domains and tasks, by changing the agents at this level.

2.4 Mind

This agent is responsible for carrying out the reasoning and learning process in order to generate the summary. Moreover, it tries to form a perception about the user and its environment in order to give more accurate results. Learning and perception are carried out by manipulation of fuzzy parameters, which essentially change the priority of data in the LTM with respect to a particular user. The user feedback and actions are utilized to perform reinforced learning. Depending upon whether the user likes or dislikes the produced summary, the fuzzy edge weights are changed in order to map the system model to the user mind model.

We maintain a separate log for each user so that, perceptions for individual users can be created through manipulations of edge weights. For example, depending upon previous user actions, the system can understand that the user perceives apple as a fruit rather than Electronics company. Reasoning is performed by spreading activations in the word-mesh and the sentence-mesh. Relations among nodes which are not directly connected are realized on the fly, depending upon the user input and environmental conditions.

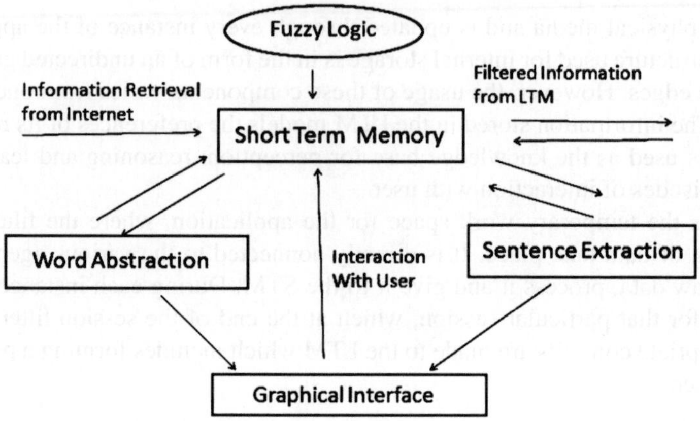

Fig. 2 Interaction between different agents in Iintelli

3 The Proposed Framework: Iintelli

To show the utilization of our cognition system, we have developed an agent based application, Iintelli, which is capable of multiple document text summarizations. The application consists of various agents, which are independent and self-sustaining in their respective fields of work. The system has two different agents to perform text summarization, one for word abstraction and another for sentence extraction. Other agents interact with the user and represent human knowledge and experience using competitive learning models, namely Maxnet and Mexican Hat. While interacting with the user, just like a human, the system tries to develop a perception about the current environment. This is the most important feature of our text summarizer since it allows the user to connect and express more accurately, and hence, produce coherent results.

Figure 2 shows the interaction between the different modules. The agent for information acquisition sends the raw data to the STM where the mind performs reasoning and learning depending upon the current perception, and passes on the filtered data to the word abstraction and sentence extraction agents. These agents perform summarization as described below and present the result to the user with the help of a GUI. The user may then perform actions and give feedback, which is then used for reinforced learning by the system.

3.1 Learning System Through Fuzzy Logic-Based Word-Mesh and Sentence-Mesh

The design followed in the system performs learning and reasoning by manipulating a set of parameters and weights, which determine the correlation of different textual units of the text. To represent these parameters, we have used fuzzy values with sigmoid membership function. There are many other membership functions, which could have been used like triangular membership function, trapezoidal membership function, etc. However, after several trials, it was found that the sigmoid membership function [8] produces the best result. The function is defined as

$$p_x^{\lambda}(x) = \frac{1}{1 + e^{(-\lambda(x-a))}} \tag{5}$$

3.2 Word Abstraction

This agent assigns high importance of to the keyword provided by the user and works on its derivatives and related keywords. The central idea to word abstraction is picking up common features among words and using these features to assign a weight to each of the related words. In the context of Iintelli, the keywords provided by the user are given the maximum weight and word abstraction is employed on the source document. This yields words with varying weights to be introduced to the knowledge base.

Basic Structure of Long Term Memory (Multi-dimensional Fuzzy Word Mesh)

The main data structure for the long-term memory is an undirected graph. The nodes of the graphs are words, which have been found after using word abstraction technique in many iterations of the system. The edges of the graph are weighted edges signifying the intensity of relationship between the connected words. Sometimes, the relationship between the words may vary according to the context; hence, the edges also store the name of the topic in which context the weight is assigned. Thus, two nodes may have multiple edges between them. The importance of semantic relationship among words is not gauged by absolute crisp numbers; instead, a fuzzy value is assigned to it based on the information gained from the training data. The fuzziness ensures the absence of crisp logic, which sometimes may render false negatives.

Knowledge Base Filtering

To prepare the word mesh, the source document is parsed through once and the sentences containing the original key words are picked up for further processing. Further processing would include word abstraction i.e. finding out syntactic and semantic relationship [9] among words in sentences and storing the words related to the given keywords in the database.

Once the word mesh has been prepared with all its associated weights in terms of fuzzy values, the whole document is passed again to include those sentences also into consideration that have any of the words in the mesh present in them. To do this each word in the sentence is checked for its presence in the word mesh and if it is found, then according to its edge in the word mesh, its weight is added to the overall score of the sentence. A point to note in this context is that the overall score is also computed as a fuzzy value. Once we have the sentences with each of them having its own "score" in the form of a fuzzy value, the sentences are put in a priority queue whose implementation is modified to give the best sentence in terms of score. The overall score is computed as

$$Score = \sum_{j=0}^{k} weight_j \tag{6}$$

k is the number of words of the database present in the sentence.
$weight_j$ is the fuzzy weight of the $word_j$.

Reinforcement Using Maxnet

The approach towards text summarization uses reinforced learning. From the interactive user interface, the user either accepts a sentence as valid or penalizes a sentence. The penalty imposed on the sentence is updated throughout the word mesh through successive activations. All the words that are present in the sentence and are a part of the word mesh are penalized by a constant factor in which its crossover point is reduced.

The user interaction leads to reinforced learning, which also includes competitive learning through the use of winner takes all strategy. This form of lateral inhibition is like that of Maxnet where the winning node is awarded whereas all the nodes, which are penalized by the user, have their weight reduced by a finite amount.

3.3 Sentence Extraction

The agent is used for extracting important and useful sentences directly from the documents based upon some predetermined features. This is a very useful technique for performing text summarization of multiple documents.

The algorithm works by creating a mesh where each node represents a sentence. The edges between two sentences represent the level of similarity of the two sentences. The level of similarity is calculated by making a feature vector for each sentence. Once, we have the feature vector for both the sentences, we calculate the similarity between these two sentences which is represented as edge weights between two given nodes. To reduce the computational load, we put use a threshold

T1 to remove those edges, which have weights less than this; they would be weak and hence, would not contribute to the result significantly.

After this, we extract those sentences, which seem to be important to all the documents. This is done by using a form of competitive learning to let the weighted graph converge, so that we can extract the top ranked sentences. Unlike the previous agent, we use Mexican Hat for performing competitive learning, which is basically the soft winner takes all approach of competitive learning. The basic idea is that, the nodes which would have a strong centrality factor associated with them would increase the rank of those nodes which share a strongly weighted connection with them. A threshold T2 is used to determine the number of epochs after which one could say that graph has converged. However, to avoid pre mature convergence of the graph and noise reduction, we add randomness in the algorithm. This randomness ensures that it does not get stuck in local minima of the state space. Rather, it allows one to walk around randomly, giving incentives to the node (and its strongly associated nodes) where it is present currently until the graph converges.

Once the graph converges, we pick the sentence with the highest centrality values and present it as the summary of the documents in question.

4 Experiments

The implementation has been done using C++ and Qt Library. To demonstrate the utility of the application, we have performed experiments on the Cran Data Set [10], which comprises of 1400 text documents from various fields and compared it with the results produced by MEAD [11] which is a natural language processing-based summarizer. We divided the data set into groups of ten documents each and used both the agents (sentence extraction and word abstraction) for summarizing. The summaries generated were then compared to calculate the percentage match. The percentage match was calculated as the ratio of the total number of matched sentences and the total number of sentences in the summary for each data set. We compared the performance of the system against MEAD, with and without the fuzzy component.

Figure 3a shows the comparison between MEAD and Iintelli without the fuzzy component. The graph shows how the output is affected as we vary the threshold, which determines the convergence of the graph. Figure 4a shows the comparison between MEAD and Iintelli with the fuzzy system component. The graph shows how the output gets affected as we vary the initial weights, which determine the relationship strength among different textual units and with the slope of sigmoid fuzzy numbers.

Figures 4b, 5a, b show how the percentage match varies among the different document sets. We have plotted bar graphs for only those constant values which give the best result in each corresponding experiment. Figure 4b shows the distribution of percentage match among the document sets with $Slope = 0.7$ and varying NN weight. Figure 5a is plotted with $T2 = 2 \times 10^{-4}$ and varying $T1$ whereas Fig. 5b is plotted with $Slope = 0.7$ and varying $T1$.

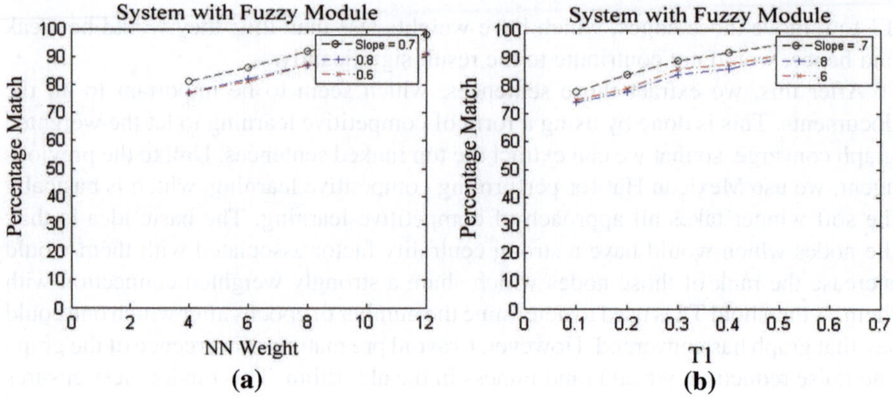

Fig. 3 Comparison between MEAD and Iintelli with fuzzy component. **a** Varying NN weight, **b** varying T1

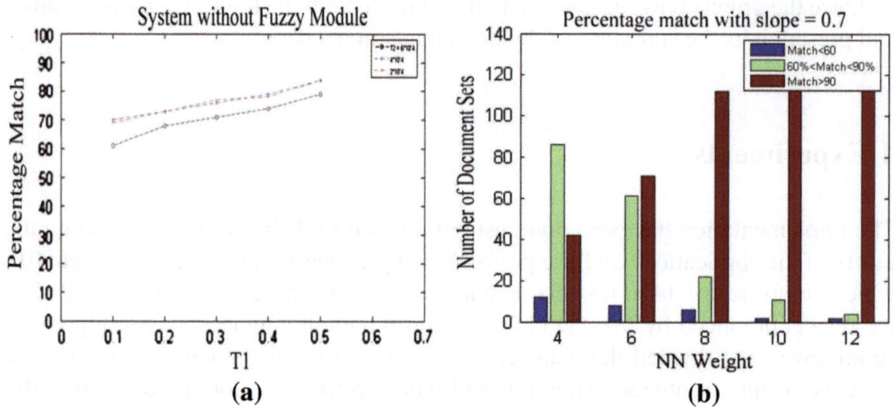

Fig. 4 **a** The comparison between MEAD and Iintelli without the fuzzy component. **b** Distribution of percentage match among the document sets with $Slope = 0.7$ and varying NN weight

Clearly, one can observe that Iintelli works better with the fuzzy component as we are able to represent human knowledge more appropriately. Also, it can be seen that the competitive models are able to output, almost all the sentences, which the MEAD does. Moreover, due to interaction with the user, the system is able to produce more coherent results. The system appears to work the best with the slope of fuzzy number set to 0.7. Moreover, we observe that as we increase the threshold T1, the results improve and eventually these saturate beyond 0.4. The system seems to work best with T2 set to 2×10^{-4}. The bar graphs shows that with $Slope = 0.7$ and $T1 = 0.5$, we are able to get greater than 90 % in over 120 document sets out of 140.

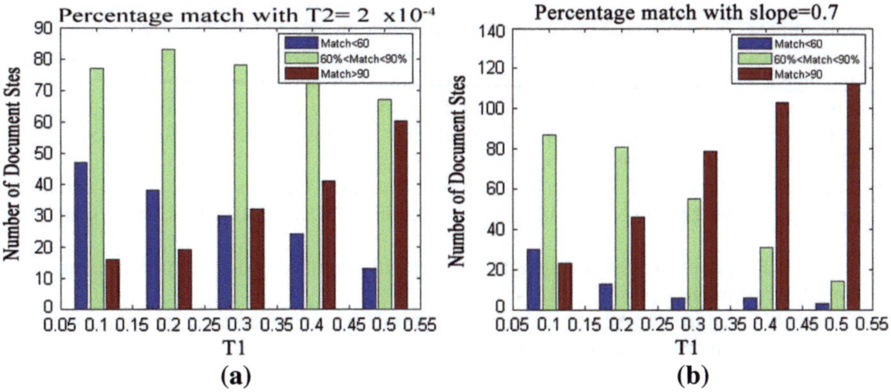

Fig. 5 Distribution of percentage match among the document sets with varying T1 **a** $T2 = 2 \times 10^{-4}$ **b** Slope = 0.7

5 Conclusion and Future Work

We have presented hybrid cognitive system which is capable of multi-document summarization through competitive learning models. We have developed an application, Iintelli, a multiple document text summarizer, which is capable of learning using competitive learning models and compared it with MEAD. The results of these experiments show that the system is able to produce similar output as MEAD. Moreover, the text summarizer allows human computer interaction, which obviously gives out better and more coherent results. The experiments also show that we get better results, if fuzzy logic is used for representation of human knowledge and experience.

To show that the framework works well with other applications also, the same system can be extended for image ranking and association with the help of different set of features. One can also develop an abstractive summarizer or other knowledge discovery and generation applications.

References

1. Skowron, M., Irran, J., Krenn, B.: Computational framework for and the realization of cognitive agents providing intelligent assistance capabilities. In: 18th European Conference on Artificial Intelligence, Cognitive Robotics Workshop, pp. 88–96 (2008)
2. Xu, F., Adolphs, P., Uszkoreit, H., Cheng, X., Li, H.: Gossip Galore: A Conversational Web Agent for Collecting and Sharing Pop Trivia. In: Filipe, J., Fred, A., Sharp, B. (eds.) Proceedings of ICAART 2009—First International Conference on Agents and Artificial Intelligence. 19–21 Jan 2009. Porto, Portugal (2009) (Forthcoming)
3. Eis, Ch.: RASCALLI platform: a dynamic modular runtime environment for agent modeling. Master's Thesis, Vienna University of Technology, Vienna, Austria (2008)

4. Kostadinov, S., Grinberg, M.: The embodiment of a DUAL/AMBR based cognitive model in the RASCALLI multi-agent platform. In: Proceedings of 8th International Conference on Intelligent Virtual Agents (IVA-08). LNCS, vol. 5208, pp. 356–363 (2008)
5. Eis, Ch., Skowron, M., Krenn, B.: Virtual agent modeling in the RASCALLI platform. In: PerMIS'08— Performance Metrics for Intelligent Systems Workshop, Gaithersburg, MD, USA, August 19–21 (2008)
6. Mihalcea, R., Tarau, P.: TextRank: bringing order into texts. In: Proceedings of the Conference on Empirical Methods in Natural Language Processing (EMNLP 2004), Barcelona, Spain, July (2004)
7. Radev, D.R., Erkan, G.: Lexrank: graph-based lexical centrality as salience in text summarization. J. Artif. Intell. Res. (JAIR) **22**, 457–479 (2004)
8. Dombi, J., Gyorbiro, N.: Addition of sigmoid-shaped fuzzy intervals using the Dombi operator and infinite sum theorems, Fuzzy Sets Syst. **157**(7), 952–963 (2006)
9. Li, Y., McLean, D., Bandar, Z.A., O'Shea, J.D., Crockett, K.: Sentence similarity based on semantic nets and corpus statistics. Knowl. Data Eng. IEEE Trans. Data Eng. **18**(8), 1138–1150 (2006)
10. Glasgow IDOM—Cranfield collection: http://ir.dcs.gla.ac.uk/resources/test_collections/cran/. Accessed 14 Apr 2011
11. MEAD: http://www.ncibi.org/gateway/mead.html. Accessed 14 Mar 2011

Breast Cancer Detection and Classification Using Support Vector Machines and Pulse Coupled Neural Network

Aboul Ella Hassanien, Nashwa El-Bendary, Miloš Kudělka and Václav Snášel

Abstract This article introduces a hybrid scheme that combines the advantages of pulse coupled neural networks (PCNNs) and support vector machine, in conjunction with type-II fuzzy sets and wavelet to enhance the contrast of the original images and feature extraction. An application of MRI breast cancer imaging has been chosen and hybridization scheme have been applied to see their ability and accuracy to classify the breast cancer images into two outcomes: cancer or non-cancer. In order to enhance the contrast of the input image, identify the region of interest and detect the boundary of the breast pattern, a type-II fuzzy-based enhancement and PCNN-based segmentation were applied. Finally, wavelet-based features are extracted and normalized and a support vector machine classifier were employed to evaluate the ability of the lesion descriptors for discrimination of different regions of interest

A. E. Hassanien (✉)
Faculty of Computers and Information, Blind Center of Technology,
Cairo University, Cairo, Egypt
e-mail: aboitcairo@gmail.com

A. E. Hassanien
ABO Research Laboratory, Cairo, Egypt

N. El-Bendary
Arab Academy for Science, Technology, and Maritime Transport, Cairo, Egypt
e-mail: nashwa_m@aast.edu

N. El-Bendary
ABO Research Laboratory, Cairo, Egypt

N. El-Bendary
VSB-Technical University of Ostrava, Ostrava, Czech Republic

M. Kudělka · V. Snášel
Faculty of Electrical Engendering and Computer Science,
VSB-Technical University of Ostrava, Ostrava, Czech Republic
e-mail: kudělka@inflex.cz

V. Snášel
e-mail: václav.snášel@vsb.cz

M. Kudělka et al. (eds.), *Proceedings of the Third International Conference on Intelligent Human Computer Interaction (IHCI 2011), Prague, Czech Republic, August, 2011*, Advances in Intelligent Systems and Computing 179, DOI: 10.1007/978-3-642-31603-6_23,

to determine whether they represent cancer or not. To evaluate the performance of presented approach, we present tests on different breast MRI images.

1 Introduction

MRI imaging techniques, in particulars MRI breast imaging are widely used in medical diagnosis [1–3]. For cancer detection, breast MRI is performed using a gadolinium-based contrast agent to increase contrast in the resulting images. When gadolinium injected into the patient bloodstream, enters the magnetic field, its temperature increases slightly. The signals created by the passing gadolinium are picked up by special radio equipment inside the MRI machine. Gadolinium contrast agent makes certain tissues, abnormalities or disease processes more clearly visible on a magnetic resonance imaging (MRI) scans. Those signals are sent to a computer, which creates images of the breast tissue. The presence of the gadolinium in the veins highlights the circulation inside your breast, and helps create a high-contrast image. Breast cancers are identified on the basis of significant early increase in signal intensity following intravenous injection of contrast agent. Contrast-enhanced breast MRIs acquired by contrast injection have been shown to be very sensitive in the detection of breast cancer, but are also time-consuming and cause waste of medical resources [4, 5]. Over the last few decades we have seen a new era of machine learning emerging that is focussing on the principles, theoretical aspects, and design methodology of algorithms gleaned from nature. Machine learning techniques are a well-established paradigm with current systems having many of the characteristics of biological computers and being capable of performing a variety of tasks that are difficult or impossible to do using conventional techniques. It is focussing on the principles, theoretical aspects, and design methodology of algorithms gleaned as well [6, 17].

In this paper we show how a combination of various machine intelligence techniques can be used to analyze MRI breast images for the purpose of diagnosing women breast cancer. Our approach utilizes type-II fuzzy algorithm for improving the quality of MRI images [4]. Then, we perform segmentation utilizing pulse coupled neural networks in order to extract regions of interest. Wavelet features are extracted from these regions and finally, support vector machine is then performed for the actual diagnosis and discrimination of different regions of interest to determine whether they represent cancer or not.

The rest of the paper is organized as follows. Section 2 gives a brief introduction to the type-II fuzzy, PCNN, wavelet and support vector machine. Section 3 presents the proposed MRI breast image analysis approach in detail. Experimental results are discussed in Sect. 4 while Sect. 5 concludes.

2 Backgrounds

Due to space limitations we provide only a brief explanation of the the type-II fuzzy, wavelet and PCNN, and support vector machine. A more comprehensive review can be found in sources such as [4, 7].

2.1 The Type-II Fuzzy Sets

Fuzzy logic systems usually employ the type-I fuzzy sets and represent uncertainty by numbers in the range [0,1] which are referred to as degrees of membership. Type-II fuzzy sets are an extension of type-I fuzzy sets with an additional dimension that represent the uncertainty about the degrees of membership. The type-II fuzzy sets are useful in circumstances where it is difficult to determine the exact membership function for a fuzzy set. The type-I membership functions are precise in the sense that once they have been chosen all the uncertainty disappears. However, type-II membership functions are fuzzy themselves. The simplest type-II fuzzy sets are interval type-II sets whose elements' degree of membership are intervals with secondary membership degree of 1.0 [8, 9].

The type-II fuzzy logic works as follows: the crisp inputs are first fuzzified into either type-0 (known as singleton fuzzification) or type-I, which then activate the inference engine and the rule base to produce output type-II fuzzy sets. These the type-II fuzzy sets are then processed by a type-reducer (which combines the output sets and then performs a centroid calculation), leading to an interval type-II fuzzy sets called the type-reduced set. A defuzzifier then defuzzifies the type-reduced set to produce crisp outputs [4, 8]. Algorithm-1 shows the general steps of the introduced Type-II, for more details referee to our published paper in [4].

2.2 Pulse-Coupled Neural Networks

The PCNN was originally presented by Eckhorn et al. [10] in order to explain the synchronous neuronal burst phenomena in the cat visual cortex. The information of the input is converted to a set of pulse images which present the objects and edges of the images, so pulse-coupled means that the image is analyzed by looking multiple times at different coupled pixels. The pulse-coupled neural network is a neural network that has the ability to extract edges, image segments and texture information from images. The PCNN is very generic. Only a few changes are necessary to effective operating on different types of data. This is an advantage over published image segmentation algorithms which generally require information about the target before they are effective. But the parameters still have to be set manually [7, 15].

Algorithm 1: Type-2 Fuzzy Image Enhancement

Input: MRI breast image
Output: Fuzzy type-I and Fuzzy type-II enhanced images of the input image
Processing:
Step-1 Fuzzy hyperbolization
For each grey level
Compute the the fuzzy type-I membership value
Step-2:Fuzzy type-I enhanced image calculation
Compute the new grey levels of the enhanced image
Step-3 Calculation of Fuzzy type-II membership function
Compute the lower membership values $\mu_{LOWER}(x)$ and the upper membership values $\mu_{UPPER}(x)$
Divide the image into sub-images
For every grey level value **Do**
Calculate a window of equal size
Compute the fuzzy Type-II membership function
Step-4 Fuzzy type-2 enhanced image calculation
Compute the new grey levels of the enhanced image

The PCNN is significantly different than other artificial neural network model in both its structure and operation. There is no training involved. Each neuron in the processing layer is directly tied to an image pixel or a set of neighboring image pixels, these are the feeding inputs, and they are also linked to nearby neurons, the linking inputs. The feeding inputs are iteratively processed and together with the linking inputs produce a pulse train. The PCNN neuron consists of three parts: the dendritic tree, the linking modulation, and the pulse generator. The role of dendritic tree is receiving the inputs from two kinds of receptive fields. Depending on the type of receptive filed, it is subdivided into two channels (the linking and the feeding). The linking receives external stimulus while the feeding receives external stimulus and local stimulus.

2.3 Wavelet Transform

Wavelet is a well-known image analysis methodology [11, 12]. It is known that the use of wavelets provides a richer feature space. It provides approximation with excellent time and frequency resolution. The produced wavelet coefficient yields a smaller set of more robust features, which can improve the probability of correct classification.

The fundamental idea behind wavelets is to analyze the signal at different scales or resolutions, which is called multiresolution [11]. Wavelets are a class of functions used to localize a given signal in both space and scaling domains. A family of wavelets can be constructed from a mother wavelet. Compared to Windowed Fourier analysis, a mother wavelet is stretched or compressed to change the size of the window. In

Table 1 Selected features [13]

No.	Name	Description
1	AMLCE	Average multi-resolution local contrast entropy
2	RMSE	Root mean square error
3	ASM	Angular second moment
4	CON	Contrast
5	DIS	Dissimilarity
6	COR	Correlation
7	IDM	Inverse difference moment
8	SD	Standard deviation
9	ADWC	Average of detailed wavelet coefficient
10	NSSDWC	Sum of square of detailed wavelet coefficient

this way, big wavelets give an approximate image of the signal, while smaller and smaller wavelets zoom in on details. Therefore, wavelets automatically adapt to both the high-frequency and the low-frequency components of a signal by different sizes of windows. Any small change in the wavelet representation produces a correspondingly small change in the original signal, which means local mistakes will not influence the entire transform. The wavelet transform is suited for nonstationary signals, such as very brief signals and signals with interesting components at different scales [11]. Wavelets mean small waves that segments data into different frequency components and transfer each component with different resolution that is matched to its scale. The main idea of wavelet analysis is to see both coarse and detail data without heavy computational penalty. The goal of most modern wavelet researches is to create a set of basis functions and transform them in order to give an information. The set of used features can be seen in Table 1.

2.4 Support Vector Machines

Support vector machine (SVM) is a promising classification technique proposed by Vapnik [14]. This technique has been widely studied for classification, regression and density estimation. It was recently used in many applications including pattern recognition, bioinformatics, cancer diagnosis, etc. SVM maps the input patterns into a higher dimensional feature space through some non-linear mapping selected a priori. SVM has given proofs of its high generalization ability. For a two-class problem, this is typically formulated as an optimization problem. The goal of SVM is to find a linear optimal hyperplane, so that the margin of separation between the two classes is maximized [14].

Suppose we have N training data points $\{(x_1, y_1), (x_2, y_2), \ldots, (x_N, y_N)\}$, where $x_i \in R^d$ and $y_i \in \{+1, -1\}$. Consider a hyperplane defined by (w, b), where w is a weight vector and b is a bias. A new object x can be classified. In practice, the data is often not linearly separable. However, one can still implement a linear model by transform the data points via a non-linear mapping to another higher dimensional space (feature space) such that the data points will be linear separable. This mapping is done by a kernel function.

3 Breast Cancer Detection and Classification System

The proposed system is using the advantages of fuzzy sets, PCNNs, and support vector machine, in conjunction with wavelet-based feature extraction to detect and classify the breast cancer images into two outcomes: cancer or non-cancer. It is comprised of the following three fundamental building phases: (1) **Pre-processing**: in the first phase of the investigation, a pre-processing algorithms based on fuzzy Type-II and pulse coupled neural networks are presented. It is adopted and used to improve the quality of the images and to make the feature extraction phase more reliable. (2) **Feature extraction-based wavelet transform**: in the second phase, wavelet-based features have been extracted, normalized and represented in a database as vector values, and (3) **Classification using support vector machine**: the last phase is the classification and prediction of new objects, it is dependent on the support vector machine.

These three phases are described in detail in the following section along with the steps involved and the characteristics feature for each phase.

3.1 Pre-Processing Phase

To increase the efficiency of the classification and prediction process, a pre-processing stage should be considered to enhance the quality of the input breast images before feature extraction and classification process. In this paper, algorithm based on fuzzy type-II is adopted and used to enhance the contrast of the MRI breast image. Fuzzy type-II set obtained by blurring a type-I membership function. It uses interval-based sets to construct the type-II fuzzy set by defining the upper and lower membership values. For more details refer to [4].

This part describes in more details how the PCNN algorithm detects the boundary of breast cancer automatically. The PCNN segmentation work as follows: an input gray-scale image is composed of $M \times N$ pixels. This image can be represented as an array of $M \times N$ normalized intensity values. Then the array is fed in at the $M \times N$ inputs of PCNN. If initially all neurons are set to 0, the input results in activation of all of the neurons at a first iteration. The threshold of each neuron, significantly increases when the neuron fires; then the threshold value decays with time. When the threshold falls below the respective neuron's potential, the neuron fires again, which again raises the threshold. The process continues creating binary pulses for each neuron. While this process goes on, neurons encourage their neighbors to fire simultaneously in a way that is supported through interconnections. The firing neurons begin to communicate with their nearest neighbors, which in turn communicate with their neighbors. The result is an autowave that expands from active regions. Thus, if a group of neurons is close to firing, one neuron can trigger the group. Due to linking between neurons, the pulsing activity of invoked neurons leads to the synchronization between groups of neurons corresponding to subregions of the image that have similar properties and produces a temporal series of binary images.

The success of the application of PCNNs to image segmentation depends on the proper setting of the various parameters of the network, such as the linking parameter, thresholds, decay time constants, and the interconnection matrices. Proper setting of the parameters is especially important when intensity significantly varies across a single segment.

3.2 Feature Extraction Phase: Wavelet Transform

In this work, the features will be defined in wavelet transform domain [13]. In the case of images, we consider the image an $N \times N$ square matrix of signal values. We decompose the image row-wise for every row using 1-D decomposition algorithm. The resulting two matrices are transposed and processed row-wise again to obtain four $\frac{N}{2} \times \frac{N}{2}$ square matrices this procedure can be repeated any number of time to represent the resolution level we want to reach. The total number of coefficients after decomposition always equals the initial input coefficient. This matrix will be considered as our co-occurrence matrix. Since cancer creates discontinuities in the tissue, in order for cancer to survive, it develops its own blood supply system which is different than the supply system of normal tissue [13]. Therefore, the variance of the edge is higher when the continuity breaks down, so cancerous tissue is expected to have more edges, and the time between the edges is smaller than non cancerous regions. The activity level of this area is high. The activity level can be measured as the entropy of this coefficient. The traditional entropy measure is global quantity calculated for the whole image, and then it is not appropriate for quantifying the distribution of the information at different scales of resolution.

3.3 Classification Using Support Vector Machine

In a third phase, the wavelet-based features were used for classification. These feature vectors were used to obtain high classification precision using SVM, which seeks the optimal separating hyperplane between two classes by focusing on the training samples that lie on the class boundaries while discarding other training samples effectively. The simplest classification problem that SVM could deal with is the linearly separable binary classification. In our case, the two classes (Cancer and non-cancer) could be distinguished by a linear separation and the line (hyperplane, in general)

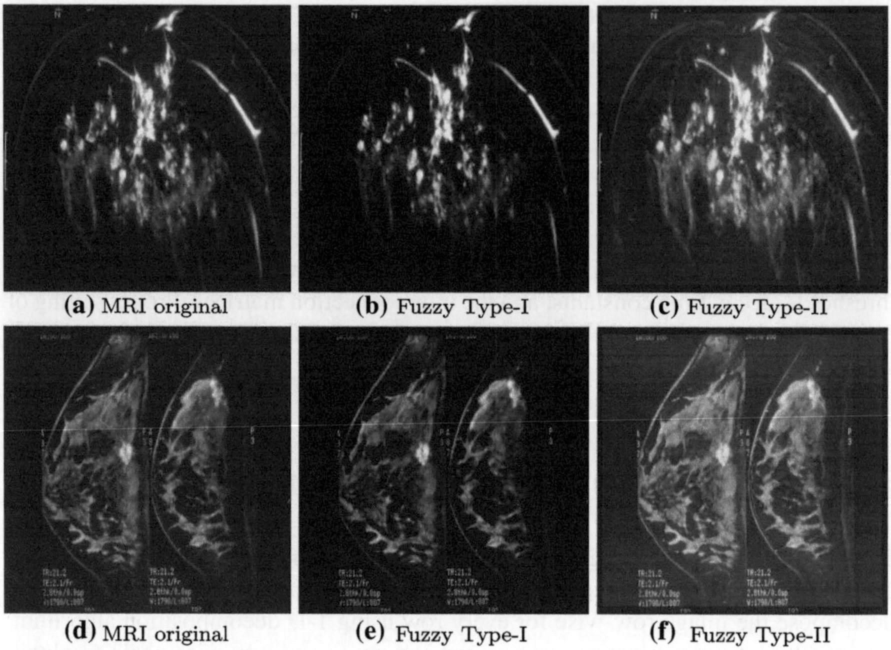

Fig. 1 Type-II fuzzy MRI breast enhancement results

would lack any curves. The basic idea is to find the optimal separating hyperplane that gives the maximum margin between the cancer and non-cancer objects. The hyperplane is determined support vectors that are closest to the decision surface which is determined by the inner product of training data, which enables us to map the input feature vectors into a higher-dimensional inner product feature space. The feature space could be implicitly defined by linear kernel.A soft margin is introduced, when two classes cannot be distinguished and separated perfectly.

4 Experimental Results and Discussion

In the following experiments, real breast MRIs acquired from patients with abnormal pathologies were used for performance evaluation. The data sets used in this paper contains 120 images, which belong to two categories: normal, abnormal. There are 70 normal images, 50 benign and malign, which are considered abnormal. Due to the differences in conditions under which acquisition occurred, the images are not at the same level of contrast and we therefore apply the pre-processing step. The various kinds of tissue found in the breast all appear differently on MRI scans. Cancer can have a unique appearance as well.

Figure 1 shows obtained results of the type-II fuzzy set-based image enhancement on MRI images. The first column of each row depicts the original MRI breast image,

Table 2 Analysis of fuzzy type-I and type-II

Image	Type	Total of pixels	Ratio	Accuracy
Image1	Type-I	238400	2306309	9.674110
Image1	Type-II	238400	−3656010	−15.335612
Image2	Type-I	214120	−299760	7.677180
Image2	Type-II	214120	2102354	−0.092700

the second column shows the result of a locally adaptive type-I fuzzy approach, while the third column gives the result of the new type-II fuzzy set approach. It shows that the results obtained by type-II fuzzy is much better than the results obtained by type-I fuzzy. Moreover, to study the accuracy of the enhancement results using type-II fuzzy compared with Type-I fuzzy, we calculate the accuracy based on the total number of pixels and the ratio of the pixels in the original and the enhanced one [4]. Table 2 shows the accuracy of samples of the MRI breast images and we observe that the experimental results indicate the promising possibilities of type-II fuzzy approach give a good better quality compared with the commonly used type-I fuzzy approach.

As for the segmentation results, the accuracy rate was used to evaluate the performance of segmentation. It was calculated based on the overlap of the standard manually labeled reference image by an expert and a collection of segmentation results obtained with the PCNN approach (refer to Fig. 2). The first row depicts the original MRI breast image, the second row shows the result of PCNN segmentation results. The accuracy rate, which represents the accuracy rate of the segmented area in class k relative to the area in the standard reference image was quantified as the overlap fraction and is defined as:

$$Accuracy = \frac{SA(k) \cap SeA(k)}{SA(k)} \tag{1}$$

where $Sa(k)$ and $SeA(k)$ are the standard and the segmented image, respectively. While $SA(k) \cap SeA(k)$ represents the intersection areas of class k between the segmented image and the standard image, and the denominator represents the area in class k in the standard image.

After the wavelet-based features have been extracted and selected. The assessment of the classification results has been made using the support vector machine SVM was applied to categorize the MRI images into two outcome: cancer or non-cancer. The performance and time cost of SVM were compared with rough sets on the same images. Experimental results showed that SVM (0.98 % accuracy) compared with the rough sets (0.92 % accuracy) and it also was faster than rough sets. The only drawback of use SVM is that the parameters such as the cost parameter which controls the trade-off between allowing training errors and forcing rigid margins, and the kernel type needs to be tuned. Moreover, the SVM mapping to higher dimension is complex and training time increases exponentially with the input data dimension.

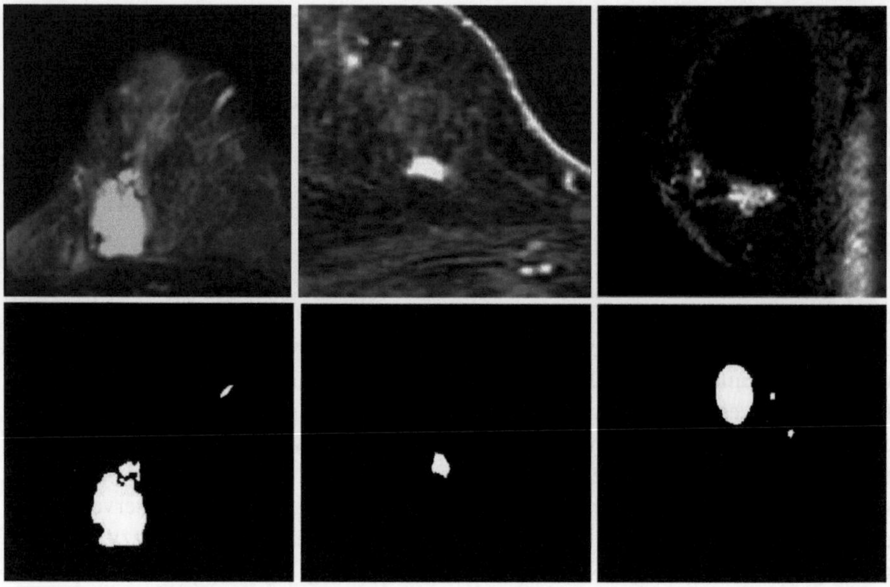

Fig. 2 MRI breast segmentation results

5 Conclusions and Future Works

In this article, we introduced a hybrid scheme that combines the advantages of PCNNs and support vector machine, in conjunction with type-II fuzzy sets and wavelet to enhance the contrast of the original images and feature extraction. An application of MRI breast cancer imaging has been chosen and hybridization scheme have been applied to see their ability and accuracy to classify the breast cancer images into two outcomes: cancer or non-cancer. Experimental results showed that SVM (0.98 % accuracy) compared with the rough sets (0.92 % accuracy) and it also was faster than rough sets. The only drawback of use SVM is that the parameters such as the cost parameter which controls the trade-off between allowing training errors and forcing rigid margins, and the kernel type needs to be tuned. Moreover, the SVM mapping to higher dimension is complex and training time increases exponentially with the input data dimension. More images and deep comparative analysis with other machine learning techniques such as neural network and decision trees will be our future works.

Acknowledgments This work has been supported by Cairo University, project Bio-inspired Technology in Women Breast Cancer Classification, Prediction and Visualization.

References

1. Wendy, D., Constance, L., Savannah, P.: Breast MRI for cancer detection and characterization: a review of evidence-based clinical applications. Acad. Radiol. **15**(4), 408–416 (2008)
2. Giger, ML.: Computerized analysis of images in the detection and diagnosis of breast cancer. Semin. Ultrasound. CT MRI 25(5), 411–418 (2004)
3. Klifa, C., Carballido-Gamio, J., Wilmes, L., Laprie, A., Shepherd, J., Gibbs, J., Fan, B., Noworolski, S., Hylton, N.: Magnetic resonance imaging for secondary assessment of breast density in a high-risk cohort. Magn. Reson. Imaging **28**(1), 8–15 (2010)
4. Hassanien, A.E., Soluiman, O., El-Bendary, N.: Contrast enhancement of breast MRI images based on Fuzzy Type-II. In: Proceeding of the 6th International Conference on Soft Computing Models in Industrial and Environmental Applications, Advances in Intelligent and, Soft Computing, vol. 87, pp. 77–83. Springer Berlin Heidelberg, Berlin, Heidelberg (2011)
5. Yuan, Y., Giger, ML., Li, H., Bhooshan, N., Sennett, C.A.: Multimodality computer-aided breast cancer diagnosis with FFDM and DCE-MRI. Acad. Radiol. **17**(9), 1158–1167 (2010)
6. Schaefer, G., Hassanien, A.E., Jiang, J.: Computational Intelligence in Medical Imaging Techniques and Applications. CRC Press, Boca Raton (2008)
7. Hassanien, A.E., Al-Qaherib, H., El-Dahshan El-Sayed, A.: Prostate boundary detection in ultrasound images using biologically-inspired spiking neural network. Appl. Soft Comput **11**(2), 2035–2041 (2011)
8. Ensafi, P., Tizhoosh, H.R.: Type-2 fuzzy image enhancement. Lecture Notes in Computer Science, vol. 3656, pp. 159–166 (2005)
9. Salama, M.A., Hassanien, A.E., Fahmy, A.A.: Feature evaluation based fuzzy C-mean classification. In: proceeding of the IEEE International Conference on Fuzzy Systems (FUZZ-IEEE)—FUZZ-IEEE 2011, Taipei, Taiwan, June 27–June 30 (2011)
10. Eckhorn, R., Bauer, R., Jordan, W., Brosch, M., Kruse, W., Munk, M., Reitboeck, H.J.: Coherent oscillations: a mechanism of feature linking in the visual cortex. Biol. Cybern. **60**, 121–130 (1988)
11. Stephane, G., Mallat, A.: Theory for multi-resolution signal decomposition: the wavelet representation. IEEE Trans. Pattern Anal. Mach. Intell. **II**(7), 674–693 (1989)
12. Hubbard, B.B.: The world according to wavelets. A K Peters Wellesley, Massachusetts (1995)
13. Own, H.S., Hassanien, A.E.: Image registration based in multiresolution local contrast entropy in wavelet transform domain. In: IEEE 14th International Conference In Digital Signal Processing, vol. 2, pp. 889–892. 1–3 July, Greece (2002)
14. Cortes, C., Vapnik, V.: Support-vector network. Mach. Learn. **20**, 273–297 (1995)
15. Hassanien, A.E.: Fuzzy-rough hybrid scheme for breast cancer detection. Image Comput. Vis. J. **25**(2), 172–183 (2007)
16. Eckhorn, R.: Neural mechanisms from visual cortex suggest basic circuits for linking field models. IEEE Trans. Neural Netw. **10**, 464–479 (1999)
17. Moon, W.K., Chang, S.-C., Huang, C.-S., Chang, R.-F.: Breast tumor classification using fuzzy clustering for breast elastography. Ultrasound Med. Biol. **37**(5), 700–708 (2011)

Author Cooperation Based on Terms of Article Titles from DBLP

Štěpán Minks, Jan Martinovič, Pavla Dráždilová and Kateřina Slaninová

Abstract Very interesting source of information about scientific publishing in computer science is database DBLP. This database allows bibliographic information about main publications from conferences, journals and books in this area. In the article we deal with strength extraction between authors based on their association. The research presented in this article is partly motivated by work of Mori et al. From this paper we have used the approach for extraction of initial metadata, and we have inspired how to take advantage from Jaccard coefficient principals for description of the strength of associations between authors. Method is usable for development of synthetic coauthors network, where as input is used the set of words, which will describe the network (the authors used these words in publication titles).

1 Introduction

Generally, *keyword* is defined as a word, which in connection with other words occurs in text more often than with other ones. After some author's text is analyzed, it could be said, that these words are unique—they are keywords. Usage is evident;

Š. Minks (✉) · J. Martinovič · P. Dráždilová · K. Slaninová
Department of Computer Science, VŠB—Technical University of Ostrava,
17. listopadu 15, 708 33Ostrava, Czech Republic
e-mail: min111@vsb.cz

J. Martinovič
e-mail: jan.martinovic@vsb.cz

P. Dráždilová
e-mail: pavla.drazdilova@vsb.cz

K. Slaninová
e-mail: katerina.slaninova@vsb.cz

M. Kudělka et al. (eds.), *Proceedings of the Third International Conference on Intelligent Human Computer Interaction (IHCI 2011), Prague, Czech Republic, August, 2011*, Advances in Intelligent Systems and Computing 179, DOI: 10.1007/978-3-642-31603-6_24, © Springer-Verlag Berlin Heidelberg 2013

searched information in specific area, based on input string, could be extended by these keywords, and even more by metadata extracted from articles of important authors.

Moreover, no object in our concern is placed in a world without interaction with other objects. At least, it always holds a part of world. The meaning of the object can differ depending on how it is viewed. This is generally called as object's context. In the following text we reduce the author's context to another person, that is with given author in relation through their articles published together. It will be very clearly seen, how another person can change author's initial set of extracted metadata. How much the set of terms will be changed depends not only on the strength of relation between authors, but also on how much initial set is relevant to the context.

The research presented in this article is partly motivated by work of Mori et al. (published in [1]). From this paper we have used the approach for extraction of initial metadata, and we have inspired how to take advantage from Jaccard coefficient principals for description of the strength of associations between objects. We also used knowledge from Information Retrieval, which is aimed to query expansion research. However, our approach differs from traditional methods in a way, that instead of using the new term for expansion query, we use information about author's context (defined by coauthors' publications in the same domain). This domain is then represented by initial keywords evaluated by these coauthors.

Very interesting source of information about scientific publishing in computer science is database DBLP.[1] DBLP database allows bibliographic information about main publications from conferences, journals and books in this area. Furthermore, there could be found references to electronic versions of presented publications. Originally, DBLP meant 'Database systems and Logic Programming', but afterwards the server's interest became wider; today, DBLP could be interpreted as 'Digital Bibliography and Library Project'.

Since we did not have the complete access to all electronic publications in DBLP at this time, we used as input text just the titles of the articles. In the future, we would like to expand the approach to abstracts from the most important sources. Following approach is not limited by the amount of input data; hence, it could be easily used to gain much more accurate results.

In the article we first discuss a way of extracting term set from DBLP and its usage (see Sect. 2). In Sect. 3 we deal with strength extraction between authors based on their association, followed by experiments described in Sect. 4 and final conclusion as Sect. 5.

2 Keyword Extraction from DBLP

Metadata could be gained by various approaches and methods; manually or with help of specific tools and ways for their extraction. Presented method shows how to extract metadata as keywords from DBLP and how to evaluate them.

[1] Online access to DBLP database is available on http://dblp.uni-trier.de/.

Term set is understood as a collection of all keywords, which are extracted from the input text. To complete such term set for DBLP and given author with afterward assessment concerned about relations between them, there is need to obtain the base input text first. It does not matter whether we have used articles from Wikipedia in the past, currently paper titles from DBLP, or all the DBLP database with full texts in the future; it is always important to do the preprocessing phase.

In the case of article titles from DBLP, there is not much work with their preprocessing. With the titles we can work as with 'pure text'. We work with sequentially XML file with all records of titles, where given author is written down as one of the authors. Then, for selected author we simply extract only appropriate publications and join them together into one plain text, which is used as the input document. On the other hand, in the case of using the articles from Wikipedia, there could be much more complex preprocessing; at least removing all HTML tags.

The extracted text is then proceed to methods, which remove words that do not carry any important information. The main issue of this paper is not to describe this kind of methods. More could be found in [2, 3].

However, it is obvious that the import expression is in some cases made by two, three or more following words, where their exact form is very important. For example words 'sql' and 'object' are very common expressions by themselves, but when they are used in collocation 'object sql', it could take our interest. For this purpose, there is commonly used simply method, were from the initial text are rid of obviously the stop words (not important words), and the rest remains in extracted form. Then, possible double and triple terms are identified, merged into one single term and added into the term set collection. That of course more stresses the machine, on which computations are done. For complex methods, we must reference readers to much appropriate literature [4].

After the method selection and term set initialization, we can order the set by term occurrences to get rid of not important terms and to save the computation time. But we should be aware that not always the higher occurrences (even approximately) means the better relevancy. The initial set could contain a thousands of terms, especially if there is not used any kind of special methods and preprocessing. It partly depends on how our algorithm for initial term set extraction is good. But even a well written advanced algorithm could give us result that most occurrences have words like 'http'. How to programmatically decide whether this word is important or not? In the following section, there is presented a way how to evaluate this kind of words to reorder their position in list, but still at the beginning it is crucial to choose right threshold, which reduce and point out terms from initial term set for computing. Unfortunately, usually the final keywords order in list does not rely on their occurrences in text [5].

Because we use Boolean model for indexation [6], where only single-word terms in base forms (without suffixes etc.) are indexed, the term made of two or three words, could be difficult to perform. Therefore we use the method, where the appropriate terms are at first splitted into single words and then put to their base form as well. Therefore, we simply make two, respectively three queries to one title, and ensure if both terms are really presented. Of course, we can obtain the positive result, even the complex term is not in fact in the title, but experiments proof that in short length titles results are not significantly distorted. We are of course aware that this is not possible in the longer text. For these purposes, we suggest for all possible terms preprocessing of their occurrences in advance.

It is obvious that we must always find compromises between accuracy and computation time. To realize how much is important to reduce very long term set collection, see the theorem in Sect. 3. For every keyword of every author (and coauthor) we must compare number of occurrences in his articles and articles in whole DBLP; it means that in more than one million articles we ensure if the keyword is in the title or not. Effectively suggested preprocessing of their occurrences could help in this phase in advance.

2.1 Coauthoring in DBLP

Besides the computation of evaluated term set itself, we can compute *association strength* between the two authors. This method is not only interesting by itself, but it is also essential for extended evaluation of the term list by selected context. Coauthors' relevancy is based on the coauthoring on the same articles. This relevancy is then approximated by Jaccard coefficient [7].

Let A be a set of all authors in DBLP. We define then single author A_i, for who we want to evaluate the strength of association with the other coauthors. In DBLP, coauthoring could be computed in a way that we go thought all the author's articles in DBLP and note all the coauthors participated. Afterwards, we can order them by frequency of together published articles. The set of coauthors of author A_i is marked as C_{A_i}. Let set P be a set all publications in DBLP and P_{A_i} be a set of all publications of author A_i.

The association strength between the authors A_i and A_j can be defined with Jaccard coefficient that reflects mainly the proximity of both authors from number of their common publications:

$$Q(A_i, A_j) = \frac{|P_{A_i} \cap P_{A_j}|}{|P_{A_j}| + |P_{A_i}| - |P_{A_i} \cap P_{A_j}|} \tag{1}$$

If this method is applied to all the authors in DBLP, we obtain weighted undirected graph that can be considered as a synthetic coauthors network (with re-weighted edges between authors). More about this research area can be found in [8, 9].

3 Searching Relations Between Authors on the Basis of Association

The keyword evaluation method was till the certain point inspired by the original article [1]. In this case, however, we cannot use the association strength, as in the calculation of the relevance between the two authors based of the articles.

If we define a set T as the set of all terms of the DBLP and T_{A_i} as the set of all the terms that could be found in the titles of articles by A_i, then t_k is the term belonging to the author A_i (t_k in T_{A_i}). Furthermore, we define (t_k in T_{A_i}) as the number of terms t_k in the headlines of articles by T_{A_i}. This number is then approximated by the number of term t_k in the titles of articles DBLP (t_k in T). The higher value, the term becomes t_k less relevant. In addition, the result is approximated by T_{A_i}, because there is an assumption that T_{A_i}, which has a high cardinality, lower the importance of the individual terms, while low cardinality indicates that the author has only one subject matter:

$$R(T_{A_i}, t_k) = \frac{(t_k\ in\ T_{A_i})}{(t_k\ in\ T) + |T_{A_i}| - (t_k\ in\ T_{A_i})} \tag{2}$$

$$R_{Norm}(T_{A_i}, t_k) = \frac{R(T_{A_i}, t_k)}{MAX(R(T_{A_i}, t_1), \ldots, R(T_{A_i}, t_{|T_{A_i}|}))} \tag{3}$$

3.1 *Method of Using Another Author as Context*

Because we have defined the relation between the authors and we can express the relevance of author's terms, we can assign to given term the best suitable coauthor. We can demonstrate the usage of significance of each coauthor as well. Our reflections were inspired by associative memory, where one is able to better recall the event, which is associated with something significant (although it was already forgotten). For a given author is significant the term, which associate him the best coauthor in the selected area.

The method extension, including the author's coauthor as context, is then constructed analogically. The context is either selected randomly, or is first calculated for a given author according to the formula 1. Afterwards, the authors are selected from the evaluated list of coauthors.

Let us define a threshold t, which indicates the term limit relevance to the author. For lower value than threshold is the result of t equal 0:

$$\begin{aligned} &ContextScore(T_{A_i}, T_{A_j}, P_{A_i}, P_{A_j}, t_k) \\ &\quad = R_{Norm}(T_{A_i}, t_k) R_{Norm}(T_{A_j}, t_k) Q(P_{A_i}, P_{A_j}) \end{aligned} \tag{4}$$

The entire procedure is shown on the Fig. 1.

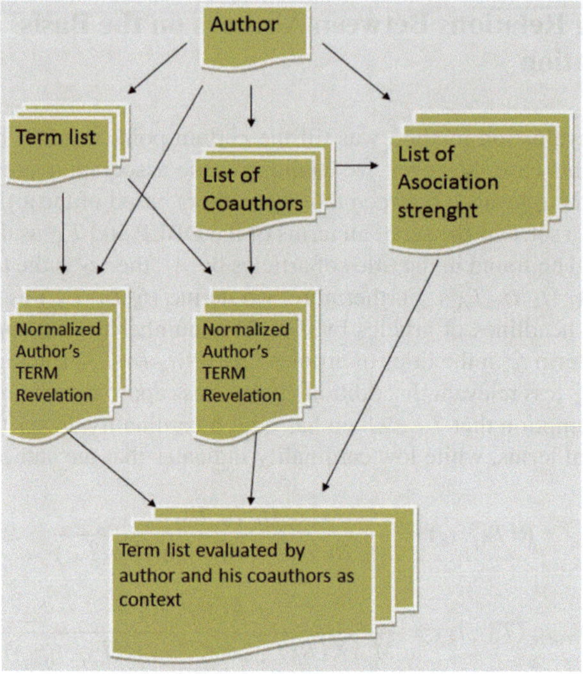

Fig. 1 Procedure of using another author as context

Table 1 Coauthorship of the Author1

Coauthors of Author1	Number of articles	Association strength
Author2	6	0,75
Author3	5	0,71
Author0	7	0,33
Author12	1	0,14
Author8	1	0,09
Author6	7	0,04

4 Experiments

For the basic computation of the coauthorship, we chose the name of our colleagues (name are anatomized) and applied them to the formula 1. The results were limited to the coauthors with whom the author has published at least once. The obtained results are showed in Table 1 for Author1 and in the Table 2 for Author0.

We can immediately notice that even though the Author1 do not share with the Author0 the most of articles, the ratio with this author is 0.33. Conversely then, although the Author0 has with the Author6 15 articles, the Author6 publishes with

Table 2 Coauthorship of the Author0

Coauthors of Author0	Number of articles	Association strength
Author1	7	0,33
Author2	7	0,33
Author3	5	0,24
Author4	4	0,10
Author5	2	0,09
Author6	15	0,08
Author7	3	0,08
Author8	2	0,08
Author9	2	0,07
Author10	1	0,05
...		
Author24	1	0,03
Author25	1	0,02
Author26	1	0,01

Table 3 Key terms for the Author0

t_k	$t_k\ in\ T_{A_{Author0}}$	Ratio
Vector model improvement	2	0,0110
Telemetric	1	0,0109
Strategy description	2	0,0105
Guardian	2	0,0104
Elearning	3	0,0077
Biotelemetric	1	0,0059
Text compression parameters	1	0,0057
dblp	1	0,0055
Compression parameters	1	0,0055
Synthetic social	1	0,0054
User segmentation based	1	0,0053
Enron	1	0,0052
Highly modular	1	0,0052
Emergent flood	1	0,0052
Based overlapping clustering	1	0,0052

many other authors. Therefore, the Author0 has not such a strong association with this Author6.

4.1 Key Terms Computation

First, we have implemented the formula 2 for calculation of the keywords for the Author0 and Author1. We have selected only the first 15 terms for illustration

Table 4 Key terms for the Author1

t_k	$t_k\ in\ T_{A_{Author1}}$	Ratio
dblp	1	0,0143
Synthetic social	1	0,0135
Based overlapping clustering	1	0,0123
Set based overlapping	1	0,0122
Computational intelligence methods	1	0,0109
Elearning	3	0,0108
Synthetic social networks	1	0,0097
Tolerance rough	1	0,0089
Overlapping clustering	1	0,0075
Finding students	1	0,0069
Intelligence methods	1	0,0055
Tolerance rough set	1	0,0050
Finding patterns	1	0,0047
Learning management system	1	0,0041

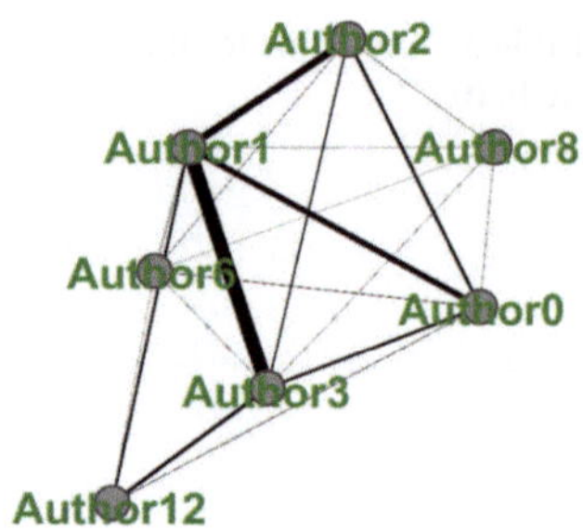

Fig. 2 Synthetic coauthors network for the terms *dblp* and *elearning*—edge weights are computed by *ContextScore*

(see Tables 3 and 4). For comparison of the Author0 and Author1 we marked by bold same terms.

Following figures show graphs of synthetic coauthors network created for the Author1. In the Fig. 2 we can see the network generated for the terms *dblp* and *elearning*. Example of the par of the bigger synthetic coauthors network is in the Fig. 3.

5 Conclusion

Research presented in this article is oriented to the strength extraction between authors based in their association. The method was presented using the data collection from DBLP database, that allows bibliographic information about main publications from scientific publishing in the computer science area.

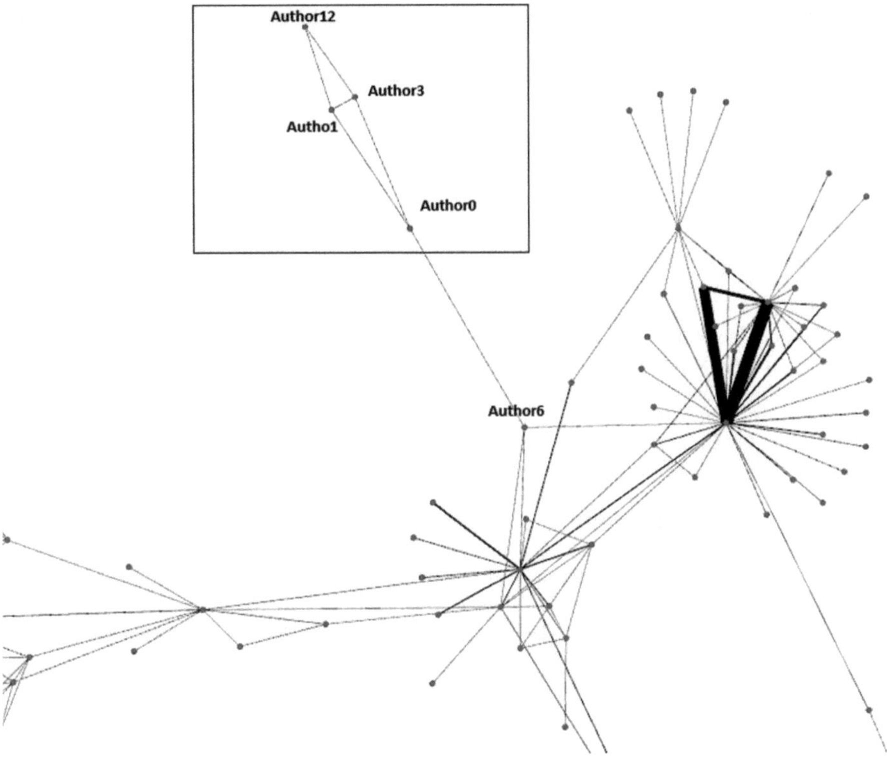

Fig. 3 Synthetic coauthors network for the *rough set*—edge weights are computed by *ContextScore*

The proposed method is usable for the development of social network. The description of this network is based on the set of terms (as the input), which are used in the publication titles by the given authors. Using this method we have obtained the new weight in synthetic coauthors network DBLP. By means of the set of selected terms, belonging to one (or more) authors, we can construct the subnetwork with only the context-related coauthors. This subnetwork can be very helpful in searching of the authors publishing in the same area, defined by selected terms.

It follows that this method can be used to a certain extent for prediction as well. In the cases, when we have more coauthors in the subnetwork, there is common that the frequency of publishing varies for different pairs of coauthors. There exists high probability, that the author will publish with the author, for which he/she has high association strength. For more detailed and more precised description of the network, we intent to use the abstracts of articles, which will be our further topic in our future experiments. From the experiments is evident, that the method could be exploitable for topic extraction in DBLP network.

Acknowledgments This work is supported by SGS, VŠB—Technical University of Ostrava, Czech Republic, under the grant No. SP2011/172.

References

1. Mori, J., Matsuo, Y., Ishizuka, M., Faltings, B.: Keyword extraction from the Web for FOAF metadata. In: Proceedings of the 1st Workshop on Friend of a Friend, Social Networking and the (Semantic) Web, (2004)
2. Konchady, M.: Text Mining Application Programming (Programming Se-ries). Charles River Media, Rockland (2006)
3. Porter, M.F.: An algorithm for suffix stripping. Program **14**, 130–137 (1980)
4. Lopez, P., Laurent, R.: HUMB: automatic key term extraction from scientific articles in GROBID, pp. 248–251. Computational Linguistics, July (2010)
5. Jacquemin, C., Didier, B.: Term extraction and automatic indexing. In: Mitkov, R. (ed.) Handbook of Computational Linguistics, pp. 599–615. Oxford University Press, Oxford (2003)
6. Lashkari, A.H., Mahdavi, F., Ghomi, V.: A boolean model in information retrieval for search engines (2009). doi:10.1109/ICIME.2009.101
7. Deza, E., Deza, M.: Dictionary of Distances, pp. 1–391. Elsevier, Amsterdam (2006)
8. Ding, Y.: Scientific collaboration and endorsement: Network analysis of coauthorship and citation networks. J. informetr. **5**(1), 187–203 (2011)
9. Santamaría, R.R.T.: Overlapping clustered graphs: co-authorship networks visualization. Lect. Notes Comput. Sci. **5166**, 190–199 (2008)

Local Community Detection and Visualization: Experiment Based on Student Data

Miloš Kudělka, Pavla Drážďilová, Eliška Ochodková, Kateřina Slaninová and Zdeněk Horák

Abstract This paper is focused on the detection of communities in social networks. We propose and describe a novel method for detecting local communities. We have used this method in an experiment on student social networks in order to prove our hypothesis about the nature of student communities. The results of the experiment rationalized our hypothesis and confirmed the effectiveness of the described method of local community detection.

1 Introduction

Networks (or graphs in general) as collections of vertices and edges connecting some subset of vertices can be used for the representation of many datasets nowadays. Typical examples of such datasets (which also pose a new challenges for informational retrieval techniques), are the World Wide Web, Internet and collaboration or social networks. Recently, there has been a considerable interest in algorithms for finding communities in networks. Communities are groups of vertices with dense internal connections, but sparser connections to other communities. In networks, the

M. Kudělka (✉) · P. Drážďilová · E. Ochodková · K. Slaninová · Z. Horák
Department of Computer Science, FEECS, VŠB—Technical University of Ostrava, Ostrava, Czech Republic
e-mail: miloš.kudělka@vsb.cz

P. Drážďilová
e-mail: pavla.drazdilova@vsb.cz

E. Ochodková
e-mail: eliska.ochodkova@vsb.cz

K. Slaninová
e-mail: katerina.slaninova@vsb.cz

Z. Horák
e-mail: zdenek.horak@vsb.cz

M. Kudělka et al. (eds.), *Proceedings of the Third International Conference on Intelligent Human Computer Interaction (IHCI 2011), Prague, Czech Republic, August, 2011*, Advances in Intelligent Systems and Computing 179, DOI: 10.1007/978-3-642-31603-6_25, © Springer-Verlag Berlin Heidelberg 2013

detection of related entity groups, i.e., communities, has a practical importance, as it helps us to understand the network structure. In the field of social networks, this problem has been studied for decades.

However, methods that require knowledge of the context of a whole network to divide the network are proving to be unusable for large and densely connected data, see Sect. 2. In our paper, we present a method, which applies the known approaches of local community detection [4] and makes use of the measurement of dependence between the network vertices. This method is based on the analysis and examination of the nearest surroundings of one or a group of vertices [11]. Therefore, the knowledge of the whole context of the network is not essential.

In order to prove the effectiveness of our method on real data, we have chosen the data about students at our university for the experiment. The students choose individual tutorials in the students information system, thus there are no administratively or formally established groups for the entire length of the study appear. Students choose tutorials for individual subjects separately. Despite this, as teachers, we tend to notice that small communities of students, who attend classes together, appear. In the context of our faculty, this means hundreds of students in every year of study. The information about the particular study progress is hidden in the data, we obtained an annonymized copy only. The hypothesis that we wanted to confirm was based on several small communities familiar to us. In nearly every case the students have known each other from high school or have commuted to the university from the same microregions. The aim of our experiment was to find out whether it is possible to detect small communities in a large network and if these are actually students with the characteristics mentioned above.

The paper is organized as follows. After the introduction, related approaches are mentioned in Sect. 2. A concept of dependency used for community detection is introduced in Sect. 3 and further on, a detection algorithm is described in Sect. 4. Section 5 focuses on the experiment with described data collection, on it's results and results visualization.

2 Related Approaches

To detect communities, graph partitioning methods or hierarchical clustering has been applied. Originally, graph partitioning methods, based on edge removal [9, 10, 15], divide the vertices of a network into a given number of (non-overlapping) groups of a given size, while the number of edges between groups is minimal. The number and size of groups are fixed and the groups acquired are typically equal-sized. The next technique currently in use is hierarchical clustering [16]. The idea behind it is to develop a measure of similarity x_{ij} between pairs (i, j) of vertices, based on the network structure.

The use of these approaches is problematic for networks which are either too large or too dynamic to be known completely. Generally, the fixed number and size of groups (usually unknown at the beginning of the process), are the constraints on community detection, as well as non-overlapping groups.

Various beneficial metric functions have been proposed to help to solve the problems. Many community-finding algorithms are based on maximizing the quantity known as modularity [8, 14, 17], but any algorithm using modularity requires complete knowledge of the entire network.

Under these circumstances, one must rely on a local method capable of finding a particular community within a network, without the knowledge of the structure outside of the discovered community. Additional motivation for local methods comes from networks that require a rather demanding generation or exploration with a crawler [16].

Several local methods exist, all of which attempt to find the community containing a particular starting node [1–3, 5–7, 12]. Typically, local community detecting techniques randomly start from a vertex v, and gradually merge neighboring vertices one-at-a-time by optimizing a measure metric.

Authors of proposed methods usually work with the terms *local community, community core, community boundary* and *community shell* similarly as is shown in Fig. 2 and communities are usually generated from a random starting vertex that could affect the community. These methods differ in the decision process which states that certain vertex (vertices) is suitable enough to be added to the existing community in the next iteration of the algorithm.

Methods mentioned above have some drawbacks. Some algorithms, such as the iterative local expansion algorithm [5], need the number of communities in a complex network to be known in advance. Otherwise, the detected communities could include many outliers [7, 12]. Finally, the result of local community detection depends on an starting vertex and a predefined threshold l, where l is the distance from the starting vertex to all shell vertices [1]. This approach works well only if the initial vertex is in the core.

In our approach, we were inspired by methods for local community detection [3]. During the detection process, we investigate how a vertex is dependent on the community. In addition, the result of our detection algorithm is a biconnected subgraph.

3 Measuring the Dependency of a Vertex on a Community

As a motivation for the explanation of the term *dependency*, let us show an example with two adjacent vertices x and y in a weighted undirected network. The strength of the relationship between two vertices of the network is expressed by the weight of their edge.

Consider the situation in Fig. 1a, b, where vertices x and y share an edge of a weight 3. The vertices in the first figure have no additional neighbours. In the second figure the vertex y is adjacent to three additional vertices and the weight of the edge between them is 1. The intuition behind the term *dependency* says, that the relation between vertices in Fig. 1a is balanced, while the situation in Fig. 1b is different—the

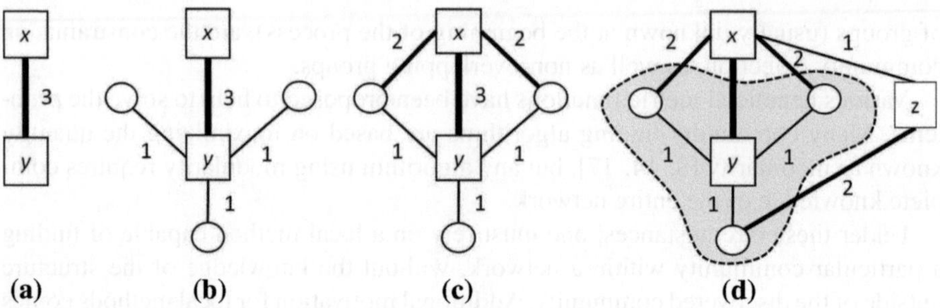

Fig. 1 Examples of dependency between two vertices

vertex y is less dependent on the vertex x, than vice versa. Figure 1c contains two additional edges (with weight 2) between the vertex x and two different vertices. In this situation, vertex x is no longer so highly dependent on vertex y because of the two new edges.

When thinking about dependency in Fig. 1c, we should also consider that the new edges include common neighbours of the vertices x and y (which mediate some part of the dependency on the vertex y). Starting from the aforementioned intuition, the dependency between two vertices of the undirected network should be understood as a generally asymmetric measure which describes relationships between two vertices using a new kind of network weights.

3.1 The Dependency Between Two Vertices

Let $E(x)$ be the set of all edges adjacent to the vertex x. Let $\text{Adj}(x, y)$ be the set of all edges between vertex x and any of the neighbours of vertex y. Clearly, $\text{Adj}(x, y) \subseteq E(x)$. Let $W(e)$ be the weight of edge e and $W(v_1, v_2)$ is the weight of an edge between vertices v_1 and v_2 ($W(v_1, v_2) = 0$, if there is no such edge).

Let x be not an isolated vertex of the network. The dependency $D(x, y)$ of vertex x on vertex y is defined as:

$$D(x, y) = \frac{W(x, y) + \sum_{e_i \in \text{Adj}(x,y)} W(e_i) \cdot R(e_i)}{\sum_{e_i \in E(x)} W(e_i)}, \tag{1}$$

$$R(e_i) = \frac{W(y, v_i)}{W(e_i) + W(y, v_i)}, \tag{2}$$

where $R(e_i)$ is the coefficient of dependency of vertex x on vertex y via the common neighbour v_i, therefore $v_i \in e_i$.

For situations depicted on Fig. 1 holds:

1a: $D(x, y) = D(y, x) = 1$,
2b: $D(x, y) = 1, D(y, x) = \frac{1}{2}$,
3c: $D(x, y) = \frac{13}{21}, D(y, x) = \frac{13}{18}$.

This dependency describes a relation of one vertex on another vertex from the point of view of their surroundings. In all presented examples the calculated dependency is greater than or equal to 0.5. Presented equations infer $D(x, y) \in \langle 0; 1 \rangle$. The dependency being equal to zero means that vertices x and y have no common edge or neighbour. The full dependency (dependency equal to one) describes a situation where vertex x has only one common edge with vertex y.

Remark 2 The dependency is non-zero, if at least one of the following conditions hold:

1. There exists an edge between vertices x and y.
2. Vertices x and y share at least one common neighbour.

3.2 Dependency of a Vertex on a Set of Network Vertices

The relation of dependency of one vertex on another could be generalized as a relation of dependency of vertex x on a set of n vertices $Y = \{y_1, y_2, \ldots, y_n\}$. In this case, two things have to be taken into consideration:

1. Vertex x can have more neighbours from set Y. Let the set of these neighbours be $N(x, Y)$, $N(x, Y) \subseteq E(x)$.
2. Let $\mathrm{Adj}(x, Y)$ be a set of all edges containing the vertex x and one of the neighbours v_i, $i = 1, \ldots, n$ of the vertices from set Y, so that $v_i \notin N(x, Y)$. Any neighbour v_i of the vertex x then can be a neighbour of more than one vertex y_j, $j = 1, \ldots, n$, $i \neq j$ from set Y at the same time. Let the set of these vertices be $Y(v_i)$.

Let x not be an isolated network vertex. Then the dependency $D(x, Y)$ of vertex x on set Y of vertices is defined as:

$$D(x, y) = \frac{\sum_{y_i \in N(x,Y)} W(x, y_i) + \sum_{e_i \in \mathrm{Adj}(x,Y)} W(e_i) \cdot R(e_i)}{\sum_{e_i \in E(x)} W(e_i)}, \tag{3}$$

$$R(e_i) = MAX_{y_j \in Y(v_i)} \left(\frac{W(y_j, v_i)}{W(e_i) + W(y_j, v_i)} \right). \tag{4}$$

The value of the dependency coefficient $R(e_i)$ in Eq. 4 is a maximum of values calculated for individual vertices of the set $Y(v_i)$, which have a mutual vertex v_i with an edge e_i, whose second vertex is vertex x. If in the situations in Fig. 1c, d (where there is also vertex z with edges of values 1 and 2) is calculated a dependency of vertex x, resp. y on sets of the vertices left V_x and V_y, while it holds: $z \notin V_x, V_y$, then it is valid:

1c: $D(x, V_x) = 1, D(y, V_y) = \frac{5}{6}$,
1d: $D(x, V_x) = \frac{23}{24}, D(y, V_y) = 1$.

4 Community Detection Based on Vertex Dependency

A local network community is a set of vertices from a network (undirected weighted graph). This set is composed of the community base and vertices in the surroundings of this base. Every vertex in the community (excluding max. one base vertex) is dependent on the rest of community vertices.

Definition 1 (Community base) A *community base* is a starting set of n vertices appropriately chosen in advance, which by definition belongs to the community and which meets the following criteria:

1. It is a biconnected subgraph.
2. At least $(n - 1)$ vertices have to be dependent on the other base vertices.

Definition 2 (Local community expansion) A *local community expansion* is an iterative process, in which only base vertices are considered to be a community, while the other network vertices are gradually examined and the vertices that follow certain criteria then progressively expand the community. *Recognized network vertices* are vertices, which are, at any point of the local expansion process, decided to belong to the community. *Unrecognized network vertices* are all the vertices left.

Definition 3 (Recognition of a network vertex) An unrecognized network vertex becomes an recognized one, if during the process of local expansion, it meets the following criteria:

1. It is adjacent to at least two different community vertices.
2. It is dependent on the other community vertices.

Remark 1

1. A community is a biconnected subgraph (in all phases of local community expansion).
2. A community can, under given circumstances, contain a cycle of any length. That means, in case of meeting the first condition, the vertex is adjacent to at least two different community vertices, which do not have an edge between them. However, in this situation, it could be assumed that a certain (though informal) weak edge exists. This assumption is based on the so called triadic closure, whilst an informal edge expresses the affiliation of vertices with the same community.

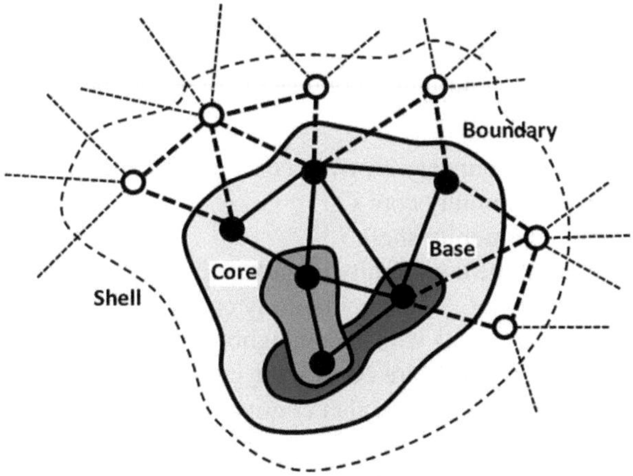

Fig. 2 Local community illustration—following terms are used: community base (a starting set of vertices), community core C, community boundary B, community shell C

Further on, the detection algorithm is described. Its result is a community with fairly strong properties, especially biconnection. Also the dependency of any vertex of the community on the other community vertices is high. This dependency appears to be essential for the algorithm. We have chosen an threshold of 0.5, which corresponds with natural intuition, for measuring the affiliation of a vertex to a community. If we further consider that vertex x is dependent on vertex y, respectively on the set of vertices Y, we have to assume that it holds: $D(x, y) \geq 0.5$, resp. $D(x, Y) \geq 0.5$.

Community L is composed of all vertices of the core and the community boundary, so that $L = C \cup B$, see Fig. 2 (in addition, community L contains all base vertices by definition). The *community core* C comprises all of the vertices which do not have any edges oriented outside of community L (out of community core C and community boundary B). *Community boundary* B is created by all vertices that belong to community L, but do not belong to community core C. Every vertex belonging to community boundary B therefore has at least one edge oriented outside community L. *Community shell* S is composed of vertices that do not belong to community L, but are adjacent to it, thus have at least one edge oriented inside the community (to community boundary B). Shell S comprises unrecognized networks vertices, which can become recognized network vertices (and would be moved to community L in the process of local expansion).

4.1 Algorithm

The algorithm is based on a gradual examination of the vertices from the shell community, and on subsequent recognition of the vertices that meet the criteria for allegiance to the community. Only vertices that do not meet the criteria remain in the shell. Then the vertices that newly belong to the shell (the chosen neighbours of the vertices moved to the community boundary) are added there, and the process is repeated.

1. Create a community boundary B, which is equal to community base.
2. Create an empty community core C.
3. Create an empty community shell S.
4. Move the vertices of the community boundary B which do not have neighbours outside the community L, to the community core C.
5. Refill the community shell with new neighbours of the vertices added to the community boundary B that are outside the community L.
6. Calculate the dependency on the other community vertices for each vertex of the community shell S.
7. Move every vertex from community shell S, which meets the criteria for recognition and move for community L, to community boundary B.
8. If at least one vertex from the shell S has been recognized (moved to community L), repeat the process from the step 4.
9. Community L is a unification of the core and the community boundary, so $L = C \cup B$.

Remark 2 In the algorithm described above, it is necessary to recalculate the dependency on a community in every iteration for every vertex of the shell. That comes with a complication (in each step, it is a complication given by the number of vertices and the number of vertices examined in the surroundings of each one of them). Therefore, a context for dependency calculation is preserved for each vertex of the shell during the calculation. The dependency is then recalculated only in case this context is changed after the iteration step.

4.2 Detection of All Network Communities

Unless it is said otherwise, a community base is considered to be two vertices connected by an edge, while at least one of the vertices is dependent on the other one. This couple should be called *the edge base*. The purpose of implementing the edge base is an effective detection of more communities. For example, there are more communities around vertex a in the Fig. 3a. We shall assume that all edges containing vertex a are edge bases, while the second vertex of every edge base is always dependent on vertex a. If we want to detect all of the communities that are based on vertex a (and are dependent on it and its surroundings), then it is essential to detect the communities for all edge bases of vertex a and to remove the duplicities. The result (three communities) is in the Fig. 3b. If we need to detect all communities of a network, it is possible to proceed in a similar way. At first, local communities are detected for every single edge base. Afterwards (or in the process), the duplications are removed.

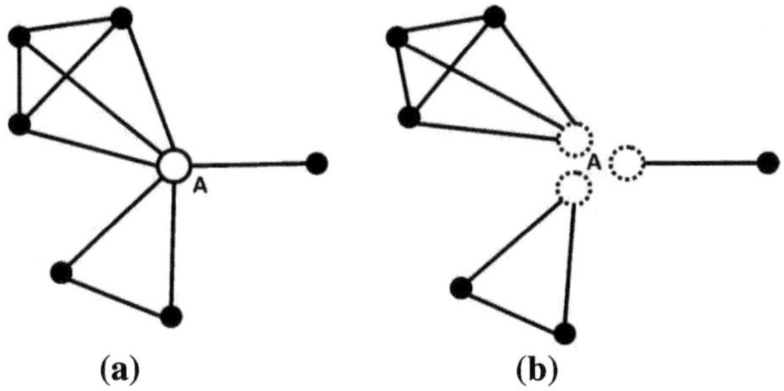

Fig. 3 Three communities around the vertex A

Remark 3 The two detected local communities L_1 and L_2 can have vertices in four different set relations:

1. $L_1 = L_2$,
2. $L_1 \cap L_2$,
3. $L_1 \subset L_2$ resp. $L_2 \subset L_1$,
4. $L_1 \cap L_2 = X, X \neq \emptyset, X \subset L_1, X \subset L_2$ (in Fig. 3b, each pair of detected communities has a mutual vertex a, but in general, this could also be vertices of a mutual sub-community).

In our approach, we focus only on communities that have at least three vertices and no sub-communities. The algorithm for detecting communities for the whole network is therefore adjusted to suit our focus, so it removes the sub-communities during the process of detection.

5 Experiment

In our experiment, we have used the data collection from the information system (called *Edison*) of our university. This system is used for the organization, planning and evidence of students' evaluation at VŠB—Technical University in Ostrava, Czech Republic. We have selected three consecutive academic years (2008/2009, 2009/2010 and 2010/2011) and explored the students' behavior during these three years. Our motivation was to verify whether there are students communities formed by students repeatedly attending the same courses and tutorials (and sharing the same schedule). The main hypothesis was that the students attending the same (or similar enough) courses and their tutorials together have known each other from the past. From the point of view of social network analysis, we can say that between such students, the relations on the basis of their similar behavior exist. We have focused on finding the reasons of such behavior and on discovering its potential influence on their study results.

5.1 Data Collection

The data collection consisted of 85, 161 records with information about students and the academic year and semester (fall, summer), the course and the study group. The study group corresponds to a particular tutorial of a course, as one course can consist of several tutorials held simultaneously for multiple study groups due to organizational reasons. Moreover, we have acquired additional information about the students such as their first year of studies, previously graduated school and the location of these schools.

A bipartite graph $|S| \times |T|$ with 85, 161 edges has been constructed from a set of students S and a set of tutorials T. In this bipartite graph, the edge represents the fact that a certain student $s_i, i = 1, 2, \ldots, |S|$ has attended the tutorial $t_l, l = 1, 2, \ldots, |T|$. A one-mode projection from the two-mode form of bipartite graph [13] was then obtained and represented by a weighted matrix $|S| \times |S|$.

The edge weight was based on a number of tutorials attended by a pair of students $(s_i, s_j), i, j = 1, 2, \ldots, |S|$ and was defined by Eq. 5:

$$w(s_i, s_j) = \frac{\sum_{k=1}^{5} \sqrt{t_{ij}^{k} t_{ij}^{k+1}}}{K_{ij}}. \tag{5}$$

The variable k denotes the number of pairs of consecutive semesters in our dataset, and K_{ij} denotes the number of consecutive semesters, when tutorials were shared by a pair of students (s_i, s_j). The obtained graph consisted of 2, 585 vertices and 319, 482 edges; 268, 197 edges were of weight 0.

5.2 Application of Detection Algorithm

In very densely connected networks the bindings (and vertices) with low importance could be regarded as noise, which complicates the analysis. When we remove bindings (edges) and vertices having low importance, we perform a reduction of the network while maintaining the substantial properties. In this situation, important properties are not the mathematical ones, but the social properties (in our case, only students and bindings that carry information essential for analysis remain in the network). For our experiment, we have rounded the weights to whole numbers, and then gradually conducted the reduction for each edge weight. For weight 1 it means that every edge of weight 0 has been removed from the network. For weight 7, all edges of weight lower than 7 (0, 1, 2, 3, 4, 5, 6) have been removed. It is evident that after this reduction the network breaks into connected components. Table 1 includes detailed statistics of how many connected components with at least three vertices emerge due to the reduction (count), including the number of vertices of the highest weight (max), and how many and how large communities have been detected (all), as well as how many of these communities are complete sub-graphs (cmpl).

Table 1 Detailed statistics of student network analysis

Reduction Weight	Components Count/max	Communities All/cmpl.	Vertices 3	4–5 All/cmpl.	6–10 All/cmpl.	11–20 All/cmpl.	21–42 All/cmpl.
7	2/4	2/2	1	1/1			
6	11/4	10/10	9	1/1			
5	22/6	20/18	14	5/4	1/0		
4	41/6	26/25	18	7/6	1/1		
3	66/9	38/36	23	13/12	2/1		
2	87/339	58/48	32	17/13	6/3	3/0	
0, 1	12 / 1,377	112/82	58	27/17	20/7	5/0	2/0

Remark 1 If an edge is of weight 1, it means that two students attend approximately one tutorial in a semester together. As a result, it can be assumed that there is no strong social relationship between them and the situation is more likely random. However, if an edge is of weight for example 3, it may be assumed that a regular collective attendance of three tutorials is not random, and it is a sign of a stronger social connection between students. In case of an edge of weight 7, there is an evident and firm social connection between students (they attend virtually the same tutorials).

As we can see in Table 1, small communities have been detected in consistent and densely connected components of the network. The largest detected community has only 42 vertices (with the second largest having 38 vertices). No other community has more than 18 vertices. That holds also for densely connected networks without reduction, whilst the major part of detected communities has no more than 10 vertices. On the other hand, we can also see that communities connected by edges of high weight can be found even without a detection algorithm (they are virtually equal to connected components that remain after a higher reduction of a network). Nevertheless, there has been a match only in case of the reduction by weight 7, where the only two found network components with at least three vertices corresponded precisely with two detected communities.

The detection algorithm is based on a relatively demanding repeated calculation of dependency of individual shell vertices on currently detected vertices of a community. Despite that, it provides results in a reasonable time. For example, the detection of the largest community (with 42 vertices) takes less than 0.1 s, while 9 iterations in total are needed for local expansion (CPU Intel i7-2720QM, 2.20 GHz, Windows 7, 64bit, .NET Framework).

5.3 Experiment Results

Using the algorithm for calculation of the relation based on the vertex dependency, we have obtained the set of communities L_x, where x gives the minimal weight of considered edges.

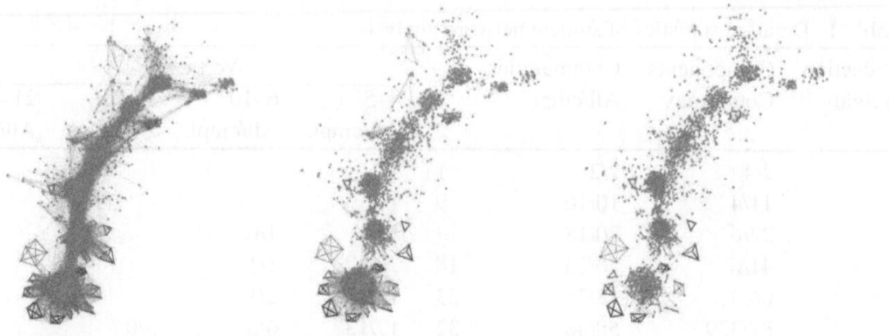

(a) (b) (c)

Fig. 4 Student network and communities visualization before and after reduction. **a** Whole network with detected communities. **b** Edge weight ≥ 1. **c** Edge weight ≥ 2

Table 2 Community description

Set	#Communities	Reason—same school	microregion	city
L_5	20	12	4	3
L_6	10	8	2	0
L_7	2	1	1	0

The graph reduction, after the exclusion of appropriate edges for the given level of the edge weight, is described in Fig. 4. At the beginning, detected communities L_0 could not be properly visualized, because the network is very dense, see Fig. 4a. The networks with L_1 and L_2 communities are presented in Fig. 4b, c.

Further on, our findings for the first three sets of communities with the highest level of edge weight are depicted. The set L_7 consists of only two communities, where one community is a complete subgraph of 4 vertices and the second community has 3 vertices. From the observation of data collection, we can say that in both these communities, students studied at the same high school, or are from the same microregion.

Set L_6 consists of communities with edge weights 6, 7. In comparison with set L_7, the number of triangles in set L_6 has increased. In 9 communities from L_6 we can find schoolmates from the same high school.

There are communities with edge weights 5, 6, 7 in set L_5. There are communities of 6, 5 and 4 vertices, which do not create complete subgraphs. All these communities comply with the fact that the students are from the same high school. We can find another 13 complete subgraphs in this set as well. Only one community differs from the others, as we are not able to find the reason for its formation with certainty. Different reasons probably exist, however we do not have enough information about them (room-mates, common hobbies, and others). The total sum of communities in set L_5 is 20.

A detailed description is presented in the Table 2:

6 Conclusion

The goal of our article was an experiment with real data focused on the detection of communities in the network of our university students. The experiment confirmed our hypothesis about the causes of natural groupings of students into relatively small communities. These communities consist, by a greater degree, of the students who are likely to know each other from high school or commute from the same microregion. To detect communities we have used our own algorithm based on property, which we call dependency. In the article we have shown that, thanks to the application of our algorithm, we are able to detect a small community in a densely interconnected network. In our future work, we will focus on the comparison of our approach with other community detection methods.

Acknowledgments This work was partially supported by SGS, VSB-Technical University of Ostrava, Czech Republic, under the grants No. SP2011/172.

References

1. Bagrow, J.P.: Evaluating local community methods in networks. J. Stat. Mech. **2008**(5), P05001 (2008)
2. Bagrow, J.P., Bollt, E.M.: A local method for detecting communities. Phys. Rev. E **72**, 046108 (2005)
3. Chen, J., Zaïane, O.R., Goebel, R.: Detecting communities in large networks by iterative local expansion. In: CASoN IEEE Computer Society, pp. 105–112. (2009)
4. Chen, J., Zaïane, O.R., Goebel, R.: Local community identification in social networks. In: ASONAM, IEEE Computer Society, pp. 237–242. (2009)
5. Chen, J., Zaiïane, O.R., Goebel, R.: Detecting communities in social networks using max-min modularity. In: SDM, SIAM, pp. 978–989. (2009)
6. Chen, Q., Wu, T.T.: A method for local community detection by finding maximal-degree nodes. In: ICMLC, pp. 8–13 (2010)
7. Clauset, A.: Finding local community structure in networks. Phys. Rev. E **72**(2), 026132 (2005)
8. Clauset, A., Newman, M.E.J., Moore, C.: Finding community structure in very large networks. Phys. Rev. E **70**, 066111 (2004)
9. Fiedler, M.: Algebraic connectivity of graphs. Czech. Math. J. **23**, 298–305 (1973)
10. Kernighan, B.W., Lin, S.: An efficient heuristic procedure for partitioning graphs. Bell Syst. Tech. J. **49**(2), 291–308 (1970)
11. Horák, Z., Snášel, V., Abraham, A., Kudělka, M.: Weighted co-authorship network based on forgetting. Future Inf. Technol. **185**, 72–79 (2011)
12. Luo, F., Wang, J., Promislow, E.: Exploring local community structures in large networks. In: IEEE/WIC/ACM International Conference on WI, pp. 233–239. (2006)
13. Newman, M.: Networks: An Introduction. Oxford University Press, Inc., New York (2010)
14. Newman, M.E.J., Girvan, M.: Finding and evaluating community structure in networks. Phys. Rev. E **69**(2), 026113 (2004). PT: J; PN: Part 2; PG: 15
15. Pothen, A., Simon, H.D., Liou, K.P.: Partitioning sparse matrices with eigenvectors of graphs. SIAM J. Matrix Anal. Appl. **11**, 430–452 (1990)
16. Schaeffer, S.E.: Graph clustering. Comput. Sci. Rev. **1**(1), 27–64 (2007)
17. Wakita, K., Tsurumi, T.: Finding community structure in mega-scale social networks: [extended abstract]. In: Proceedings of the 16th International Conference on World Wide Web, WWW '07, pp. 1275–1276. ACM, New York, NY, USA (2007)

Index

M. Kudělka et al. (eds.), *Proceedings of the Third International Conference on Intelligent Human Computer Interaction (IHCI 2011), Prague, Czech Republic, August, 2011*, Advances in Intelligent Systems and Computing 179, DOI: 10.1007/978-3-642-31603-6,

MIX
Papier aus verantwortungsvollen Quellen
Paper from responsible sources
FSC® C105338
FSC
www.fsc.org

If you have any concerns about our products,
you can contact us on
ProductSafety@springernature.com

In case Publisher is established outside the EU,
the EU authorized representative is:
Springer Nature Customer Service Center GmbH
Europaplatz 3, 69115 Heidelberg, Germany

Printed by Libri Plureos GmbH
in Hamburg, Germany